Advances in
ATOMIC, MOLECULAR, AND OPTICAL PHYSICS

VOLUME 37

Editors

BENJAMIN BEDERSON
New York University
New York, New York

HERBERT WALTHER
Max-Plank-Institut für Quantenoptik
Garching bei München
Germany

Editorial Board

P. R. BERMAN
University of Michigan
Ann Arbor, Michigan

M. GAVRILA
F. O. M. Institute voor Atoom-en Molecuulfysica
Amsterdam, The Netherlands

M. INOKUTI
Argonne National Laboratory
Argonne, Illinois

W. D. PHILIPS
National Institute for Standards and Technology
Gaithersburg, Maryland

Founding Editor

SIR DAVID BATES

ADVANCES IN
ATOMIC, MOLECULAR, AND OPTICAL PHYSICS

Edited by

Benjamin Bederson
DEPARTMENT OF PHYSICS
NEW YORK UNIVERSITY
NEW YORK, NEW YORK

Herbert Walther
UNIVERSITY OF MUNICH AND
MAX-PLANK INSTITUT FÜR QUANTENOPTIK
MUNICH, GERMANY

Volume 37

ACADEMIC PRESS

San Diego   London   Boston
New York   Sydney   Tokyo   Toronto

This book is printed on acid-free paper. ∞

Copyright © 1996 by ACADEMIC PRESS

All Rights Reserved.
No part of this publication may be reproduced or transmitted in any form or by any means, electronic or mechanical, including photocopy, recording, or any information storage and retrieval system, without permission in writing from the publisher.

Academic Press, Inc.
525 B Street, Suite 1900, San Diego, California 92101-4495, USA
http://www.apnet.com

Academic Press Limited
24-28 Oval Road, London NW1 7DX, UK
http://www.hbuk.co.uk/ap/

International Standard Serial Number: 1049-250X

International Standard Book Number: 0-12-003837-4

PRINTED IN THE UNITED STATES OF AMERICA
96 97 98 99 00 01 QW 9 8 7 6 5 4 3 2 1

Contents

CONTRIBUTORS   vii

Evanescent Light-Wave Atom Mirrors, Resonators, Waveguides, and Traps

Jonathan P. Dowling and Julio Gea-Banacloche

I. Introduction	2
II. Atom Mirrors: A Brief Theoretical Review	10
III. Atom Resonators: Fabry–Pérot Type	23
IV. Atom Waveguides	39
V. Blue-Detuned Concave Atom Traps	59
VI. Red-Detuned Convex Atom Traps and Guides	75
VII. Conclusions and Summary	88
References	90

Optical Lattices

P. S. Jessen and I. H. Deutsch

I. Introduction	95
II. The 1D Lin $\perp$ Lin Model System	97
III. Crystallography of Optical Lattices	104
IV. Laser Cooling in Optical Lattices: Theory	109
V. Spectroscopy	119
VI. New Developments	128
References	136

Channeling Heavy Ions through Crystalline Lattices

Herbert F. Krause and Sheldon Datz

I. Introduction	139
II. Channeling Trajectories and Interactions Potentials	144
III. Planar Channeling	145
IV. Axial Channeling	146
V. Experimental Methods	150
VI. Charge Changing Collisions	152
VII. Radiative Electron Capture	153
VIII. Electron Impact Ionization	158

IX. Dielectronic Excitation and Recombination and Resonant Transfer
   with Excitation 161
X. Resonant Coherent Excitation 166
   References 176

Evaporative Cooling
Wolfgang Ketterle and N. J. van Druten

I. Introduction 181
II. Theoretical Models for Evaporative Cooling 184
III. The Role of Collisions for Real Atoms 201
IV. Experimental Techniques 209
V. Summary of Evaporative Cooling Experiments 227
VI. Outlook 229
    References 231

Nonclassical States of Motion in Ion Traps
J. I. Cirac, A. S. Parkins, R. Blatt, and P. Zoller

I. Introduction 238
II. Model 244
III. Sideband Cooling: Preparation in the Ground State 252
IV. Preparation of Fock States 258
V. Preparation of Squeezed States 274
VI. Preparation of Schrödinger Cat States 278
VII. Analysis of the Nonclassical States of Motion 283
VIII. Conclusions 292
    References 292

The Physics of Highly Charged Heavy Ions Revealed by Storage / Cooler Rings
P. H. Mokler and Th. Stöhlker

I. Introduction 297
II. The Physics of Highly Charged Heavy Ions 300
III. Storage and Cooler Rings for Heavy Ions 316
IV. Charge Changing Processes 332
V. Atomic Structure Studies 348
VI. Future Developments 362
    References 365

SUBJECT INDEX 371
CONTENTS OF VOLUMES IN THIS SERIAL 379

Contributors

Numbers in parentheses indicate the pages on which the authors' contributions begin.

R. BLATT (237), Institut für Experimentalphysik, Universität Innsbruck, A-6020 Innsbruck, Austria

J. I. CIRAC (237), Departamento de Fisica Aplicada, Universidad de Castilla–La Mancha, 13071 Ciudad Real, Spain

SHELDON DATZ (139), Physics Division, Oak Ridge National Laboratory, Oak Ridge, Tennessee 37831

I. H. DEUTSCH (95), Center for Advanced Studies, University of New Mexico, Albuquerque, New Mexico 87131

JONATHAN P. DOWLING (1), Reseach, Development, and Engineering Center, Weapons Sciences Directorate, AMSMI-RD-WS-ST, U. S. Army Missile Command, Redstone Arsenal, Alabama 35898

JULIO GEA-BANACLOCHE (1), Physics Department, University of Arkansas, Fayetteville, Arkansas 72701

P. S. JESSEN (95), Optical Sciences Center, University of Arizona, Tucson, Arizona 85721

WOLFGANG KETTERLE (181), Department of Physics and Research Laboratory of Electronics, Massachusetts Institute of Technology, Cambridge, Massachusetts 02139

HERBERT F. KRAUSE (139), Physics Division, Oak Ridge National Laboratory, Oak Ridge, Tennessee 37831

P. H. MOKLER (297), G.S.I.-Darmstadt, 64220 Darmstadt, Germany, and University of Geisen, Germany

A. S. PARKINS (237), University of Waikato, Hamilton, New Zealand

TH. STÖHLKER (297), Institute für Kernphysik, Universität Frankfurt, Germany, and G.S.I.-Darmstadt, 64220 Darmstadt, Germany

N. J. VAN DRUTEN (181), Department of Physics and Research Laboratory of Electronics, Massachusetts Institute of Technology, Cambridge, Massachusetts 02139

P. ZOLLER (237), Institut für Theoretische Physik, Universität Innsbruck, A-6020 Innsbruck, Austria

EVANESCENT LIGHT-WAVE ATOM MIRRORS, RESONATORS, WAVEGUIDES, AND TRAPS

JONATHAN P. DOWLING

Research, Development, and Engineering Center
Weapons Sciences Directorate
U.S. Army Missile Command
Redstone Arsenal, Alabama

JULIO GEA-BANACLOCHE

Physics Department
University of Arkansas
Fayetteville, Arkansas

I. Introduction	2
II. Atom Mirrors: A Brief Theoretical Review	10
A. Optical Forces on a Neutral Atom	10
B. The Evanescent-Wave Atom Mirror	13
C. Effective Potential Treatment	16
D. Surface-Plasmon-Enhanced Evanescent Wave	18
E. Dielectric-Waveguide-Enhanced Evanescent Wave	19
F. Other Electromagnetic Mirror Schemes	21
III. Atom Resonators: Fabry–Pérot Type	23
A. Cavities with Two Mirrors	23
B. Gravitational Cavity with Parabolic Mirror	29
C. Red–Blue Pushme–Pullyou Resonator	36
IV. Atom Waveguides	39
A. Red-Detuned, Propagating Light-Wave, Hollow Fiber Guides	39
B. Blue-Detuned, Evanescent Light-Wave, Hollow Fiber Guides	45
C. Parallel-Mirror Waveguide	55
IV. Blue-Detuned Concave Atom Traps	59
A. Particle-in-a-Box with Gravity	60
B. Pyramidal Gravitational Trap	65
C. Conical Gravitational Trap	68
D. Evanescent-Wave Cooling in Gravitational Traps	72
VI. Red-Detuned Convex Atom Traps and Guides	75
A. Microsphere Whispering-Gallery Trap	76
B. External Solid Fiber Guide	81
C. Inverted Cone Yukawa-Potential Trap	86
VII. Conclusions and Summary	88
References	90

I. Introduction

For many years, it has been known that light can be used to trap and manipulate small dielectric particles and atoms (Ashkin, 1970, 1978; Minogin and Letokhov, 1987; Kazanstev *et al.*, 1985, 1990; Marti and Balykin, 1993; Meystre and Stenholm, 1985; Wallis, 1995). In particular, the intense coherent light of lasers has been used to cool neutral atoms down to the micro-Kelvin (Chu and Wieman, 1989) and now even the nano-Kelvin regimes (Reichel *et al.*, 1995). At such low temperatures, the de Broglie wavelike character of the atoms becomes pronounced, making it necessary to treat the atoms as wave phenomena. To this end, the study of atom optics has recently developed, in which atom optical elements are fabricated in order to manipulate atoms, while utilizing and preserving the coherence and superposition properties inherent in their wavelike propagation (Balykin and Letokhov, 1989b, 1990; Mlynek *et al.*, 1992; Adams, 1994; Adams *et al.*, 1994b; Sigel and Mlynek, 1993; Pillet, 1994). For example, there has been a concerted effort to study theoretically and produce experimentally the atom optic analogs of photonic optical elements, such as atom beam splitters (Glasgow *et al.*, 1991; Murphy *et al.*, 1993, 1994, 1995; Pfau *et al.*, 1993a, b; Deutschmann *et al.*, 1993b), atom diffraction gratings (Keith *et al.*, 1988, 1991b; Hajnal and Opat, 1989; Baldwin *et al.*, 1990; Zhang and Walls, 1993; Deutschmann *et al.*, 1993a; Christ *et al.*, 1994; Feron *et al.*, 1994; Stenlake *et al.*, 1994), atom lenses (Carnal *et al.*, 1991; Keith *et al.*, 1991b; Ketterle and Pritchard, 1992; Averbukh *et al.*, 1994), atom interferometers (Kasevitch and Chu, 1991; Keith *et al.*, 1991a; Mlynek *et al.*, 1992; Scully and Dowling, 1993; Wilkens *et al.*, 1993; Pillet, 1994; Adams *et al.*, 1994a), and—last but not least—atom mirrors. It is light-induced atom mirrors, and their application to making atom resonators, waveguides, and traps, that we shall focus on in this chapter.

As with all the other atom optical elements just mentioned, a good atom mirror must reflect the atoms specularly while preserving the coherence of the de Broglie wavefunction. Otherwise, the mirrors would be useless for making interferometers and resonators that depend on interference to function. Curved atom mirrors can also be used as focusing devices that—unlike atom lenses—are free of the large "chromatic" aberration that is associated with atomic beams, the velocity spreads of which are often substantial (Balykin and Letokhov, 1989b). Atoms cannot easily be scattered coherently off a crystal of ordinary matter. At thermal velocities, surface roughness effects will cause the atom wave to scatter diffusely, whereas at low velocities van der Waals forces between the atom and the material surface cause them to stick. However, if the surface is suitably prepared—say by ultrafine polishing—then the diffuse scattering can be minimized. This was shown experimentally by Anderson *et al.* (1986), where about a 50% specular reflection was achieved for rather fast thermal cesium atoms grazing a highly polished glass surface. However, this technique will not affect the van der Waals sticking that will tend to dominate the reflection process for slow, ultracooled atoms. Nearly 100% coherent reflection of hydrogen atoms has been achieved also, when the surface was composed of a single quantum state of liquid helium (Berkhout *et al.*, 1989; Berkhout and Walraven, 1993). In this experiment, the liquid helium was rotated to form a paraboloid of revolution, thereby reflecting the atoms from a hydrogen beam into a focus. The experimental difficulties in dealing with a spinning, liquid helium mirror, however, are clear. Another recent idea is to make a coherent magnetic atom mirror, suggested by Hinds and collaborators (Roach *et al.*, 1995). The mirror here consists of a periodically magnetized ferromagnetic surface, and the reflection process utilizes the Stern–Gerlach effect and results in an experimentally observed specular reflectance of nearly 100% for cold atoms.

It was first suggested by Cook and Hill in 1982, that an evanescent light-wave field might be used to form a specular atom mirror (see Fig. 1). Such an evanescent field can easily be formed on the vacuum side of a vacuum–dielectric interface, when light is undergoing total internal reflection at this interface from the dielectric side (Born and Wolf, 1985). A discussion of this atom mirror will be given in Section II; suffice it to say that such an evanescent field can provide quite a sharp and high potential barrier for atoms that are nearly resonant with a field that is blue-detuned. As long as the laser intensity is sufficiently high, the atoms interact only with the evanescent field and not the actual dielectric surface, minimizing diffuse scattering and van der Waals sticking. In addition, such a light-wave mirror is much easier to fabricate and control than the alternatives previously mentioned.

An interesting point to note is that an evanescent light wave exerts a classical force on small dielectric particles, such as spheres (Ashkin, 1970), as well as a quantum mechanical force on atoms. The quantum gradient dipole force on an atom can either be repulsive or attractive—away or toward the region of high field, respectively—depending on whether the laser field is red-detuned below or blue-detuned above the atomic resonance, respectively (Balykin *et al.*, 1987; Minogin and Letokhov, 1987; Balykin and Letokhov, 1989b, 1990; Kazantsev *et al.*, 1990). Until recently, the conventional wisdom seems to have been that the classical evanescent light-wave dipole force on, say, small dielectric spheres cannot be repulsive —in part due to the lack of a resonance with respect to which one could tune above (Marti and Balykin, 1993; Almaas and Brevik, 1995). However, recent experiments by Kawata and Sugiura (1992)—as well as theoretical calculations by Chang *et al.* (1994)—show that a classical repulsive force on a dielectric microsphere is possible, provided that the evanescent laser field is blue-detuned just above one of the classical *Mie resonances* of the sphere.

After the suggestion by Cook and Hill in 1982 that an evanescent light wave could be used to specularly reflect atoms, numerous theoretical and experimental investigations of this effect have occurred. The first observation of such a reflection of atoms was made by Balykin *et al.* (1987, 1988), who used a beam of thermal sodium atoms at grazing incidence to the atom mirror. Since the potential presented to the atoms by the evanescent field is large, but not infinite, the need for grazing incidence is apparent; if the atoms have too large a velocity component normal to the mirror surface, they would penetrate the barrier and scatter diffusely or stick. The first experiment showing the reflection of supercooled atoms from a mirror at normal incidence was performed by Kasevich *et al.* (1990). In this experiment, to which they gave the whimsical moniker "the atomic trampoline," the atoms were first supercooled in a magneto-optical trap and then dropped from a very small height onto a planar, evanescent field mirror. About two bounces of the atoms were observed before they were lost due to lack of confinement in the transverse direction. (Due to the transverse Gaussian intensity profile, the mirror surface is slightly convex, even if the dielectric surface is planar. Spontaneous emission events can quickly "kick" the atom transversely off the mirror.) After the second bounce, small perturbations transverse to the mirror normal would cause the atoms to wander and impact a point on the dielectric mirror not "coated" by the evanescent field, due to the small spot size of the totally internally reflecting laser beam. Cook and Hill (1982) and Kasevitch *et al.* (1990) suggested that a curved or parabolic—rather than planar—evanescent mirror might help to alleviate this problem. The Schrödinger modal structure of such a parabolic, evanescent light-wave, gravitational trap was first calculated by Wallis *et al.* (1992). Their results (reviewed in Section III.B) indicated that such a trampoline would tend to confine the atoms at the focus of the paraboloid of revolution—qualifying the device as an actual atom trap. The group of Phillips at NIST (Helmerson *et al.*, 1992) was the first to attempt to confine atoms in such a parabolic gravitational atom trap, but they were *still* plagued by a low level of transverse confinement and a loss of atoms after a couple of bounces (K. Helmerson, private communication, 1993). Similar problems also occurred in the early experimental work of Aminoff *et al.* (1993a). Improvements, however, finally led to the observation of a spectacular 10 bounces, by Aminoff *et al.* (1993b). (See Fig. 6.)

One problem with parabolic gravitational traps—in terms of transverse confinement—is that they must be very shallow so that the total internal reflection condition is everywhere satisfied across the laser spot at the vertex (see Fig. 5). A way around this problem was suggested by Dowling (1993), by Dowling and Gea-Banacloche (1994, 1995); and independently by Ovchinnikov *et el.* (1995a) and Söding *et al.* (1995), in the form of pyramidal and conical evanescent light-wave gravitational traps (see Fig. 17). In these geometries, the laser beam is not incident on the dielectric–vacuum interface at a large incident angle—as in the planar and parabolic trampolines—but is incident precisely normal to the interface. Instead of a shallow parabola, a sharp and narrow conical or pyramidal fissure is etched out of the dielectric (see Pangaribuan *et al.*, 1992), and now the laser beam totally internally reflects off the sides of this feature, coating the vacuum side with a repellent evanescent potential in the form of a cone or pyramid. The modal structure of these conical and pyramidal traps was first worked out by Dowling (1993) and Dowling and Gea-Banacloche (1994, 1995), as will be discussed in Section V. Ovchinnikov *et al.* (1995a, b) and Söding *et al.* (1995) have demonstrated theoretically that such traps can be used in conjunction with Sisyphus-like cooling and a geometric cooling mechanism to bring alkali atoms to the recoil limit. In particular, the predicted phase-space density at the vertex of the pyramidal trap is on the order of that required for Bose condensation, as will also be discussed in Section V. Preliminary experimental results, utilizing pyramidal and conical evanescent light-wave traps, have been reported by Lee *et al.* (1996).

In all of the geometries just discussed, these traps can be thought of as atom resonators with the evanescent field supplying the restoring force from below, and gravity from above. Of course, gravity need not be a consideration for trapping if two or more evanescent mirrors are used. In fact, the idea of an enclosed evanescent light-wave resonator was first put forth by Cook and Hill in 1982, in the same paper in which they first described the mirror concept. Balykin and Letokhov (1989a) later proposed large atom cavities made from two or more light-wave mirrors, in which they considered the longitudinal and transverse modes in an atom-optical analogy to the theory of large laser cavities (see Fig. 4). These large cavities will be discussed in Section III.A. Ovchinnikov *et al.* (1991) then proposed a simple, one- or two-dimensional, resonator or trap with only one dielectric–vacuum interface, but two totally internally reflecting laser beams with different angles of incidence and detunings of opposite sign—the "pushme–pullyou" trap, discussed in Section III.C. The disadvantage of this trap is that the atom spends its time in a region of high (red-detuned) field intensity—and hence will be prone to spontaneous emission and subsequent decoherence and heating, limiting the usefulness of the trap as a quantum resonator.

In 1993, Wilkens *et al.* discussed the theory of an actual Fabry–Pérot resonator for atoms. Unlike the schemes of Balykin and Letokhov (1989a) —in which the atoms move more or less ballistically on classical trajectories between widely separated mirrors—this resonator was to operate in the quantum regime, where the partially transparent mirrors were to have a separation on the order of the de Broglie wavelength of the atoms (see also Balykin, 1989). This scheme will be reviewed in Section III.A. The modal structure of three-dimensional, boxlike resonators—including gravity—was first worked out by us (Dowling, 1993; Dowling and Gea-Banacloche, 1994, 1995) and will be discussed in Section V.A. Consideration will be given to a fully quantum, particle-in-a-box resonator—or quantum atom "dot"—in the context of its use as a gravimeter or a qubit for quantum computation (Feynman, 1982, 1985, 1986; Deutsch and Joza, 1992; Ekert, 1995).

Another type of quantum resonator we will discuss in Section IV is the evanescent light-wave atom waveguide. The idea of actually using an optically pumped hollow glass fiber as an atom waveguide was proposed theoretically by Ol'Shanii, Ovchinnikov, and Letokhov (OOL) (1993) and independently by Savage, Marksteiner, and Zoller (SMZ) (1993) see also Marksteiner *et al.*, 1994). In the OOL scheme, a hollow glass fiber without cladding guides a J_0 transverse Gaussian optical EH_{11} field mode that has an intensity maximum along the fiber hollow axis (see Fig. 8). The field is red-detuned *below* the atomic resonance, so that the atoms to be guided are attracted to the region of high field intensity and hence will tend to coast down the axis—held in place by the approximately harmonic potential of the transverse part of the propagating optical EH_{11} mode. An experiment at JILA (Renn *et al.*, 1995) has demonstrated that such a waveguide is operable (see Fig. 9). Unfortunately, with this OOL scheme —as with any red-detuning scheme—the atoms are localized in a region of high field intensity and hence are prone to excitement and consequent spontaneous decay with subsequent heating and loss of coherence. Hence, the OOL guide will have limited usefulness as a coherent waveguide for use, say, in the arms of an atom interferometer. It seems destined to function more as a "garden hose" for the atoms as they bounce ballistically and essentially classically down the fiber axis. For this reason the SMZ scheme is more appealing (see Fig. 10). Here, the field is blue-detuned above the atomic resonance and localized primarily in the fiber shell wall, undergoing total internal reflection off the shell–vacuum and the shell–cladding interfaces. This field then coats the walls of the hollow of the fiber with a blue-detuned, repulsive, evanescent fuzz that guides the atoms. However, now the atoms move primarily in a vacuum—interacting only very briefly on each bounce with the exponentially thin layer of evanescent field. For this reason—if the field is sufficiently far detuned—spontaneous emission can be minimized and coherence of the atom's de Broglie wavefunction preserved. Although this SMZ scheme is apparently more difficult to implement experimentally (Renn *et al.*, 1996; see Fig. 13), its fundamental appeal has stimulated a series of theoretical investigations of this and related waveguides (Marksteiner *et al.*, 1994; Jhe *et al.*, 1994; Harris and Savage, 1995; Ito *et al.*, 1995), which shall be discussed further in Section IV.B. The experimental observation of rubidium atoms guided by blue-detuned evanescent waves in a hollow fiber has been made by the JILA group of Renn *et al.* (1996) and independently by the Japanese–Korean collaboration of Ito *et al.* (1996). In addition, Ito *et al.* performed two-step laser photoionization spectroscopy on the guided atoms and utilized quantum state selectivity of the guiding potential to separate the two stable isotopes of rubidium from each other. We will discuss the JILA experiment is some detail in Section IV.C.

We also introduce in Section IV.C, for the first time, the idea of a parallel plane-mirror atom waveguide. In this setup, two blue-detuned evanescent fields are made on the glass surfaces bounding a two-dimensional atom waveguide (see Fig. 14). Unlike the Fabry–Pérot scheme of Wilkens *et al.* (1993), discussed in Section III, here we think of the atoms moving parallel to the mirror surfaces, rather than normal. Hence, the atoms are confined coherently to a two-dimensional space, leading to the possibility of investigating with a neutral vapor of atoms such two-dimensional quantum phenomena as anyonic statistics (Iengo and Lechner, 1992) and the Kosterlitz–Thouless Coulomb-gas phase transition (Minnhagen, 1987).

In Section VI, we discuss several convex, evanescent light-wave traps or guides in which at least one field is red-detuned and hence attractive, but a centrifugal force or a blue-detuned field provides a repulsive counterforce to allow the atoms to remain confined in stable orbits around the convex, dielectric, optical resonator. Prototypical of these is the dielectric microsphere trap of Mabuchi and Kimble (1994), where blue- and red-detuned optical whispering-gallery modes of dielectric microspheres are pumped by a laser. Since these modes propagate around the equator of the sphere by total internal reflection, they produce large, evanescent fields at the dielectric–vacuum interface (Treussart *et al.*, 1994). The idea is to red-detune one field so that, on the one hand, the atom is attracted to the sphere, but on the other hand it is repelled by centrifugal force (or a blue-detuned field) and hence orbits the microsphere (see Figs. 22 and 23). A new scheme to use this same mechanism to guide atoms along the outside of solid fibers—suggested here for the first time by us—will be presented in Section VI.B. A similar trapping mechanism has also been proposed to trap an orbiting atom around the strong evanescent Yukawa-type field that is emitted when a light-guiding fiber terminates in a sharp point or inverted cone (Pangaribuan *et al.*, 1992; Hori *et al.*, 1992).

So far, we have introduced some of the interesting things one can do with evanescent light-wave mirrors, without going much into the theory and development of the mirror concept itself. We rectify this now. As mentioned, the fact that an optical field exerts a gradient (dipole) force on an atom has been known for a long time (Ashkin, 1970, 1978). It was Cook and Hill who in 1982 suggested the evanescent light-wave atom mirror in its simplest form: a single laser beam totally internally reflecting off a dielectric interface. A simple modification is to have two counterpropagating beams reflecting instead. In this case, the beams interfere and produce a corrugation or sinusoidal oscillation in the evanescent potential field, which can be used to make a reflection atom-diffraction grating or a beam splitter, as first pointed out by Hajnal and Opat (1989), who predicted diffraction orders for grazing incidence thermal sodium atoms to be separated by relatively large angles of about 5 mrad. Hajnal *et al.* (1989) and Baldwin *et al.* (1990) searched unsuccessfully for these diffracted beams in an experimental setup similar to the original one of Balykin *et al.* (1987, 1988), also using sodium atoms at grazing incidence, but here off counterpropagating corrugated evanescent waves instead. The problem in observing this effect is attributable to the large Doppler shift in atomic velocities relative to the stationary grating, which weakens the diffraction pattern, as pointed out by Deutschmann *et al.* (1993a) in their dressed-state theoretical model of the diffraction process. Basically, in order to see a significant amount of diffraction, the Rabi frequency of the standing evanescent field—which is proportional to the laser intensity I—must be much larger than the Doppler shift. This criterion is difficult to satisfy for fast thermal atoms at grazing incidence, leaving the experimentalist three options: (1) Crank up the intensity, (2) slow down the atoms, or (3) translate the grating along with the atoms to reduce the Doppler shift. Since the production of the evanescent wave from cw dye lasers is already quite power consuming, the amount of power increase to be had is limited. The first successful solution was to move the grating, as was done by Stenlake *et al.* (1994). This they accomplished by detuning the two counterpropagating beams slightly from each other so that the beat frequency caused the grating to move transversely along the surface in the same direction as the incident atoms, thereby reducing the Doppler shift dramatically. The observation of different orders of diffraction other than the first was not conclusive in this experiment, however. Very clear second diffraction orders were obtained by Christ *et al.* (1994), who proceeded by slowing down the atoms rather than speeding up the grating. Using a beam of metastable neon atoms—slowed to a velocity of 7000 cm/sec by the large magnetic field gradient of a Zeeman-slower—this group observed clear signatures of both the first and second diffraction orders, separated by about 50 mrad. Since then, there have been several quantum theoretical investigations of the corrugated evanescent mirror, functioning as a diffraction grating (Tan and Walls, 1994; Murphy *et al.*, 1993, 1994; Feron *et al.*, 1994; Savage *et al.*, 1995) and as a beam splitter (Deutschmann *et al.*, 1993b).

In addition to making a mirror grating that oscillates periodically in space—to produce ordinary spatial diffraction—one can also consider the complementary scenario in which the evanescent mirror is spatially uniform but periodically modulated in time to produce a temporal "diffraction." Precisely this was done in the atom phase modulation experiment of Steane *et al.* (1995), in which they used an acousto-optic modulator to harmonically oscillate the evanescent-mirror potential height. The initially monochromatic atom de Broglie waves were reflected with quantized sidebands, now introduced in the time of flight signals. [The phase shift of the atom's de Broglie wave upon reflection has been calculated by Henkel *et al.* (1994).]

The evanescent-wave mirror can also be used as a quantum state selector, reflecting atoms in one quantum state specularly while scattering identical atoms in a different quantum state diffusely. This was demonstrated experimentally early on by Balykin *et al.* (1988), who observed a ratio of nearly 100 between the reflection coefficients of sodium atoms in the $F = 2$ versus the $F = 1$ ground state sublevels. A fully quantum theory of this and related phenomena was developed by Zhang and Walls (1992, 1993). Ito *et al.* (1996) have used this idea—in conjunction with a blue-detuned, evanescent light-wave, hollow fiber atom waveguide—to selectively guide two stable isotopes of rubidium. Only one isotope is guided by the fiber, yielding an efficient isotope separator.

One of the principal experimental drawbacks of the evanescent light-wave mirror—as originally envisioned by Cook and Hill (1982)—is that it requires quite high laser power to produce a sufficiently large potential barrier to reflect atoms with any realistic component of velocity normal to the surface, while not introducing an unacceptable degree of spontaneous emission probability (see Section II). Hence, there has been a search for methods to enhance the evanescent field produced at the dielectric interface for a given laser power, so that low cost, low power diode lasers could be used instead of high power dye lasers. Quite a bit of success in this goal has been achieved by the use of surface-plasmon-enhanced evanescent fields, produced when a thin metallic coating is placed on the vacuum side of the usual vacuum–dielectric interface, as is discussed in Section II. This idea was first suggested and demonstrated experimentally by the group of Hänsch at the Max-Planck-Institut für Quantenoptik (Esslinger *et al.*, 1993). They were able to obtain large grazing incidence reflection angles of 2.5 mrad for thermal rubidium atoms, using only 6 mW of diode laser power. Further experiments on surface-plasmon-enhanced atom reflection were carried out with metastable neon by Feron *et al.* (1993), and with metastable argon by Seifert *et al.* (1994a).

A second approach to enhance the evanescent field, advanced by Kaiser *et al.* (1994) and Seifert *et al.* (1994b), is to coat the dielectric surface with a multilayer dielectric thin-film structure in order to form a waveguide (see Fig. 2). The laser beam tunnels into the waveguide, in which large fields then build up, presenting an enhanced evanescent wave to the vacuum. This approach has some advantages over surface plasmons in the degree that spontaneous emission probabilities can be kept low by careful design of the thin film, as discussed in Section II.E.

We now have finished a rather exhaustive historical introduction to the evanescent light-wave atom mirror and its many possible applications in atom optics. For the rest of this chapter, we shall focus on the use of the evanescent field for making atom mirrors, resonators, waveguides, and traps.

II. Atom Mirrors: A Brief Theoretical Review

A. OPTICAL FORCES ON A NEUTRAL ATOM

The force produced by an evanescent light field on an atom is just a special case of the so-called dipole or gradient force that acts on an atom in a spatially inhomogeneous field. There are a number of ways to derive this force, of which probably the simplest is a semiclassical approach due to Cook (1979).

Let the interaction potential term $V(\mathbf{R})$ between the field and the atom be, in the dipole approximation,

$$V(\mathbf{R}) = -\mathbf{d} \cdot \mathbf{E}(\mathbf{R}) \tag{1}$$

where $\mathbf{R}$ is the position operator for the atomic center of mass and $\mathbf{d}$ is the atomic dipole moment operator. Ehrenfest's theorem allows one to write the expectation value of the force on the atom as

$$\mathbf{F} = \langle \nabla(\mathbf{d} \cdot \mathbf{E}) \rangle \tag{2}$$

where the gradient operator is taken with respect to **R**. Since **d** does not depend on **R** (it is only a function of the atomic electron relative coordinate operator), Eq. (2) may be rewritten as

$$\mathbf{F} = \langle (\mathbf{d} \cdot \hat{\mathbf{e}}) \nabla E \rangle \tag{3}$$

where $\hat{\mathbf{e}}$ is a unit polarization vector (assumed to be also independent of **R**) and E is the field amplitude. Assuming that the dimensions of the atomic wavepacket are small compared with the characteristic lengths over which E varies [such as λ_{opt} the optical wavelength, or, for an evanescent wave, the penetration depth, $\delta \approx \lambda_{\text{opt}}/(2\pi)$], we may approximate Eq. (3) by

$$\mathbf{F} = \langle \mathbf{d} \cdot \hat{\mathbf{e}} \rangle \nabla E \tag{4}$$

and treat **R** as a classical variable. It is not clear how good an approximation this may be for some of the atom cavities to be considered in this chapter, where the atom wavepacket is delocalized over a distance large compared with the field's penetration depth.

The expectation value appearing in Eq. (4) involves the atomic dipole **d** induced by the field and, as such, depends on **R**. It can be calculated in a straightforward way from the optical Bloch equations under a number of assumptions. If only two levels, $|1\rangle$ and $|2\rangle$, are involved in the process, the matrix element $\langle 1|\mathbf{d} \cdot \hat{\mathbf{e}}|2\rangle \equiv \mu$ can be taken to be real, and the expectation value

$$\langle \mathbf{d} \cdot \hat{\mathbf{e}} \rangle = \mu(\rho_{12} + \rho_{21}) \tag{5}$$

where ρ_{12} and ρ_{21} are atomic density-matrix elements.

If $\omega_0 \equiv (E_2 - E_1)/\hbar$ is the natural frequency of the $|1\rangle \to |2\rangle$ transition, then in the absence of interaction, we have $\rho_{12} \propto e^{i\omega_0 t}$ and $\rho_{21} \propto e^{-i\omega_0 t}$. Similarly, assuming the field $E(t, \mathbf{R})$ is monochromatic, it can be written as a sum of two rapidly varying complex exponentials,

$$E(t, \mathbf{R}) = \tfrac{1}{2} E(\mathbf{R})(e^{-i\omega t + i\theta} + e^{i\omega t - i\theta}) \tag{6}$$

where the phase factor θ may be a function of **R** (e.g., for a standing wave $\theta = 0$, and for a running wave $\theta = \mathbf{k} \cdot \mathbf{R}$), and $E(\mathbf{R})$ is a real amplitude. Making the rotating wave approximation in Eq. (4), i.e., neglecting the terms that oscillate at or near twice the optical frequency, yields

$$\mathbf{F} = \tfrac{1}{2} \mu \rho_{12} (\nabla E + iE \nabla \theta) e^{-i\omega t + i\theta} + \text{c.c.} \tag{7}$$

The optical Bloch equations for a two-level atom at the point **R**, with the interaction of Eq. (1) and the field of Eq. (6), are readily written down. For an atom at rest, and for times longer than $1/\gamma$ (where γ is the decay rate of the upper level), one can substitute into Eq. (7) the steady state solution,

$$\rho_{12} = -\frac{\Omega}{\gamma^2 + 4\Delta^2 + 2\Omega^2}(2\Delta + i\gamma)e^{i\omega t - i\theta} \qquad (8)$$

where $\Delta = \omega - \omega_0$ is the laser detuning, and $\Omega(\mathbf{R}) = \mu E(\mathbf{R})/\hbar$ is the (position dependent) atomic Rabi frequency. Substituting this into Eq. (7) yields the force:

$$\mathbf{F} = \frac{\hbar}{\gamma^2 + 4\Delta^2 + 2\Omega^2}(-\Delta \nabla \Omega^2 + \gamma \Omega^2 \nabla \theta) \qquad (9)$$

The first term in this expression is the dipole or gradient force, which acts in the direction of the gradient of the field intensity, that is, either toward low field intensities, if the detuning Δ is positive (blue-detuned), or toward high field intensities if Δ is negative (red-detuned). The second term is the radiation pressure force, as can be seen from the fact that, for a traveling plane wave where $\theta = \mathbf{k}_{opt} \cdot \mathbf{r}$, it points in the direction of the wavevector $\mathbf{k}_{opt}$. In the present context, this is also often called the "spontaneous" force, since it is proportional to the spontaneous emission rate γ (see Gordon and Ashkin, 1980).

An interesting interpretation of the dipole force has been given by Dalibard and Cohen-Tannoudji (1985), who show that in the limit of very strong fields ($\Omega \gg \gamma$), or very large detunings ($\Delta \gg \gamma$) this force equals the gradient of the dressed-state energies, weighted by the relative populations of the dressed states at a given point in the field. They have also shown that the work done against the dipole force equals the change in energy of both the atom *and* the field, and that the latter is not negligible —even in the quasi-static limit. This explains why it is impossible to obtain the dipole force semiclassically from naive energetic considerations, e.g., as the gradient of the expectation value of the energy in Eq. (1). This difficulty is circumvented by the Ehrenfest theorem approach used here. (Of course, semiclassical theories are notorious for violating energy conservation anyway.) In addition to yielding a number of valuable insights, the dressed-state approach of Dalibard and Cohen-Tannoudji is probably the one best-suited to deal with transient regimes and moving atoms.

The expression of Eq. (9) is, strictly speaking, valid only for an atom at rest. Motion complicates matters in at least two ways: There is the Doppler shift of the atomic lines, and the finite response time of the atoms (of the order of $1/\gamma$), which causes the density-matrix elements ρ_{ij} to have, in general, at any point in space and time, values that are somewhat different from the steady state values corresponding to the instantaneous field at that point. A fully quantized field treatment—including atom motion effects—has been given by Gordon and Ashkin (1980); see also Dalibard and Cohen-Tannoudji (1985) for a careful study of velocity dependent corrections to the dipole force.

The velocity-dependent forces are in fact extremely important in the physics of atomic cooling, since they may lead to damping (or, conversely, to a "heating up") of the atomic velocity distribution. For small detuning, which is typically the case for cooling experiments, the first-order velocity-dependent corrections are of the order of $v/(\gamma\lambda)$. (Although, in general, instead of the wavelength λ one should consider any characteristic length scale over which the field changes.) For large detuning ($\Delta \gg \gamma$), which is typically the situation for atomic mirror experiments, the corrections are of the order of $v/(\Delta\lambda)$, and may be more easily neglected provided the detuning is high enough.

Quantized-field treatments show that the dipole force is associated with the process of absorption followed by stimulated emission, whereas the spontaneous force is associated with absorption of a photon that is traveling in the direction of $\mathbf{k}$ (or $\nabla\theta$) followed by spontaneous emission with equal probability in all directions. These treatments also show that, in addition to the average forces calculated, there are random fluctuations of both the dipole and the spontaneous force due to the discrete nature of the emission and absorption processes. Spontaneous emission, in particular, results in the atom being "kicked" around in random directions by the recoil of the emitted photons and heated up. It is interesting to note that, for the case of an atom interacting with thermal radiation, all these effects were originally calculated by Einstein (1917) in the classic paper in which he speculated for the first time on the quantized momentum carried by a light quantum.

B. THE EVANESCENT-WAVE ATOM MIRROR

Cook and Hill (1982) appear to have been the first to suggest the use of the dipole force, due to an evanescent light wave, to make a mirror for atoms. Assume that a plane wave, coming from the $y < 0$ half-space, is totally internally reflected at a dielectric–vacuum interface at $y = 0$ (see Fig. 1). The electric field of the transmitted evanescent wave is of the form

$$\mathbf{E} = \hat{\mathbf{e}} E_0 e^{-\alpha y} \cos(\omega t - k_{\text{opt}} x) \qquad (10)$$

which is of the form of Eq. (6) with $E = E_0 e^{-\alpha y}$ and $\theta = k_{\text{opt}} x$. The resulting dipole force is along the y axis, i.e., perpendicular to the

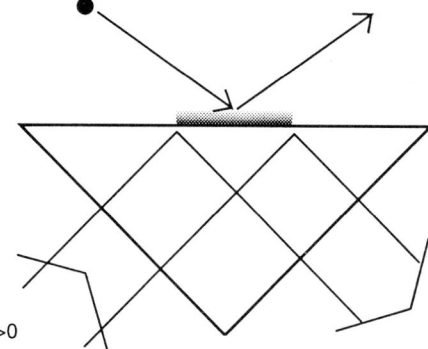

FIG. 1. Evanescent-wave atom mirror. The atom reflects off the strong field gradient on the vacuum side of the interface, which is due to the evanescent field of the internally reflected laser beam.

interface, and is given by

$$F_y = \frac{2\hbar \alpha \Delta \Omega_0^2}{\gamma^2 + 4\Delta^2 + 2\Omega_0^2 e^{-2\alpha y}} e^{-2\alpha y} \tag{11}$$

where $\Omega_0 = \mu E_0/\hbar$ is the Rabi frequency associated with the original (incoming) field. In particular, if the incoming field, propagating in the dielectric and undergoing total-internal reflection, has a dielectric electric field amplitude E_i associated with a Rabi frequency $\Omega_i = \mu E_i/\hbar$, then applying the Fresnel laws (Jackson, 1975) for transmission at the index n to the vacuum–dielectric interface gives a value of (Kaiser et al., 1994)

$$\Omega_0^2 = \frac{4n^2 \cos^2\theta_i}{(n^2 - 1)[(n^2 + 1)\sin^2\theta_i - 1]^p} \Omega_i^2 \tag{12}$$

where θ_i is the angle of incidence of the incoming beam in the dielectric, where $p = 0, 1$ for TE or TM polarizations, respectively.

The force of Eq. (11) by itself would result in specular reflection of an atom at the interface. There is, however, also the spontaneous force to consider. For the field of Eq. (10), the spontaneous force, the second term of Eq. (9), yields

$$F_x = \frac{\hbar k_{\text{opt}} \gamma \Omega_0^2}{\gamma^2 + 4\Delta^2 + 2\Omega_0^2 e^{-2\alpha y}} e^{-2\alpha y} \tag{13}$$

To have $|F_y| \gg |F_x|$, and hence nearly specular reflection, one therefore needs a large detuning, $\Delta \gg \gamma$ (compared with the natural width of the levels). Note that for an evanescent wave of the type described, i.e., produced by total internal reflection at a dielectric–air interface, the penetration depth $\delta \equiv 1/\alpha$ is essentially of the order of the wavelength, i.e., $\alpha \approx k_{opt} = 2\pi/\lambda_{opt}$. This follows from (see, e.g., Jackson, 1975),

$$\frac{1}{\delta} \equiv \alpha = \frac{2\pi}{\lambda_{opt}} \sqrt{n^2 \sin^2\theta_i - 1} \tag{14}$$

where θ_i is the angle of incidence.

As mentioned, there is also a fluctuating part to the spontaneous force, due to the recoil of the atom, every time a photon is scattered in a random direction. (There is also a fluctuating part to the dipole force, which we shall ignore here, although evidence suggests that it may be important in causing some diffusion of the reflected atom beam.) For atoms cooled to the recoil limit, these random kicks result in changes in velocity comparable with the atom's original velocity itself; i.e., they can substantially alter the trajectory of either the reflected or the incoming atoms. Thus, a further requirement for near-specular reflection to occur is that the atom should not spontaneously scatter a photon during the time τ it spends in the field, that is the product of $\gamma\tau$ with the probability to find the atom in the excited state should be small. The latter probability scales as $\Omega^2/(4\Delta^2 + 2\Omega^2)$ for large detunings, and hence we require

$$\frac{2\gamma}{\alpha v_\perp} \frac{\Omega_0^2}{4\Delta^2 + 2\Omega_0^2} \ll 1 \tag{15}$$

where the time τ spent by the atom in the field has been estimated as $\tau \approx 2/(\alpha v_\perp)$, with $v_\perp$ the component of the atom velocity perpendicular to the interface. As discussed above, for the velocity dependent forces to be negligible we require $\gamma \gg v/\lambda_{opt}$; hence, the only way to reduce the expression in Eq. (15) is to consider very large detunings, $\Delta \gg \gamma, \Omega_0$.

Finally, deviations from specular reflection may also arise from the dipole force itself, when the field incident on the surface is not a pure plane wave. Suppose, for instance, that it is a Gaussian laser beam and that the evanescent field has an additional x and z dependent envelope

$$\mathbf{E} = \hat{\mathbf{e}} E_0 e^{-\alpha y - x^2/w_x^2 - z^2/w_z^2} \cos(\omega t - k_{opt} x) \tag{16}$$

with beam widths w_x and w_z along the surface dimensions x and z (the value of w_x will depend also on the angle of incidence, if the plane of incidence is the xy plane, as assumed here). Equation (9) then yields a component of the dipole force parallel to the surface, in addition to the forces of Eqs. (11) and (13):

$$\mathbf{F}_{\text{dip}\parallel} = \frac{4\hbar\,\Delta\Omega^2(\mathbf{r})}{\gamma^2 + 4\Delta^2 + 2\Omega^2(\mathbf{r})}\left(\frac{x^2}{w_x^2}\hat{\mathbf{x}} + \frac{z^2}{w_z^2}\hat{\mathbf{z}}\right) \tag{17}$$

where $\hat{\mathbf{x}}$ and $\hat{\mathbf{z}}$ are unit vectors. The force of Eq. (17) is zero at the center of the laser beam and grows away from it. In essence, it would cause a reflected atomic beam to diverge slightly; i.e., the flat surface acts as a slightly convex, and probably somewhat astigmatic, mirror for the incoming atoms.

We may take the total vector $\mathbf{F}_{\text{dip}}$ as defining the effective "normal" to the atom mirror at any given point. This normal is tilted relative to the y axis by an angle of about $\tan(F_\parallel/F_y)$. Assuming that an atom coming in at normal incidence, a distance x off the axis is reflected specularly about the effective normal, the angle of reflection would equal $2F_\parallel/F_y = 4x/(\alpha w_x^2)$, for $z = 0$. This immediately yields an equivalent focal length for the divergent mirror along the xy plane of $f_x = \alpha w_x^2/4$. Since the penetration depth $\delta \equiv 1/\alpha \approx \lambda_{\text{opt}}/(2\pi)$ is of the order of the wavelength or smaller, this defocusing effect is typically quite small.

C. Effective Potential Treatment

Equation (11) shows that the dipole force can, in general, be derived from an effective potential:

$$V(\mathbf{R}) = \frac{\hbar\Delta}{2}\ln\left(1 + \frac{2\Omega^2(\mathbf{R})}{\gamma^2 + 4\Delta^2}\right) \tag{18a}$$

$$\underset{\Delta \gg \Omega}{\approx} \hbar\Delta\,\frac{\Omega^2(\mathbf{R})}{\gamma^2 + 4\Delta^2} \tag{18b}$$

$$\underset{\Delta \gg \Omega,\gamma}{\approx} \frac{\hbar\Omega^2(\mathbf{R})}{4\Delta} \tag{18c}$$

(chosen here so that it will vanish when the field intensity $\Omega = 0$). The usefulness of this effective potential for the evanescent-wave mirror application is that (among other things) it allows one to figure out easily what the maximum velocity of an atom can be in the direction perpendicular to the wall, if it is to be reflected. That is, the evanescent wave may be regarded as providing a potential hill for the atom whose maximum height $V_{\max}$ (at the wall itself) is given by Eqs. (18) with $\Omega = \Omega_0$, as determined by Eq. (12). If the incoming atom has a component of velocity perpendicular to the wall, $v_\perp$, such that $\frac{1}{2} M v_\perp^2 < V_{max}$, it will be reflected; otherwise, it will reach the wall and stick there because of the van der Waals forces, or scatter diffusely.

Because a large detuning is required to overcome the spontaneous force, the height of the barrier can be estimated from the approximation

$$V_{max} \approx \frac{\hbar \Omega_0^2}{4\Delta} \quad (19)$$

which holds provided $\Delta \gg \Omega_0, \gamma$, as per Eq. (18c).

As discussed earlier, in order to achieve near-specular reflection, it is important that essentially no spontaneous emission events occur during the time τ that the atom spends in the field, which results in the condition of Eq. (15). For a given value of the incoming atom's normal velocity, $v_\perp$, we can eliminate Ω_0 from this expression by setting it equal to the minimum value necessary to reflect the atom, i.e., by applying Eq. (19),

$$\Omega_0^2 \approx \frac{2 M v_\perp^2 \Delta}{\hbar} \quad (20)$$

Substituting Eq. (20) in the low spontaneous emission condition of Eq. (15) yields the condition

$$\frac{\gamma M v_\perp}{\alpha \hbar \Delta} \ll 1 \quad (21)$$

which shows that for a given detuning it is actually advantageous to have slowly moving atoms, even though they spend more time in the field overall, provided Ω_0 is reduced accordingly, since the probability of having the atom reach the upper state (and hence of being able to decay by spontaneous emission) decreases as Ω_0^2, but only linearly with Δ. Note that the condition of Eq. (21) is automatically satisfied, under the assumption that $\Delta \gg \gamma$, if the atom is cooled to the recoil limit, i.e., if $Mv \approx \hbar k_{opt}$, since α and k_{opt} are typically of the same order of magnitude.

Clearly, a purely geometric way to reduce $v_\perp$ is to try to reflect a beam of atoms at grazing incidence. This is, in fact, how the evanescent-wave atom mirror was first demonstrated, by Balykin *et al.* (1987, 1988). The reflection of normally incident atoms (cooled to about 25 μK in a magneto-optical trap) was first demonstrated by Kasevich *et al.* (1990).

Another obvious use of the effective potential concept, of particular relevance for the gravitational traps that we discuss here, is to estimate the height from which an atom can be dropped onto the atom mirror (as in the experiments by Kasevich *et al.*) and still be reflected.

Since the potential of Eq. (18) drops very fast (essentially exponentially) away from the wall, to a somewhat crude approximation it may be replaced by an infinite barrier—provided the incoming particle's energy is well below the actual barrier height, Eq. (19). For the calculation of the modal structure of gravitational traps, we shall use this approximation later. We expect it to be justified so long as the wavefunctions that we calculate have negligible tails inside the potential barrier. (In fact, the justification for this approximation has been worked out in detail by Chen and Milburn, 1995.) Nonetheless, it is still in reality a finite barrier, and tunneling to the wall (which exerts an attractive, very short range, van der Waals force on the atoms) cannot be ruled out, although tunneling times can be reduced exponentially by increasing the height of the barrier as needed. (An estimate of the tunneling probability can be found in Marksteiner *et al.*, 1994.)

Useful as it is, however, the effective potential is still only an approximation valid in the quasi-static regime of very slowly moving atoms. Precise theoretical calculations of the reflection of atoms off evanescent-wave mirrors must include velocity dependent effects and generally must be done numerically, typically as Monte Carlo (quantum trajectory) calculations. An example of such calculations is provided by the work of Seifert *et al.* (1994a), which makes use of the dressed-state approach mentioned earlier (see also Deutschmann *et al.*, 1993a; Tan and Walls, 1994; Savage *et al.*, 1995). Detailed calculations of the de Broglie wave phase shift upon reflection have been given by Henkel *et al.* (1994).

D. SURFACE-PLASMON-ENHANCED EVANESCENT WAVE

One problem with the original evanescent light-wave scheme of Cook and Hill (1982) is that the evanescent field strength, measured by Ω_0, is not very large. Hence, to reflect atoms with even modest velocities normal to the interface requires high power dye lasers. In addition, to reduce spontaneous emission, one requires a large detuning Δ, which reduces the potential $V(\mathbf{R})$ even further, as can be seen from Eq. (18c). Hence, it is extremely desirable to enhance the evanescent-wave field strength Ω_0, at a fixed detuning Δ and incident laser field strength, as measured by Ω_i, Eq. (12). If sufficient enhancement can be obtained, then low power diode lasers—rather than dye lasers—could be used, and spontaneous emission minimized. One way to do this is with surface plasmons.

Surface plasmons are electromagnetic charge-density waves propagating along a metallic surface (we quote here a standard reference, the book by Raether, 1988). If the dielectric surface, used in the atom-mirror scheme previously discussed, is covered by a thin metallic layer, the evanescent wave can excite a surface plasmon provided that it has the right polarization (electric vector in the plane of incidence, which yields a transverse magnetic (TM) polarized plasmon), and the right (i.e., resonant) wavelength. The propagation wavenumber k_{opt} that appears in the expression (10) for the evanescent wave is simply, by continuity, the x component of the wave inside the dielectric, i.e.,

$$k_{opt} = \frac{\omega}{c} n \sin \theta_i \qquad (22)$$

where n is the dielectric index of refraction. The surface-plasmon propagation wavenumber, on the other hand, is given by

$$k_p = \frac{\omega}{c} \sqrt{\frac{\varepsilon_1'}{1 + \varepsilon_1'}} \qquad (23)$$

where $\varepsilon_1 = \varepsilon_1' + i\varepsilon_1''$ is the relative permittivity of the metal. Equating Eqs. (22) and (23), so that $k_{opt} = k_p$, yields the resonance condition and defines the optimum angle of incidence. Under these conditions, a large fraction of the incoming light energy (possibly higher than 95%) is transferred to the plasmon wave. There is a correspondingly large evanescent wave on the air side of the metallic film, the intensity of which may exceed the simple dielectric case by as much as two orders of magnitude (limited somewhat by surface imperfections such as corrugations).

Plasmon-enhanced evanescent-wave mirrors have been demonstrated by Esslinger *et al.* (1993), Feron *et al.* (1993), and Seifert *et al.* (1994a). They might be very useful to achieve high reflectivity with low laser powers (such as, e.g., diode lasers).

E. DIELECTRIC-WAVEGUIDE-ENHANCED EVANESCENT WAVE

Another promising enhancement technique involves stratified dielectric media (Kaiser *et al.*, 1994; Seifert *et al.*, 1994b). In this idea, the original dielectric of index n is coated with two additional thin-film dielectric layers, first one of index $n_1 < n$ and then a second of index $n_2 > n_1$. Hence, since the top layer of index n_2 is bounded above by vacuum (of index 1), and below by index $n_1 < n_2$, it can act as an infinite planar dielectric slab waveguide (see Fig. 2). Now, light incident from the original dielectric of index n can still undergo total internal reflection at the $n \rightarrow n_1$ interface, since $n_1 < n$. However, the evanescent field produced in the n_1 region is now used to pump the n_2 waveguide, rather than to reflect atoms directly. Radiation that tunnels through the n_1 layer finds itself trapped in the n_2 waveguide layer, where it will undergo many total

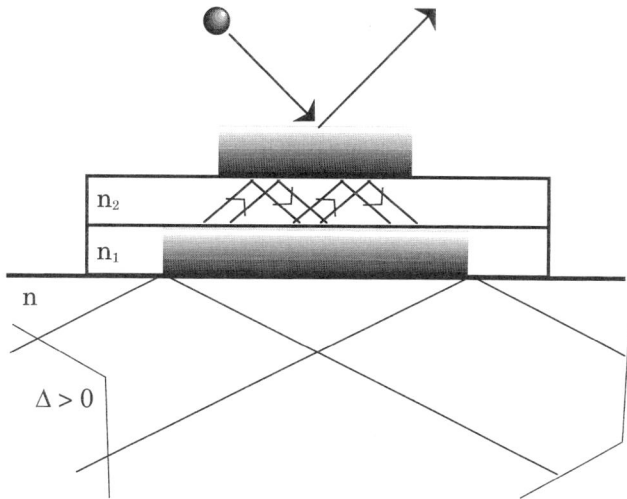

FIG. 2. Dielectric-waveguide-enhanced evanescent-wave atom mirror. Incident laser light (large arrows) from dielectric of index n reflects off a dielectric layer of index $n_1 < n$. This produces an evanescent wave (gradation) that tunnels through the n_1 layer into the n_2 layer ($n_2 > n_1$). Here, the light builds up to large intensities by multiple internal reflections and, hence, produces an enhanced evanescent wave in the vacuum, $n = 1$. The atom reflects off this blue-detuned field.

internal reflections back and forth between the $n_2 \to n_1$ and $n_2 \to$ vacuum, interfaces. For this reason, a very large field E_2 can build up in this dielectric layer—much larger than incident field E_i, or the field E_0, given implicitly by Eq. (12). Hence, the evanescent light wave that protrudes into the vacuum region has a much larger equivalent Rabi frequency, $\Omega_2 = \mu E_2 / \hbar$, than the $\Omega_0 = \mu E_0 / \hbar$ of before.

A detailed theoretical and experimental study of the dielectric-waveguide-enhanced evanescent mirror has been presented by Kaiser *et al.* (1994) and Seifert *et al.* (1994b), in which they carefully studied theoretically the minimization of spontaneous emission loss and experimentally observed the reflection from such a waveguide of metastable argon atoms with a normal velocity component of 300 cm/sec. The waveguide-enhanced scheme has two major advantages over the surface-plasmon method: (1) The decay length of the evanescent wave is not a material parameter—as in the case of surface plasmons—but a design parameter of the thin-film guide, and hence it can be chosen as small as possible to keep the atom–field interaction time τ (and thus the probability of spontaneous emission) very low. (2) There is much loss of power in the surface-plasmon technique due to ohmic heating of the metallic thin film, whereas the dielectric thin-film waveguide has a much lower loss tangent, defined by $\varepsilon''/\varepsilon'$, cf. Eq. (23), and hence much more of the incident field is coupled into the evanescent wave. Enhancement factors of several thousands seem possible using this technique.

F. OTHER ELECTROMAGNETIC MIRROR SCHEMES

The evanescent-wave-based atom mirrors have the potential advantage of trapping neutral atoms in the very large gradient of the evanescent wave, whose exponential decay away from the wall, over the dimensions of an optical wavelength, means that the atoms are for the most part unperturbed in the trap, except for brief impulse-like collisions with the mirrors. On the other hand, the presence of the glass surface complicates things somewhat, because of the van der Waals forces that can cause the atoms to stick to the walls or produce energy shifts that would affect high precision measurements. Several groups have proposed alternative atomic mirrors, also based on the gradient force, but exploiting instead the kind of intensity gradients that can be produced by focusing laser beams appropriately in free space.

One could envision, for instance, using cylindrical lenses to focus laser beams into "sheets" of light. A single such beam, with its elliptic equal-intensity contours, would naturally be a convex mirror. This is not necessarily an obstacle to the realization of resonant atom cavities (to be considered in the next section), since one could still use an unstable-resonator design (analogous to the ones used for some kinds of lasers) to confine the atoms for relatively long times (Wilkens *et al.*, 1993). Moreover, it has been shown by Davidson *et al.* (1995) that by combining two such sheets of light appropriately, one can build the equivalent of a concave atom mirror in free space.

The simplest geometry for this optical dipole trap (the trapping in the experiment is assisted by gravity) is as shown in Fig. 3. Two cylindrical laser beams, traveling along a common axis y, overlap with their long axes at right angles. Clearly, for an atom such as shown in the figure, and the blue-detuned case, there is always an upward vertical component to the dipole force, pushing it away from the region of high laser intensity below it. Moreover, for an atom sufficiently high above the xy plane, the figure also suggests that there will be a confining force in the x direction as well. Finally, it is somewhat less obvious, but nonetheless true, that a restoring force is also provided along the y axis itself by the diffractive spreading of the beam: For an atom that is essentially outside the beam at the focal plane, moving in the y direction results in its encountering a somewhat

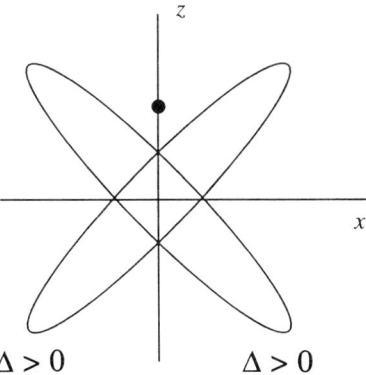

FIG. 3. A gravitationally assisted atom trap formed by two blue-detuned ($\Delta > 0$) cylindrical laser beams (the elliptical cross sections are shown). Gravity pulls the atom downward and helps trap it.

greater light intensity because of the beam spreading out in the vertical dimension on either side of the focal plane.

A simple mathematical analysis of this goes as follows. The (incoherent) superposition of the two Gaussian cylindrical beams results in an intensity

$$I \propto \frac{1}{\sqrt{w_a w_b}} \left[e^{-(x-z)^2/(2w_a^2)} e^{-(x+z)^2/(2w_b^2)} + e^{-(x+z)^2/(2w_a^2)} e^{-(x-z)^2/(2w_b^2)} \right] \quad (24)$$

where the beam waists w_a and w_b are functions of y given by $w_a^2 = a^2(1 + y^2/y_a^2)$ and $w_b^2 = b^2(1 + y^2/y_b^2)$, where a and b are the beam radii (the $1/e$ intensity points) on the focal plane, and $y_a = \pi a^2/\lambda_{\rm opt}$ and $y_b = \pi b^2/\lambda_{\rm opt}$ are the corresponding Rayleigh ranges. If $b \gg a$, we may as a lowest order approximation neglect the diffractive spreading along the long axes, i.e., treat w_b as a constant equal to b. Moreover, if we consider only values of x and z small compared to b, we may replace the exponentials involving w_b by unity in Eq. (24). Then, expanding in powers of x and y yields, to lowest order, yields

$$I \propto e^{-z^2/(2a^2)} \left[1 - \left(\frac{1}{4y_a^2} - \frac{z^2}{2a^2 y_a^2} \right) y^2 \right] \left[1 - \left(\frac{1}{2a^2} - \frac{z^2}{2a^4} \right) x^2 \right] \quad (25)$$

This explicitly shows that there is a crossover height z beyond which the atom is sufficiently outside the beams for the dipole force to act as a restoring force, both in the x (if $z > a$) and in the y (if $z > y_a/\sqrt{2}$) directions. The quadratic form of Eq. (25) shows that the restoring force is approximately harmonic in the x and y directions (but not, for a non-zero height, in the z direction, where it is given by a sharply rising Gaussian).

The result is somewhat like a concave mirror for the atoms, at least for sufficiently small x and y, albeit a highly astigmatic one since its focal length in the y direction is greater than in the x direction by about the square of the ratio of the Rayleigh range, y_a, to the beam width, a; this is typically a very large number unless the beam is brought to an extremely tight ($a \approx \lambda_{\text{opt}}$) focus. A more symmetric mirror would be achieved by a three-beam arrangement proposed by Davidson *et al.* (1995), in which three light sheets would be made to intersect at 90° to each other.

III. Atom Resonators: Fabry–Pérot Type

A. Cavities with Two Mirrors

The possibility of fabricating high reflectivity mirrors for atoms immediately suggests trying to make the equivalent, for matter waves, of the Fabry–Pérot interferometer for light. There have been at least a couple of proposals along these lines, by Balykin (1989), Balykin and Letokhov (1989a), and Wilkens *et al.* (1993). Apart from their fundamental interest, such cavities might be useful as velocity selectors for atoms, in the same way as the optical étalons may act as filters for light of a particular frequency and propagation vector. This is because of de Broglie's relationship,

$$\mathbf{k}_{\text{dB}} = \frac{M}{\hbar} \mathbf{v} \qquad (26)$$

between the atomic velocity $\mathbf{v}$ and the wavevector k_{dB} of the associated atom wave.

It is tempting to speculate about whether one might not take the analogy a step further and build something like an "atom laser" or "boser" (at least for bosonic atoms). Indeed, there has been a recent (and somewhat controversial) proposal for such a device (Wiseman *et al.*, 1995; see also Lenz *et al.*, 1994). In the absence, however, of the equivalent of stimulated emission for atoms, it seems that these atom cavities would be relegated essentially to the role of the so-called "passive" cavities in optics; i.e., they would act essentially as filters.

The equation to be solved to determine the spatial dependence of the atom energy eigenfunctions, in a cavity with perfectly reflecting mirrors, is (neglecting gravity) the same as for the optical case, i.e., the Helmholtz equation. Hence, one finds the same modes for the de Broglie matter waves as for the optical waves. For an elongated cavity (Fig. 4), such as considered by Balykin and Letokhov (1989a), this means a longitudinal mode structure involving the length L of the cavity as a characteristic

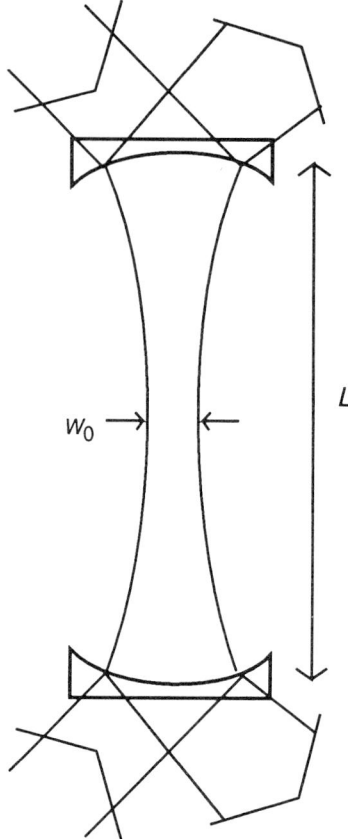

FIG. 4. Sketch of a conventional (stable-resonator) Fabry–Pérot atom cavity. The cavity uses blue-detuned, evanescent-wave atom mirrors (the coupling prism for the external laser beams is not shown).

length, as well as a transverse mode structure determined by diffraction and by geometrical factors involving the curvature of the mirrors. The characteristic transverse length scale is the waist of the principal mode, on the order of

$$w_0 \approx \sqrt{\frac{L\lambda_{dB}}{4\pi}} \qquad (27)$$

where $\lambda_{dB} = 2\pi/k_{dB} = h/(Mv)$ is the de Broglie wavelength of the atom wave. Equation (27) ignores geometrical factors that depend on the type of cavity considered (hemispherical, confocal, etc.). To within such factors, of the order of unity, Eq. (27) also yields the order of magnitude of the spot size at the mirrors for the lowest transverse mode; that is, the fundamental mode of the cavity would have an angular divergence of the order of

$$\Delta \phi \approx \frac{w_0}{L} \approx \sqrt{\frac{\lambda_{dB}}{L}} \qquad (28)$$

For a cavity of macroscopic dimensions on the order of 1 cm, such as the one considered by Balykin and Letokhov (1989a), this is necessarily a very small number. For example, for a sodium atom cooled to the Doppler limit, the corresponding velocity $v_\gamma = \hbar\gamma/(Mc) = 300$ cm/sec yields a de Broglie wavelength $\lambda_{dB} \approx 5 \times 10^{-7}$ cm. For a 5-cm cavity, this yields $\Delta\phi \approx 3 \times 10^{-4}$. Assuming that the atoms are somehow injected into the cavity with velocity components $v_\parallel$ along the axis of the cavity and $v_\perp$ perpendicular to the cavity axis, only the atoms satisfying $v_\perp/v_\parallel < \Delta\phi$ would go into the lowest transverse mode. To maximize the number of atoms satisfying this inequality, one might envision injecting the atoms with an appreciable velocity component along the cavity axis and then rapidly cooling them, using lasers, in the transverse directions; Balykin and Letokhov (1989a) discuss some schemes for doing this. The maximum longitudinal velocity that can be allowed in the cavity, however, is limited by the reflectivity of the atomic mirrors (as explained in the previous section); also, the minimum achievable transverse velocity is limited by the cooling technique (at most, one could hope for something of the order of the recoil velocity, $v_r = \hbar k_{opt}/M$, where $k_{opt} = 2\pi/\lambda_{opt}$ is the optical wavenumber. For the sodium example, v_r is of the order of 2.5 cm/sec). This would make it quite difficult to excite only the lowest transverse mode; for instance, for the sodium atom with $v_\parallel = v_\gamma$ and $v_\perp = v_r$ one still gets only $v_\perp/v_\parallel \cong 10^{-2}$, about 100 times larger than $\Delta\phi$. Thus, such a cavity would naturally be filled in a superposition of a large number of transverse modes.

Turning to the longitudinal modes, the atom wave quantization condition that $k_\parallel L = n\pi$, where $n \in \{1, 2, 3, \ldots\}$, as in the optical case, yields, together with Eq. (26), the allowed values for the longitudinal velocity,

$$v_\parallel = \frac{nh}{2ML} = n\left(\frac{\lambda_{opt}}{2L}\right)v_r \qquad (29)$$

where v_r is the recoil velocity (λ_{opt} is the wavelength of the cooling laser). Successive longitudinal modes therefore correspond to atomic velocities differing by $\Delta v = (\lambda_{opt}/L)v_r$, i.e., a very small fraction of the recoil velocity. For a macroscopic cavity, the velocity resolution necessary to operate in a single longitudinal mode may therefore be impossible to achieve with current techniques. Remarkably, however, Balykin and Letokhov (1989a) conclude that it should nonetheless be possible to achieve a large degree of degeneracy in some modes (that is, a large number of atoms all in the same mode) with some of the injection techniques they discuss. Essentially, what they show is that it might, in principle, be possible to increase the density of very cold atoms to the point where there are many more atoms than occupied modes (which would naturally result in some modes having a large amount of atoms).

It is interesting to note that the velocity resolution necessary to operate in a single longitudinal mode would, if extended to the transverse dimensions, lead to fairly large transverse mode sizes through the uncertainty principle. Specifically, if x is the transverse coordinate, an uncertainty in velocity $\Delta v_x = h/(2ML)$ implies $\Delta x \geq \hbar/(2M\Delta v_x) = L/(2\pi)$. That is, if the longitudinal temperature of the atoms is small enough to have a single longitudinal mode and the transverse temperature is as small or smaller, the transverse mode size muse be of the same order of magnitude as the cavity length itself.

From these considerations, experimentally, atoms in a multimode cavity would probably be in something close to a ballistic regime—i.e., they would form wavepackets of small dimensions compared with the cavity mode (similar to the pulses of light in a multimode laser). These packets would bounce off the mirrors in more or less classical-particle fashion. The foregoing arguments indicate that cavities in which the atoms would be stored in a true quantum regime (i.e., with their center of mass wavefunction spread coherently all over the cavity mode) would probably have to be quite small and with aspect ratios, for the cavity modes, not very far from unity. The cavity proposed by Wilkens *et al.* (1993) is of this type.

Instead of evanescent-wave mirrors, Wilkens *et al.* consider light-induced mirrors, as produced by the dipole force in strongly focused laser beams. As discussed in the previous section, such mirrors are naturally convex, leading to unstable resonator geometries, although crossed laser beams could be arranged to produce the concave light mirrors needed for stable resonator operation. Wilkens *et al.*, however, point out that an unstable cavity might have some advantages, such as allowing for an easy way to extract the atom waves, and also for the large transverse dimensions required for single longitudinal mode operation.

A laser beam focused to the diffraction limit, i.e., to a spot size of a diameter on the order of a wavelength, would provide a convex mirror of a radius of curvature $R \approx \lambda_{opt}$. For their estimates, Wilkens *et al.* assume a cavity formed by two such mirrors about $L \approx 10\lambda_{opt}$ apart. For such an unstable resonator, Eq. (27) would probably not be a very good estimate for the beam waist, since the geometrical factors neglected in this equation could be quite large. Wilkens *et al.* assume that the resonator is arranged so that $w_0 \approx L/(2\pi)$, the transverse position uncertainty calculated just for operation in a single longitudinal mode. Of course, one does not want the resonator to be so unstable that the atoms leave it after only a few bounces off the mirrors. Assuming the atom beam waist increases as it propagates according to the formula $w^2 = w_0^2(1 + z^2/z_R^2)$, with $z_R = \pi w_0^2/\lambda_{dB}$ for a Gaussian beam, one finds that the Rayleigh range z_R, over which the beam width increases appreciably, is of the order of

$$z_R \approx \frac{L^2}{4\pi\lambda_{dB}} \qquad (30)$$

A large number of bounces would be ensured if z_R could be made much larger than the cavity length. This clearly requires $L \gg \lambda_{dB}$, i.e., operation in a high longitudinal mode. Recalling also Eq. (29) for the nth longitudinal mode, we see that with $L \approx 10\lambda_{dB}$, a large value of n implies an atomic velocity $v \gg v_r$, the recoil velocity. Wilkens *et al.* estimate the largest longitudinal velocities that could still be reflected by the light mirrors, assuming a laser spectral power of $(10 \text{ mW})/\gamma$. They conclude that longitudinal mode numbers in the neighborhood of $n \approx 100$ would be feasible for a number of atomic species, and as high as $n \approx 200$ for sodium; this means that the cavity length L in Eq. (30) could be as large as $100\lambda_{dB}$, suggesting that at least tens of bounces could be achieved before the atom leaves the resonator. This is a relatively low finesse cavity, but it is certainly a good start.

It is worth wondering what might be achieved with a *stable* resonator configuration of the dimensions considered by Wilkens *et al.* The beam waist, as given by Eq. (27), can be rewritten

$$w_0 \approx \left(\frac{L\lambda_{dB}}{4\pi}\right)^{1/2} \left(\frac{v_r}{v}\right)^{1/2} \qquad (31)$$

With $L \approx 10\lambda_{dB}$, this becomes $w_0 \approx \lambda_{dB}\sqrt{v/v_r}$, which would be compatible with the condition $w_0 \approx L/(2\pi)$ for single longitudinal mode operation only if $v \approx v_r$. This would also imply, by Eq. (29), a relatively low order longitudinal mode. Such an atom cavity would begin to resemble some of the "particle-in-a-box" configurations to be discussed later. (See also the parallel-mirror waveguide, Section IV.C.)

There are a number of potential difficulties, in addition to the ones already discussed, regarding injection and cooling that need to be considered in a study of the feasibility of atom cavities. One of them is the possibility of losing the atoms as they tunnel through the light barrier (either to the glass surface, in the case of an evanescent-wave mirror, or simply to the other side of the focused laser beam). This is generally not viewed as a serious concern because the tunneling probability is an exponential function of the barrier energy, so that a relatively small increase in the light intensity can reduce the tunneling probability substantially. An estimate of this loss mechanism is given in Marksteiner *et al.* (1994).

More serious is the threat of spontaneous emission. Typically, the atom would spend most of its time in the field-free region between the mirrors and hence in the ground state; spontaneous emission, therefore, would only be possible during the time the atom is actually being reflected (estimates of this effect are given in Deutschmann *et al.*, 1993a; Seifert *et al.*, 1994b; Liston *et al.*, 1995b). Equation (29) indicates that, for any cavity bigger than the laser wavelength, the recoil from the emission of a single spontaneous photon would be enough to put the atom in another longitudinal mode. The transverse mode situation is less clear-cut: The maximum change in transverse angle upon emission of a photon, $\Delta \phi_e \approx v_r/v$, is to be compared with the acceptance angle $\Delta \phi$ of the lowest transverse mode, Eq. (28). To have $\Delta \phi_e < \Delta \phi$ requires $v_r/v < \lambda_{dB}/L$, a condition easily achieved in the small cavity considered by Wilkens *et al.* but not in the macroscopic cavities.

In any event, it is clear that spontaneous emission is a significant threat to single mode operation of an atom cavity. For their macroscopic cavity, Balykin and Letokhov conclude that spontaneous emission would be the main factor limiting the lifetime of the atom in the cavity in the single transverse mode regime (see also Liston *et al.*, 1995b), whereas in a multimode regime, with additional transverse cooling of the atoms, the main factor would likely be collisions with background gas. We discussed in the previous section ways to minimize the probability of a spontaneous emission event during the interaction of the atom with the light-induced mirror. Wilkens *et al.* (1993) provide detailed estimates of the detunings necessary to limit spontaneous emission rates to tolerable levels for their proposed cavities, for different atomic species; they conclude that detunings as large as 10 Rabi frequencies might be necessary for the strongest transitions, such as the sodium *D* line.

Finally, we have ignored so far the effect of gravity, but this would clearly be a factor for very slow atoms, especially if they have to travel over macroscopic distances. Accordingly, Balykin and Letokhov (1989a) propose arranging the cavity vertically (to prevent the atoms from dropping out in transit from one mirror to the other) and point out that in order for the atoms to reach the upper mirror their velocity must exceed $v_{min} = \sqrt{2gL}$. For a 5-cm cavity, $v_{min} = 100$ cm/sec, which is only slightly smaller than the Doppler-limited velocity for sodium, namely, $v_\gamma = 300$ cm/sec, discussed earlier, and is substantially larger than the recoil velocity. With this constraint, $v > v_{min}$, they include the effects of gravity on the cavity mode structure by using a WKB-type approximation, i.e., treating it as an effective refractive index for the de Broglie waves.

The restriction $v > v_{min}$ is naturally much less severe for the much smaller cavity of Wilkens *et al.*, although here, too, a detailed study of the atom wavefunctions would require the inclusion of gravity if the atoms are going to be kept bouncing back and forth between the mirrors for any substantial length of time.

B. Gravitational Cavity with Parabolic Mirror

In 1992, Wallis *et al.* suggested turning the gravitational constraint, just discussed, into an asset by simply removing the upper mirror in the vertical cavity considered by Balykin and Letokhov (1989a) and letting gravity confine the atoms in the vertical direction (Fig. 5). The length L of the cavity is then determined effectively by the energy of the atom, i.e., by how high it can rise; for an atom of velocity v (at the mirror), this classical turning point is

$$L = \frac{v^2}{2g} \tag{32}$$

Thus, as before, for an atom moving at 100 cm/sec, we find an effective cavity length of $L = 5$ cm, whereas the values of $z_E = 5$ mm, discussed by Wallis *et al.*, correspond to atom velocities of roughly 30 cm/sec.

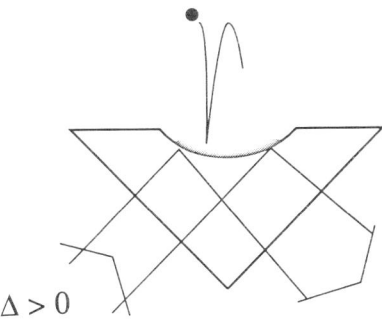

FIG. 5. A gravitational trap with a parabolic blue-detuned, evanescent-wave atom mirror.

For their discussion, Wallis *et al.* assumed the lower mirror to be a paraboloid of revolution, with an equation

$$z = \frac{x^2 + y^2}{2R} \tag{33}$$

The constant R equals the radius of curvature of the mirror at its center. The parabolic shape has the advantage that the equations of motion become separable in parabolic coordinates. The problem of the particle bouncing, under the influence of gravity, onto a mirror that is a surface of revolution of arbitrary shape has in general two constants of the motion. These are the total energy E and the z component of the angular momentum, L_z (the former because the reflection off the mirror is an elastic collision, the latter because of the rotational symmetry around the z axis). For the motion of a particle in a linear potential, there also exists (much like the Runge–Lenz vector for the Kepler problem) a third constant of motion, K, in addition to the energy and angular momentum, given by

$$K = \left[\mathbf{p} \times \mathbf{L} + \frac{M^2 g}{2} \mathbf{r} \times (\mathbf{r} \times \hat{\mathbf{z}}) \right] \cdot \hat{\mathbf{z}} \tag{34}$$

where $\hat{\mathbf{z}}$ is a unit vector in the z direction. For a parabolic mirror, and only for that case (Korsch and Lang, 1991), this quantity is also conserved upon reflection. One has then three constants of the motion, which lead to complete separability of the problem in the appropriate coordinates—in this case, parabolic coordinates.

Although a sufficiently large parabolic mirror could in principle trap atoms of any initial transverse velocity, for most of their analysis, Wallis *et al.* (1992) concerned themselves only with paraxial trajectories, i.e., those confined to a small region near the center of the mirror and having a small ratio of transverse to longitudinal velocity. There is a good experimental reason to try to keep the particle bouncing near the center of the mirror: If the mirror is to be realized by shining light from the glass side and using the evanescent-wave force, its actual size is limited by the laser spot size, and moreover, the total internal reflection condition could not be satisfied arbitrarily far away from the vertex of the mirror since the slope of the parabola, Eq. (33), increases as w/R (where w is either x or y). It is thus essential to keep w/R small. It seems reasonable to assume that for the paraxial modes it would not make a difference, in practice, whether the mirror is parabolic or spherical.

For paraxial motion, one has again a relatively clear separation between the transverse and longitudinal characteristics of the atom de Broglie waves. The longitudinal part of the problem has an energy spectrum essentially given by the exactly solvable (see, e.g., Sakurai, 1985, Section 2.4) one-dimensional problem of a particle bouncing on a flat mirror under the influence of gravity. We discuss the exact solutions to this problem in terms of Airy functions in Section V.A, but we point out here that the WKB approximation to the spectrum,

$$E_n = Mgl_g \left[\frac{3\pi}{2} \left(n - \frac{1}{4} \right) \right]^{2/3} \quad (35)$$

is excellent even for small values of the longitudinal mode number n, and essentially exact for the large values of n required for the paraxial motion. Here we have introduced the *characteristic gravitational length* l_g (z_0 in the notation of Wallis *et al.*), defined by

$$l_g \equiv \left(\frac{\hbar^2}{2gM^2} \right)^{1/3} \quad (36)$$

The quantity l_g gives the order of magnitude of the energy of the ground state wavefunction of a gravitationally bound particle resting against a flat surface, as can be seen by minimizing the total energy $Mgz + p_z^2/(2M)$ with the constraint $zp_z = \hbar/2$. Typical values of l_g for several atomic species are in the range of micrometers to tenths of micrometers (see Table I in Section V). Equation (36) shows that an atom dropped a distance L above the mirror would have a longitudinal quantum number of the order of $n \approx (L/l_g)^{3/2}$; moreover, the spacing between successive energy levels of the longitudinal motion decreases as $n^{-1/3}$, so that an uncertainty Δz in the initial height of the atom above the surface translates into a superposition of

$$\Delta n \geq n^{1/3} \left(\frac{\Delta z}{l_g} \right) \approx \sqrt{\frac{L}{l_g}} \left(\frac{\Delta z}{l_g} \right) \quad (37)$$

longitudinal modes. This can be a large number if the atoms are dropped, e.g., a few millimeters above the surface.

Equation (37) is only a rough estimate that neglects the effect of the initial atom momentum distribution. This can be accounted for by assuming some initial distribution of positions and momenta, calculating the corresponding rms energy spread [with the energy given by $E = Mgz + p_z^2/(2M)$] and setting Δn equal to the energy spread divided by the density of modes, $\partial E/\partial n$, around the energy E. For a Gaussian distribution of position and momenta, Wallis et al. obtain, in this fashion,

$$\Delta n = \frac{1}{\pi} \sqrt{\frac{L}{l_g}} \sqrt{2\left(\frac{l_g \Delta p_z}{\hbar}\right)^4 + \left(\frac{\Delta z}{l_g}\right)^2} \quad (38)$$

Equation (38) shows explicitly that it is not really advantageous to attempt to reduce Δz below the value l_g; this would introduce a large momentum uncertainty in the original distribution, a correspondingly large energy uncertainty, and hence again a large spread in n.

The transverse length scale for the atomic modes in the parabolic trap is somewhat less obvious than the longitudinal one. For paraxial modes, it can be estimated from the following approach. Consider a classical particle bouncing on the mirror; assume that at some time it has just left the mirror surface from a point a horizontal distance x from the center, with a total speed v and at an angle θ to the right of the vertical. It will land on the mirror again at a point $x' = x + (v^2/g)\sin 2\theta \approx x + 2v^2\theta/g$ (the paraxial approximation has been used to neglect the vertical rise of the curved mirror at that point; this would result in a correction of order x^2). The local normal at the point x' makes an angle $\alpha = -x'/R$ with the vertical (negative, i.e., to the left of the vertical, if x and θ are both positive); the atom lands at an angle $-\theta$ to the vertical and is then reflected at an angle $\theta' = -(2\alpha - \theta)$. The coordinate change from (x, θ) to (x', θ') can be written as

$$\begin{pmatrix} x' \\ \theta' \end{pmatrix} = \hat{M} \begin{pmatrix} x \\ \theta \end{pmatrix} \quad (39)$$

with a matrix $\hat{M}$ defined by

$$\hat{M} = \begin{pmatrix} 1 & 2v^2/g \\ -2/R & 1 - 4v^2/(Rg) \end{pmatrix} \quad (40)$$

The eigenvalues of Eq. (40) are

$$\lambda_\pm = 1 - \frac{2v^2}{Rg} \pm i \sqrt{\frac{4v^2}{Rg} - \left(\frac{2v^2}{Rg}\right)^2} \quad (41)$$

assuming that the condition

$$v^2 < Rg \quad (42)$$

holds, in which case both eigenvalues have unit modulus and the iteration of Eq. (39) is stable. For an atom dropped with essentially zero transverse velocity from a distance L above the mirror, the speed v with which it hits the mirror is $\sqrt{2Lg}$. The condition of Eq. (42) amounts to $L < R/2$; i.e., the atom should be dropped below the focus of the mirror.

Unnormalized eigenstates of $\hat{M}$ are given by

$$\mathbf{u}_\pm = \begin{pmatrix} -\dfrac{v}{g}\left(v \pm i\sqrt{Rg - v^2}\right) \\ 1 \end{pmatrix} \qquad (43)$$

Consider now an atom dropped a distance L above the mirror and a distance x_0 to the right of the mirror axis, with a small transverse velocity v_0. It first hits the mirror at $x = x_0 + v_0\sqrt{2L/g}$ and $\theta = -v_0/\sqrt{2Lg}$, where again the paraxial approximation has been used. If we express this initial condition as a linear combination $A\mathbf{u}_+ + B\mathbf{u}_-$ of the eigenstates of Eq. (43), we can easily find an expression for $x^{(N)}$ after N bounces; it will be the first component of the vector

$$\begin{pmatrix} x^{(N)} \\ \theta^{(N)} \end{pmatrix} = A\lambda_+^N \mathbf{u}_+ + B\lambda_-^N \mathbf{u}_- \qquad (44)$$

The result is

$$x^{(N)} = \left(v_0\sqrt{\dfrac{2L}{g}} + x_0\right)\cos N\phi$$

$$+ \left(x_0\sqrt{\dfrac{2L}{R-2L}} - v_0\sqrt{\dfrac{R-2L}{g}}\right)\sin N\phi \qquad (45)$$

where $\phi = \arg \lambda_+$, and where we have set $v^2 \approx 2Lg$ in Eq. (43). The maximum transverse displacement on the mirror, x_M, is now easily obtained from Eq. (45) by combining the sine and cosine terms into an overall sinusoidal function of amplitude:

$$x_M^2 = \dfrac{x_0^2}{R-2L} + \dfrac{Lv_0^2}{g} \qquad (46)$$

The maximum transverse width of the bundle of trajectories generated by $\hat{M}$ at the height L of the turning points can be estimated along similar lines. Consider an atom starting at the mirror at point x and projected at an angle θ. When it reaches its maximum height, its transverse position is given by $x + v^2\theta/g$, in the small-angle approximation. Hence, we now use the solution of Eq. (44) to calculate the maximum value of $x^{(N)} + v^2\theta^{(N)}/g$, and the result is

$$x_S^2 = x_0^2 + \frac{v_0^2}{g}(R - 2L) \qquad (47)$$

Equations (46) and (47) agree exactly with the results of Eqs. (37) and (39) in the paper of Wallis *et al.* (1992), obtained from a more complicated analysis of the classical motion using parabolic coordinates and the constants of the motion. Our derivation shows explicitly that Eqs. (46) and (47) hold in the paraxial limit, regardless of the specific shape of the mirror (i.e., whether it is parabolic or spherical).

This classical analysis can be used to estimate the transverse width of the corresponding quantum wavefunction for the atom. If we now minimize Eqs. (46) and (47), under the constraint that the transverse position and velocity cannot be effectively smaller than allowed by the uncertainty principle, i.e., $x_0 v_0 = \hbar/(2M)$, we then obtain the transverse width at the mirror,

$$x_M = \left(\frac{2 l_g^3 R^2}{R - 2L} \right)^{1/4} \qquad (48)$$

and at the height L above the mirror (about at the classical turning point), we have

$$x_S = \left[2 l_g^3 (R - 2L) \right]^{1/4} \qquad (48)$$

This estimate for the characteristic transverse length scale of the quantum problem agrees with the result of the fully quantum analysis of Wallis *et al.* In fact, they show that in the paraxial limit, and in parabolic coordinates, the transverse equation is that of a harmonic oscillator with a characteristic length scale (width of the ground state wavefunction) given by Eq. (48).

With l_g of the order of micrometers and mirror radii of the order of centimeters, Eqs. (48) yield characteristic transverse length scales of the order of 10^{-3} cm or so. This happens to be of the same order of magnitude as the transverse mode width for the Fabry–Pérot cavity of Balykin and Letokhov (1989a), Eq. (27). Hence, the problems of operating the cavity in a single transverse mode configuration remain the same for both schemes. Suggestions for increasing the degeneracy factor (number of atoms per mode) proposed by Wallis *et al.* include a combination of repeated cavity fillings with the introduction of mode-selective elements.

There are a number of idealizations in the preceding analysis. Perhaps the most important one is the neglect of spontaneous emission. This emission is only possible when the atom is interacting with the evanescent wave, since only then can the atom become excited. The time spent by the atom in the field of the evanescent wave is of the order of $2\delta/v \approx 2\delta/\sqrt{2gL}$, (where $\delta \approx 1/\alpha$, Eq. (14), is the penetration depth of the evanescent wave). The probability of spontaneous emission over this time is, for a far-detuned atom,

$$\frac{2\gamma}{\alpha\sqrt{2Lg}} \frac{\Omega_0^2}{4\Delta^2} \tag{49}$$

which should be compared with Eq. (15). Dividing this by the time between bounces, $\Delta t = 2\sqrt{2L/g}$, one obtains something like an effective decay rate Γ for the atom cavity modes (since spontaneous emission events typically result in the loss of the atom), namely,

$$\Gamma \approx \frac{\gamma \Omega_0^2}{8\alpha L \Delta^2} \cong \frac{\gamma Mg}{2\alpha\hbar\Delta} \tag{50}$$

making use of the fact that the maximum effective potential due to the evanescent wave at the mirror—$\hbar\Omega_0^2/(4\Delta)$, from Eq. (19)—should at least equal MgL, the gravitational potential energy for the atom to be reflected. An interesting consequence is that this decay rate Γ turns out then to be independent of the atom's energy. These and other effects due to the finite width of the evanescent wave (including corrections to the mode structure) have been discussed by Liston et al. (1995a, b).

An experiment with a gravitational cavity has been reported by Aminoff et al. (1993b), in which a cloud of cold cesium atoms was dropped onto the mirror and up to 10 successive bounces observed (see Fig. 6). (Preliminary reports of a similar experiment had also been given earlier by the NIST group of Helmerson et al., 1992, and Aminoff et al., 1993a.) In this experiment of l'Ecole Normale, the setup was similar to that of the atom trampoline of Kasevitch et al. (1990). The atoms are cooled in a magneto-optical trap and dropped from a height of 3 mm. About 39% of the atoms in the cloud are lost after each bounce, with losses due primarily to recoil from photon scattering and the finite mirror spot size (see Fig. 6). One drawback in this and related experiments is that the atoms are detected using probe lasers that monitor the atoms by resonance fluorescence or absorption—processes that contribute to heating and spontaneous emission decoherence. A recent proposal by Aspect et al. (1995) is to monitor the atom bounces by phase changes they induce in the laser beam that is producing the evanescent mirror. This method of detection is reminiscent of quantum nondemolition techniques, in that information such as atom number is obtained without collapsing the atom's de Broglie wavefunction.

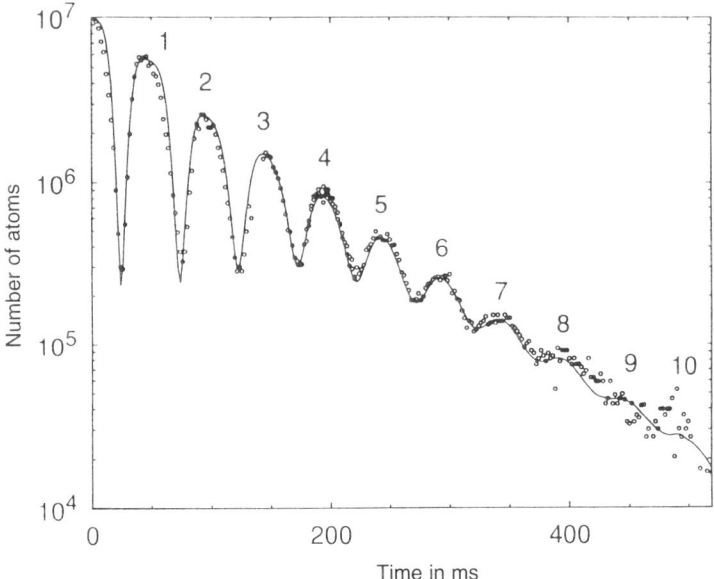

FIG. 6. Experimental observation of 10 bounces of supercooled cesium in a parabolic evanescent mirror cavity, made by the *Ecole Normale* group. The mirror consisted of 800 mW of laser power with a blue-detuning of $\Delta = 1.9$ GHz. The atoms were cooled in a magneto-optical trap to 4 μK and dropped onto the mirror from a height of 2.91 mm. The curve shows about a 39% loss rate of atoms from the trap per bounce (from Aminoff et al., 1993b, reprinted with permission).

C. RED–BLUE PUSHME–PULLYOU RESONATOR

In 1991, Ovchinnikov et al. proposed an atom resonator that combined red- and blue-detuned evanescent light waves at a dielectric interface, as shown in Fig. 7. Since the same physics in this scheme is found in the microsphere whispering-gallery trap of Mabuchi and Kimble (1994)—and other convex atom resonators—discussed in Section VI, we review it lightly here. The basic idea is to totally internally reflect a red-detuned ($\Delta < 0$) laser beam off the dielectric vacuum interface at a very small angle of incidence $\theta_r \gtrsim \theta_c$, where θ_c is the critical angle for total internal reflection, $\theta_c = \arcsin(1/n)$ where n is the dielectric index of refraction. Then, at the same point on the interface, a blue-detuned ($\Delta > 0$) beam is reflected at a much larger angle of incidence, $\theta \gg \theta_c$. From Eq. (14), the penetration depth δ of the evanescent field into the vacuum is

$$\delta \equiv \frac{1}{\alpha} = \frac{\lambda_{\text{opt}}}{2\pi\sqrt{n^2 \sin^2 \theta_i - 1}} \qquad (51)$$

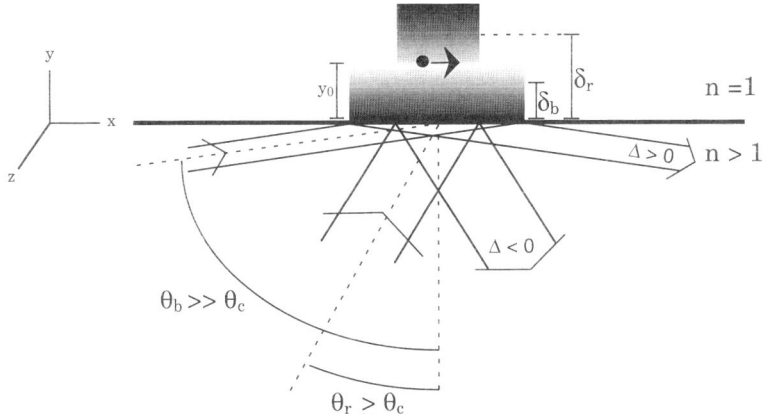

FIG. 7. The "pushme–pullyou" planar waveguide or trap Ovchinnikov et al. (1991). A red-detuned laser ($\Delta < 0$) totally internally reflects at a small incidence angle $\theta_r \gtrsim \theta_c$, where θ_c is the critical angle for total internal reflection. In addition, a blue-detuned beam ($\Delta > 0$) reflects at a large incident angle $\theta_b \gg \theta_c$. From Eq. (51), the red-detuned evanescent field has a much larger vacuum penetration or "skin" depth than the blue-detuned one, $\delta_r \gg \delta_b$. The atom is pulled by the red-detuned field and pushed by the blue-detuned one and hence remains confined to a plane of height $y = y_0$, given by Eq. (53).

where $\theta_i \in \{\theta_r, \theta_b\}$. For $\theta_r \gtrsim \theta_c$, the denominator of Eq. (51) is nearly singular, and hence this gives a δ that is very large—in practice, several optical wavelengths λ_{opt}. For $\theta_b \gg \theta_c$, the corresponding δ is a much smaller proper fraction of a wavelength. Hence, a layer of attractive red-detuned evanescent fuzz becomes sandwiched between the vacuum above and a layer of repulsive blue-detuned fuzz below, forming a two-dimensional Fabry–Pérot cavity (Fig. 7). Therefore, the atom will be *pulled* into the red-detuned layer and *pushed* away from the dielectric by the blue-detuned layer. Hence, we dub this the "pushme–pullyou" mechanism, after the fictional two-headed, pushing-pulling, llama-like beast from the Dr. Dolittle stories (Lofting, 1922). In the large detuning limit ($\Delta \gg \gamma, \Omega_r, \Omega_b$), we may approximate the vertical potential from Eq. (18c) as

$$V(y) = \frac{\hbar}{4}\left\{ \frac{\Omega_b^2}{|\Delta_b|} e^{-2\alpha_b y} - \frac{\Omega_r^2}{|\Delta_r|} e^{-2\alpha_r y} \right\} \quad (52)$$

where Ω_b and Ω_r are the blue- and red-detuned Rabi frequencies, respectively, computed from the incident field intensity via Eq. (12). (We tacitly assume here that the atom is at least a three-level system with two different resonant transitions near λ_b and λ_r, as per Mabuchi and Kimble, 1994.) Here, the inverse skin depths, $\alpha_b \equiv 1/\delta_b$ and $\alpha_r \equiv 1/\delta_r$, are computed in Eq. (51), whereas Δ_b and Δ_r are the blue and red detunings, respectively, and we have assumed that beat interference effects between the red- and blue-detuned fields average out. Assuming Δ_b and Δ_r to be large enough so that we may neglect gravity and van der Waals forces, the potential has a minimum at a height y_0, given by

$$y_0 = \frac{1}{2(\alpha_b - \alpha_r)} \ln \left[\frac{\alpha_b \Omega_b^2 |\Delta_r|}{\alpha_r \Omega_r^2 |\Delta_b|} \right] \qquad (53)$$

where we have assumed $\alpha_r < \alpha_b$, as described in the setup. Hence, atoms incident from above, with a velocity $\mathbf{v} = -v_y \hat{\mathbf{y}}$, will see the potential as a sort of reflection Fabry-Pérot if their kinetic energy exceeds the potential depth. Atoms with a velocity $\mathbf{v} = v_x \hat{\mathbf{x}} + v_z \hat{\mathbf{z}}$ will be guided along the $y = y_0$ plane in a fashion similar to the parallel waveguide to be discussed in Section IV.C. As mentioned in the introduction, this pushme-pullyou scheme has the drawback that the atoms spend a lot of time in the high intensity red-detuned region and hence will suffer much spontaneous emission heating and decoherence, as in the red-detuned hollow fiber guides to be discussed in Section IV. This effect can be offset somewhat by using very high power lasers or enhanced evanescent waves, which would allow for larger detunings and hence a lower photon absorption probability.

The pushme-pullyou resonator can also act as a trap, as well as a waveguide and Fabry-Pérot, as emphasized by Ovchinnikov *et al.* (1991). If the reflecting laser beams have transverse Gaussian profiles, then if the lasers are focused to a very small spot size on the dielectric, with a diameter on the order of λ_{opt}, there will be a large transverse gradient force of the form $\mathbf{F}_\parallel$, Eq. (17). From this equation, we see that $\mathbf{F}_\parallel \propto \hat{\boldsymbol{\rho}} \Delta$, where Δ is the detuning and $\hat{\boldsymbol{\rho}}$ is a transverse radial unit vector with origin at the spot center. Since the supercooled atoms spend most of their time in the high intensity, red-detuned region, the red-detuned transverse force will be the dominant one, and hence $\mathbf{F}_\parallel$ will be attractive toward the spot center axis, since $\Delta_r < 0$, as is clear from Eq. (17). Thus, confined below by blue-detuned pushes, and confined above and transversely by red-detuned pulls, the atom finds itself in a three-dimensional pushme-pullyou trap. For sodium atoms in such a trap, with blue- and red-detuned laser powers of $P_b = 640$ mW and $P_r = 40$ mW, respectively, Ovchinnikov *et al.* (1991) have estimated trapping times of about one second in such traps before spontaneous emission diffusion knocks them out. From the discussion of diffusion in Section IV.A, and taking Eqs (63) and (64), we can see that this time can be lengthened by increasing laser power (or enhancing the evanescent waves) and by increasing the detuning. The basic trapping mechanism of the Mabuchi–Kimble (1994) whispering-gallery microsphere trap—discussed in Section VI.A—operates on this same pushme–pullyou principle.

IV. Atom Waveguides

A. Red-Detuned, Propagating Light-Wave, Hollow Fiber Guides

The first scheme for guiding atoms we shall discuss is one originally proposed by Ol'Shanii et al. (1993) and recently demonstrated by Renn et al. (1995) at JILA. Important to this idea is the fact that a hollow glass fiber—in the form of a thin-walled cylindrical glass shell (no cladding)—can support propagating, red-detuned, leaky, grazing incidence, electromagnetic modes whose field intensities reside primarily in the fiber hollow (Marcatile and Schmeltzer, 1964; Solimeno et al. 1986). (See Fig. 8.) In particular, the lowest order of such propagating modes has a zeroth order Bessel function, $\mathbf{J}_0(\chi\rho)$, transverse profile, so that the propagating light-field intensity is maximal along the hollow fiber axis. (The radial vector $\boldsymbol{\rho}$ has norm $\rho \equiv |\boldsymbol{\rho}|$ in cylindrical coordinates.) Hence, if the field is red-detuned with respect to an atomic transition, then atoms will find themselves attracted to the axis of the hollow by the optical potential of the gradient force, $V(\rho)$, Eq. (18), and kept away from the fiber wall, where they would stick. There is also a spontaneous force, given by the second term of Eq. (9), that acts only in the z direction along the fiber axis (see

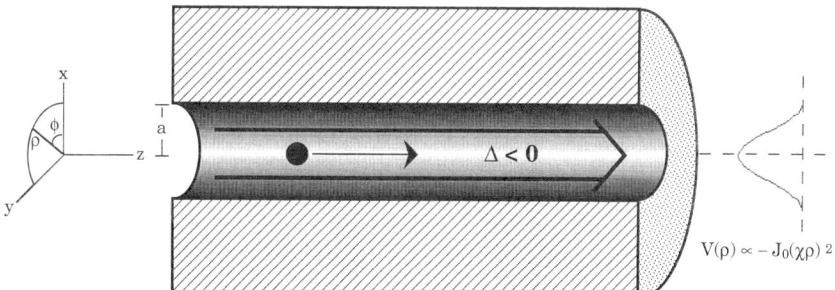

FIG. 8. Hollow fiber guide scheme of Ol'Shanii et al. (1993). Red-detuned grazing incidence EH_{11} propagating (but attenuated) laser mode (large arrow) confines atoms transversely in a negative potential of the form shown. Atoms are guided along the z axis by an approximately harmonic radial potential.

Gordon and Ashkin, 1980). For the large longitudinal atomic velocities we shall consider here, this force has a negligible effect on the atomic motion.

This scheme has the same drawbacks as any red-detuned procedure for optical atom manipulation, as mentioned in the introduction, namely, that atoms are localized in a region of high optical field strength and hence undergo many spontaneous emission events (Ovchinnokov et al., 1991). This leads to heating and diffusion of the atoms as they propagate (Balykin and Letokhov, 1989b), which will lead to their eventual ejection from the confining potential. Spontaneous emission also decoheres the atom's de Broglie wavefunction (Seifert et al., 1994b), limiting the usefulness of such an atom guide for such applications as atom interferometry. Nevertheless, this scheme is the first to be implemented experimentally (Renn et al., 1995), as it is apparently easier to carry out than the blue-detuned scheme to be discussed (Renn et al., 1996). In addition, there are applications where the precise delivery of atoms is desired—such as lithography—and where the lack of coherence does not matter.

The most general guided optical mode in a hollow glass cylindrical shell has an electric field profile given by

$$\mathbf{E}(\rho, z, t) = \frac{1}{2} \sum_{m=1}^{M} \mathbf{E}_m(\rho) e^{i(\beta_m z - \omega t)} \qquad (54)$$

where $\mathbf{E}_m(\rho)$ has the form

$$\mathbf{E}_m(\rho) = \hat{\mathbf{e}} E_m^0 J_0(\chi_m \rho) \qquad (55)$$

where $\hat{\mathbf{e}}$ is a unit transverse polarization vector. This solution is independent of the azimuthal coordinate ϕ (see Fig. 8). The wavenumbers β_m and χ_m are complex, in general, and can be found from the finite number of solutions, $m \in \{1, \ldots, M\}$, of the characteristic equation

$$\kappa_m J_0(\chi_m a) K_1(\kappa_m a) + \chi_m J_1(\chi_m a) K_0(\kappa_m a) = 0 \qquad (56)$$

where the J_i and K_i are Bessel functions and modified Bessel functions, respectively, of integer order i (Abramowitz and Stegun, 1965); a is the radius of the fiber hollow (Fig. 8); and β_m, χ_m, and κ_m are complex wavenumbers related by the two dispersion relations

$$k_{\text{opt}}^2 = \beta_m^2 + \chi_m^2 \qquad (57a)$$

$$nk_{\text{opt}}^2 = \beta_m^2 - \kappa_m^2 \qquad (57b)$$

where n is the index of refraction of the shell wall, with the hollow assumed to have index 1. Here, $k_{\text{opt}} \equiv \omega/c$, with ω the optical frequency and c the vacuum speed of light.

In general, if the guided light is multimode, $M > 1$ in Eq. (54), then each of the modal transverse fields, Eq. (55), will tend to confine the atoms along the hollow axis as long as $\Delta < 0$, but the analysis will be somewhat complicated. In the experiment of Renn *et al.* (1995), they consider the case of only single-mode optical propagation, $M = 1$, the so-called EH_{11} mode, in the limit where $ka \gg 1$ and the optical ray approximation holds. (Here, $k_{opt} = 2\pi/\lambda_{opt}$, with λ_{opt} the free space optical wavelength.) With these assumptions, the characteristic equation (56) reduces to

$$2k\sqrt{n^2 - 1}\, J_0(\chi a) - i\chi(n^2 + 1)J_1(\chi a) \tag{58}$$

where $\chi \equiv \chi_1$, and χ is related to the propagation wavenumber, $\beta \equiv \beta_1$, by the dispersion relation, Eq. (57a). The relation (57b) has been used to eliminate $\kappa \equiv \kappa_1$. The complex z-propagation wavenumber, $\beta = \beta' + i\beta''$, has an imaginary attenuation coefficient β'', which in the $k_{opt}a \gg 1$ regime has the approximate form

$$\beta'' \approx \frac{(\chi a)^2 (k_{opt}a)}{2a} \frac{n^2 + 1}{\sqrt{n^2 - 1}} \tag{59}$$

In this single EH_{11} optical mode of operation, only the $m = 1$ term in the sum Eq. (54) for $\mathbf{E}(\rho, z, t)$ survives. Hence, the guiding potential $V(\rho)$, Eq. (18b), will have in the large-detuning limit the form

$$V(\rho) \underset{\Delta \gg \Omega_0}{\approx} \hbar\Delta \frac{\Omega_0^2}{\gamma^2 + 4\Delta^2} J_0^2(\chi\rho) \tag{60}$$

where $\Omega_0 \equiv \mu E_1^0/\hbar$ is the Rabi frequency associated with the free space field, as per Eq. (55). Clearly, then, $V(\rho)$ is an attractive potential for $\Delta < 0$ with its minimum at $\rho = 0$ (see Fig. 8).

In the JILA experiment (Renn *et al.*, 1995), approximately a 3 cm length of hollow fiber is used, with an outer diameter of 144 μm and hollow core diameter of $a = 40$ μm. Light of wavelength $\lambda_{opt} = 780$ nm, resonant with the rubidium $D2$ resonance, is coupled from a Ti:sapphire laser into the EH_{11} mode of the fiber. The light has a bandwidth of 2 GHz and tunability of about ± 50 GHz around the $D2$ resonance. The solution of the characteristic equation (58) for χa for these parameters yields $\chi a = 2.405 + 0.22i$. The dispersion relation, Eq. (57), yields a z direction propagation attenuation length of $1/\beta'' \approx 6.2$ cm, where $\beta = \beta' + i\beta''$, as before. Hence, these EH_{11} optical modes are fairly strongly attenuated, or "leaky". Defining the "mode diameter" as the diameter d in which the transverse intensity function $J_0^2(\chi\rho)$ falls to $1/e$ of its original value gives a value of $d = 22$ μm, which is much less than the actual hollow diameter, $a = 40$ μm. Hence, the rubidium atoms will tend to be localized in a region near the axis by a transverse confining potential that is approximately harmonic, as can be seen for small values of $\chi\rho$ in the Taylor expansion of $J_0^2(\chi\rho) \approx 1 - (\chi\rho)^2$ in the potential $V(\rho)$, Eq. (60). Thus, the atoms glide down the hollow, oscillating about the axis.

To find the maximum transverse velocity, v_ρ^{max}, that atoms may have to be so guided, we may simply set the kinetic energy $Mv_\rho^2/2 = V(0)$, where $V(0)$ is the potential $V(\rho)$ evaluated at the axis, $\rho = 0$. [In the large detuning limit ($\Delta \gg \Omega$), we may use the approximate expression for $V(\rho)$, Eq. (60); otherwise, we need to revert to Eq. (18a).] For typical experimental numbers (Renn *et al.*, 1995), we have an optical intensity of $I = 45$ mW and a detuning of $\Delta = -6$ GHz. This yields a potential depth corresponding to a transverse capture temperature $T_\rho = V(0)/k_B = 71$ mK, when divided by Boltzmann's constant, k_B. This yields a maximum transverse capture velocity of $v_\rho^{max} = 37$ cm/sec. The fiber end opens into a thermal vapor-filled chamber of Rb atoms, and those that randomly strike the fiber with $v_\rho < v_\rho^{max}$ will be guided.

For a classical ballistic motion in straight fiber, no constraint exists on the longitudinal velocity v_z. However, if the fiber is curved, then the potential $V(\rho)$ must be able to overcome the centrifugal force rounding the bend, or else the atom will hit the wall and stick. For a given potential $V(\rho)$, the average gradient force $\langle F \rangle$ on a typical guided atom can be estimated by

$$\langle F \rangle = -\frac{2}{d} \int_0^{d/2} \left(\frac{dV}{d\rho}\right) d\rho$$
$$= \frac{2}{d}\left[V(0) - V\left(\frac{d}{2}\right)\right]$$
$$\equiv 2\frac{\Delta V}{d} \cong 2\frac{V(0)}{d} \quad (61)$$

where d is the mode diameter defined previously. Setting $\langle F \rangle$ equal to the on-axis centrifugal force $F_c = Ma_c = Mv_z^2/R$, where $a_c \equiv v_z^2/R$ is the centripetal acceleration for a radius of curvature R, yields a minimum allowable curvature radius, namely

$$R_{min} \cong \frac{Mv_z^2 d}{2V(0)} \quad (62)$$

below which atoms of longitudinal velocity v_z will not be guided. Here, d is again the mode diameter. This fact can be used to design a velocity selector for atoms—by curving the fiber, the faster tail of the velocity distribution can be removed. For the JILA experiment, R_{min} is about 18 cm.

As mentioned, red-detuned trapping and guiding schemes such as these are subject to decoherence and heating due to numerous spontaneous emission events, since the atoms are localized near a maximum in the optical field intensity. This effect is estimated in Ol'Shanii et al. (1993) and Balykin and Letokhov (1989a). (A good semiclassical discussion of the role of spontaneous emission in light-wave mirrors is given in Seifert et al., 1994b. A fully dressed-state, quantum analysis is found in Deutschmann et al., 1993a). In the limit $\Delta \gg \gamma$, a transverse diffusion coefficient D describing the random walk undergone by an atom undergoing multiple spontaneous emission events can be defined by (Kazantsev et al., 1990)

$$D \underset{\Delta \gg \gamma}{\approx} \frac{(\hbar k_{opt})^2 \gamma}{20} \left(\frac{p}{1+p} \right) + \frac{\hbar^2}{\gamma} \left[\frac{d\Omega}{d\rho} \right]^2 \left(\frac{p}{1+p} \right)^3 \quad (63a)$$

$$\underset{\Delta \gg \gamma, \Omega_0}{\approx} \frac{(\hbar k_{opt})^2 \gamma}{20} p \quad (63b)$$

where $p = 2\Omega^2/(\gamma^2 + 4\Delta^2)$. Hence, the time τ it takes for spontaneous emission heating to enlarge v_ρ sufficiently to escape the potential is

$$\tau \approx \frac{MV(0)}{\langle D \rangle} \quad (64)$$

where $\langle D \rangle$ is D averaged over $\rho \in (0, d)$. For the JILA experiment, $\langle p \rangle \approx p(0) \approx 10$ for the aforementioned experiment parameters. Hence, $\langle D \rangle \approx (\hbar k_{opt})^2 \gamma/2$, and $\tau \approx 0.12$ sec. A simple heuristic physical picture of this number is to realize that in an integration time τ, a total of $N = \tau\gamma$ spontaneous emission events will occur, each imparting a random momentum kick of $\hbar k_{opt}$ to the atom, accumulating a total energy $E = N(\hbar k_{opt})^2/(2M)$. Setting $E \approx V(0)$, the potential depth that has to be overcome, yields the same approximate expression for τ as just derived. For the JILA experiment, $\tau \cong 0.12$ sec and $\langle v_z \rangle \approx 3.3 \times 10^4$ cm/sec for the Boltzmann distribution of thermal rubidium atoms coming from the capor cell, which yields a large mean free path of guiding distance, namely, $\tau \langle v_z \rangle = 4 \times 10^3$ cm. Since the fiber used is only 3 cm long and the optical attenuation length is around 6 cm, we see that this spontaneous emission heating is not a limiting factor in this experiment. However, for guides where supercooled atoms with large de Broglie wavelengths and coherence lengths will be required, this loss factor will be a serious drawback.

The number of atoms guided per second (atom flux), as a function of detuning (from the JILA experiment) is shown in Fig. 9. Most notable is the sharp turn-on as the detuning goes from blue to red, with the flux falling off like $1/\Delta$ in the limit of large detuning, as expected from the form of $V(\rho)$, Eq. (18c), valid for $\Delta \gg \gamma, \Omega_0$. The experiment also measured the guided-atom flux as a function of the curvature radius R_{min} and found qualitatively good agreement with Eq. (62), except when $R_{min} \lesssim 1/\beta'' \approx 6$ cm, the optical fiber attenuation length. In this regime, the optical modes are severely distorted and $\mathbf{E}(\rho, z, t)$ in Eq. (54) is no longer a good solution. The actual optical intensity profile becomes asymmetric about the axis and guides the atoms into the fiber wall (Marcatile and Schmeltzer, 1964).

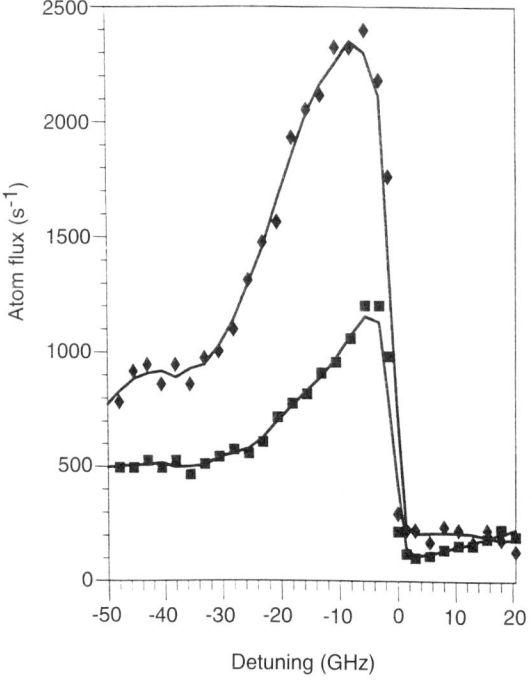

FIG. 9. Experimental realization of the red-detuned hollow fiber scheme of Ol'Shanii et al. performed by the JILA group. Thermal rubidium atoms are guided through 3 cm of hollow fiber with a hollow diameter of 40 μm. The guided atom flux is shown here as a function of detuning, Δ, for two EH_{11} mode laser intensities, $I_{top} = 3.6 \times 10^3$ and $I_{bottom} = 1.3 \times 10^3$ W/cm². Notice the sharp turn-on of flux as the detuning becomes negative. (From Renn et al., 1995, reprinted with permission).

B. Blue-Detuned, Evanescent Light-Wave, Hollow Fiber Guides

As mentioned, although guiding atoms in red-detuned, grazing incidence $J_0^2(\chi\rho)$ potentials of the EH_{11} optical modes has been demonstrated successfully (Renn et al., 1995), it nevertheless has significant drawbacks for atom optical applications. In particular, as discussed in the introduction, the atoms are guided along an attractive, red-detuned, light-potential intensity maximum and hence undergo heating and decoherence by multiple spontaneous emission events. In addition, the guiding, J_0, propagating EH_{11} modes, Eq. (55), do not propagate without loss by total internal reflection, but rather in a loose, grazing incidence confinement that is very leaky, and hence they decay away exponentially fast as they propagate, with an attenuation length $1/\beta''$, where $\beta'' = \text{Im}\{\beta\}$, as given in Eq. (59) above (see Marcatile and Schmeltzer, 1964; Snyder and Love, 1983; Solimeno et al., 1986).

Much more appealing, then, is to use a blue-detuned HE_{11} optical mode for the guide instead. This optical mode propagates practically without any attenuation by total internal reflection inside the annular "core" region between the hollow and the vacuum or cladding, with the optical intensity mostly confined to the annular core—as in the case of a dielectric waveguide (see Fig. 10). However, as in the theory of the evanescent mirror, Section II, an exponentially steep evanescent field extends into the vacuum of the hollow, to coat the surface with a blue-detuned, repulsive, evanescent fuzz that guides the atoms (see Fig. 11).

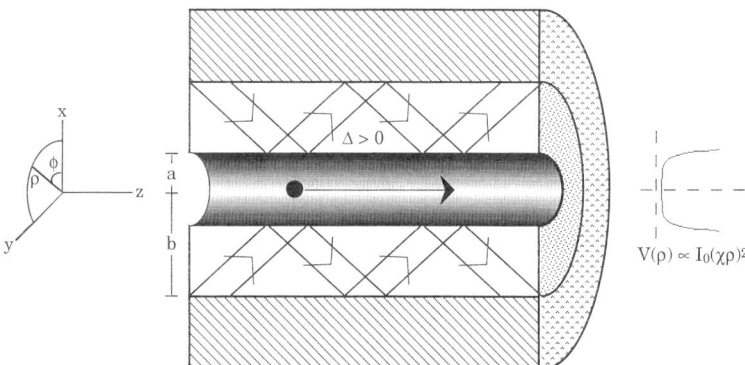

FIG. 10. Blue-detuned evanescent-wave, hollow fiber atom guide scheme of Marksteiner et al. (1994). Here, light propagates by total internal reflection in a HE_{11} mode in the annular core region, $\rho \in [a, b]$. This produces a blue-detuned, repulsive, evanescent potential on the walls of the hollow, as shown. Atoms move in a mostly field-free region with brief collisions with the mirrored walls.

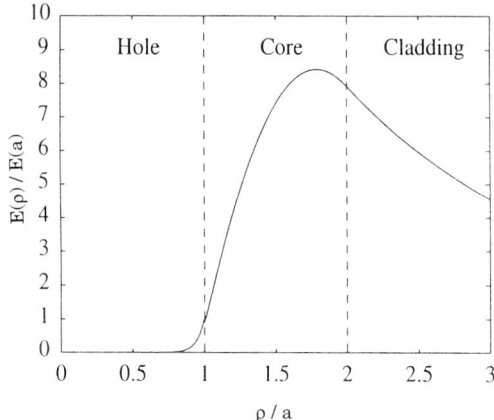

FIG. 11. The electric field amplitude of blue-detuned light propagating in a HE_{11} mode by total internal reflection in the annular core of a hollow fiber. Here, the dimensionless amplitude $E(\rho)/E(a)$ is plotted against the dimensionless radial coordinate, ρ/a. The core and cladding have indices of refraction $n_1 = 1.500$ and $n_2 = 1.497$, respectively, and the optical wavelength λ_{opt} is chosen here so that $a/\lambda_{opt} = (b - a)/\lambda_{opt} = 2.889$, where a is the hollow radius and b the outer radius of the annular core. The propagating optical field is single mode (from Marksteiner et al., 1994, reprinted with permission).

This scheme, first suggested by Savage et al. (1993), and elaborated on by Marksteiner et al. (1994) and Ito et al. (1995), has been demonstrated by the same group at JILA that carried out the red-detuned scheme, namely, Renn et al. (1996; see also Ito et al., 1996).

The primary advantage of this scheme is that the atom is guided through a region of zero optical field intensity, except for its brief reflection episodes at the evanescent potential. If the detuning is large, $\Delta \gg \Omega_0, \gamma$, then the probability of spontaneous emission is minimal, as discussed in Section II. In addition, the exponentially thin potential can be approximated by an infinite potential step, in which case the quantum propagation of a supercooled atom with a large de Broglie wavelength—comparable with the fiber diameter—will be well described by the mathematics similar to that required to model scalar microwaves propagating in a cylindrical metal waveguide. Hence, for these reasons, this blue-detuned evanescent mirror scheme is most promising for applications to atom interferometry and also to quantum atom gravitational cavities made from hollow fiber atom waveguides, as proposed by Harris and Savage in 1995.

1. Theory of Unattenuated Optical Propagation in a Hollow Waveguide

We shall follow the theoretical development of Savage *et al.* (1993) and Marksteiner *et al.* (1994), with some notational changes to maintain consistency within this chapter. We first would like to consider the form of the blue-detuned guided HE_{11} optical field that propagates down the annular core walls by total internal reflection (see Fig. 10). We take the fiber to have a transverse index of refraction profile $n(\rho)$, given by

$$n(\rho) = \begin{cases} 1, & \rho \in [0, a) \quad \text{(I)} \\ n_1, & \rho \in [a, b) \quad \text{(II)} \\ n_2, & \rho \in [b, \infty) \quad \text{(III)} \end{cases} \quad (65)$$

where region I is the hollow, II the annular core, and III the cladding. Hence, the core index n_1 applies to the annular region, $a \leq \rho < b$, and the cladding of index n_2 extends to infinity. In reality, the cladding (if there is any) extends far from the core, and hence this is a good approximation. If there is no cladding and the fiber is a simple glass capillary, as discussed in Savage *et al.* (1993) and used in the JILA experiments of Renn *et al.* (1995, 1996), then we can include that fact by setting $n_2 = 1$, i.e., a vacuum. In this latter case, if the condition $b - a \gg \lambda_{\text{opt}}$ holds, then the optical guide can be treated in the ray approximation as a dielectric slab guiding light by total internal reflection, as discussed in Savage *et al.* (1993) and Renn *et al.* (1996). That result, of course, can be obtained as a limiting case of the wave description presented here by taking $n_2 \to 1$ and replacing the Bessel function solutions, to be given, with their asymptotic form at large argument (Abramowitz and Stegun, 1965). From Snyder and Love (1983), Marksteiner *et al.* (1994), and Ito *et al.* (1995), the annular field modes that represent unattenuated propagation have electric and magnetic field profiles given in cylindrical coordinates by

$$\mathbf{E}(\rho, \phi, z, t) = \mathbf{e}(\rho, \phi) e^{i(\beta z - \omega t)} + \text{c.c.} \quad (66a)$$

$$\mathbf{H}(\rho, \phi, z, t) = \mathbf{h}(\rho, \phi) e^{i(\beta z - \omega t)} + \text{c.c.} \quad (66b)$$

where β is the propagation wavenumber, and $\omega \equiv k_{\text{opt}} c$ is the frequency of the incident field. For a particular polarization of unit direction, $\hat{\rho}$, the field envelopes $\mathbf{e}$ and $\mathbf{h}$ can be decomposed as

$$\mathbf{e} = e_\rho \hat{\rho} + e_z \hat{z} \equiv \mathbf{e}_\rho + \mathbf{e}_z \quad (67a)$$

$$\mathbf{h} = h_\rho \hat{\rho} + h_z \hat{z} \equiv \mathbf{h}_\rho + \mathbf{h}_z \quad (67b)$$

where the longitudinal components $\mathbf{e}_z$ and $\mathbf{h}_z$ obey the transverse Helmholtz equations,

$$[\nabla_\rho^2 + n_i^2 k_{\text{opt}}^2 - \beta] \begin{Bmatrix} \mathbf{e}_z \\ \mathbf{h}_z \end{Bmatrix} = 0 \qquad (68)$$

where $n_i \in \{1, n_1, n_2\}$ depending on whether we are in region I, II, or III, respectively, as per Eq. (65). The transverse components $\mathbf{e}_\rho$ and $\mathbf{h}_\rho$ can be obtained from the solutions to Eq. (68) for $\mathbf{e}_z$ and $\mathbf{h}_z$ by application of Maxwell's equations, via

$$(n_i^2 k_{\text{opt}}^2 - \beta^2)\mathbf{e}_\rho = i\beta\, \nabla_\rho \mathbf{e}_z - ik\hat{\mathbf{z}} \times \nabla_\rho \mathbf{h}_z \qquad (69\text{a})$$

$$(n_i^2 k_{\text{opt}}^2 - \beta^2)\mathbf{h}_\rho = i\beta\, \nabla_\rho \mathbf{h}_z + kn_i^2\hat{\mathbf{z}} \times \nabla_\rho \mathbf{h}_z \qquad (69\text{b})$$

(see, for example, Jackson, 1975, Section 8.2). Hence, the solutions of Eq. (68) give enough information to construct the complete electromagnetic modal solutions $\mathbf{E}$ and $\mathbf{H}$, Eq. (66). The Helmholtz equation (68) is separable, and it has physical solutions of the form

$$\begin{Bmatrix} \mathbf{e}_z \\ \mathbf{h}_z \end{Bmatrix} = \begin{cases} \begin{Bmatrix} A_1 \\ B_1 \end{Bmatrix} I_m(\chi\rho)e^{im\phi} & \text{(I)} \\[1em] \left[\begin{Bmatrix} A_2 \\ B_2 \end{Bmatrix} J_m(\kappa\rho) + \begin{Bmatrix} A_3 \\ B_3 \end{Bmatrix} Y_m(\kappa\rho)\right] e^{im\phi} & \text{(II)} \\[1em] \begin{Bmatrix} A_4 \\ B_4 \end{Bmatrix} K_m(\alpha\rho)e^{im\phi} & \text{(III)} \end{cases} \qquad (70)$$

where J_m, Y_m, I_m, and K_m are Bessel functions of integer order m (see Abramowitz and Stegun, 1965). The transverse propagation constant κ and the transverse attenuation constants χ and α obey

$$k_{\text{opt}}^2 = \beta^2 - \chi^2 \qquad (71\text{a})$$

$$n_1^2 k_{\text{opt}}^2 = \beta^2 + \kappa^2 \qquad (71\text{b})$$

$$n_2^2 k_{\text{opt}}^2 = \beta^2 - \alpha^2 \qquad (71\text{c})$$

where α is an attenuation wavenumber (and not the vacuum inverse penetration depth). The eight unknown constants $A_1, \ldots, A_4, B_1, \ldots, B_4$ are determined by demanding that $\mathbf{h}, \mathbf{e}_z$ and $n^2 \mathbf{e}_\rho$ are everywhere continuous, in particular at the interfaces (see, for example, Jackson, 1975, Section 7.3).

Details of the explicit solutions of these can be found in Marksteiner et al. (1994), Ito et al. (1995), Marcatile and Schmeltzer (1964), Snyder and Love (1983), and Solimeno et al. (1986). Several limiting cases are of interest. When the radius of the hollow, a, is on the order of the optical wavelength or smaller, $a \leq \lambda_{\text{opt}}$, the lowest order unattenuated propagating mode is an HE_{11}. The HE_{11} has azimuthal dependence on ϕ, but for circularly polarized light this averages out on the atomic motion time scale, yielding an azimuthally symmetric potential. The HE_{11} single mode electric field $E(\rho)$ is plotted in Fig. 11, with permission from Marksteiner et al. (1994), with $a = 1.65$ μm, $b = 2a$, $n_1 = 1.500$, and $n_2 = 1.497$. We can see that most of the field is in the annular core or the cladding—but the evanescent coating of the fiber hollow is clear. From Eq. (70), we see that this coating gives rise to a potential $V(\rho) \propto I_0^2(\chi\rho)$, as indicated in Fig. 10.

Another limiting case is the regime when $a \gg \lambda$. In this limit, the curvature of the fiber around the z axis has little effect on the optical propagation, and the hollow fiber may be treated like an infinite planar dielectric waveguide with a slab of index n_1 sandwiched between slabs of $n = 1$ and $n = n_2$. In this limit, the lowest order guided modes are the well-known TE_1 and TM_1 modes. For a fixed hollow radius a, one can also vary the thickness of the core annulus, namely $\Delta\rho \equiv b - a$. As a rule of thumb, when this is done, one has that (1) $\Delta\rho \ll \lambda_{\text{opt}}$, no unattenuated modes; (2) $\Delta\rho \sim \lambda_{\text{opt}}$, single mode; (3) $\Delta\rho \gg \lambda_{\text{opt}}$, multimode.

2. Schrödinger Modal Structure of a Cylindrical Atom Waveguide

Now that we have established that blue-detuned evanescent fields can in fact be made to coat the hollows of the annular fiber waveguide, we can again make the approximation, as in Section III, and Section V following, that we have a large detuning, $\Delta \gg \Omega_0, \gamma$, and neglect spontaneous emission. If, in addition, we assume that the atoms are sufficiently cooled below the recoil limit such that their de Broglie wavelength λ_{dB} is much larger than the optics evanescent penetration depth, namely $\lambda_{\text{dB}} \gg \lambda_{\text{opt}}/(2\pi)$ (see Reichel et al., 1995), then the evanescent wave potential is very well described by an infinite step potential, as depicted in Fig. 10 (see Chen and Milburn, 1995). Hence, the calculation of the atom wave modal structure reduces to solving a scalar wave equation in a cylindrical waveguide. We shall neglect the effect of gravity here, but it has been considered by Harris and Savage (1995). With these approximations, the Schrödinger equation becomes (Jackson, 1975, Section 3.7)

$$\frac{\hbar^2}{2M}\left(\frac{\partial^2}{\partial\rho^2} + \frac{1}{\rho}\frac{\partial}{\partial\rho} - \frac{l^2}{\rho^2} - k_{\text{dB}}^2\right)\psi + [E - V(\rho)]\psi = 0 \quad (72)$$

where $\psi(\rho, \phi, z) = \Re(\rho)e^{\pm il\phi}e^{\pm ik_{dB}z}$, where $l \in \{0, 1, 2, \ldots\}$ and

$$V(\rho) = \begin{cases} 0, & \rho \in [0, a) \\ \infty, & \rho \in [a, \infty) \end{cases} \quad (73)$$

where $k_{dB} = 2\pi/\lambda_{dB}$ is the atom wave propagation de Broglie wavenumber.

In Savage et al. (1993), they assume that the atom may have arbitrary velocity in the z direction, but that its transverse motion is quantized. This is an inconsistent approximation, as transverse quantization necessarily implies quantized values for k_{dB} and hence selects only particular longitudinal modes of propagation. We do not make this approximation here. The equation for $\Re$ becomes

$$\frac{d^2\Re}{d\rho^2} + \frac{1}{\rho}\frac{d\Re}{d\rho} + \left[\varepsilon - k_{dB}^2 - \frac{l^2}{\rho^2}\right]\Re = 0 \quad (74)$$

with the boundary condition $\Re(a) = 0$ at the hollow wall, and $\varepsilon \equiv 2ME/\hbar^2$ is the scaled energy. Performing the change of variable $u = \rho\sqrt{\varepsilon - k_{dB}^2}$, assuming $\varepsilon - k_{dB}^2 > 0$ for a propagating unbound atom, yields the transformed equation for $\Re$ (Jackson, 1975, Section 3.7):

$$\frac{d^2\Re}{du^2} + \frac{1}{u}\frac{d\Re}{du} + \left[1 - \frac{l^2}{u}\right]\Re = 0 \quad (75)$$

where the boundary conditions $\Re(a) = 0$ and $\Re(0) < \infty$ imply

$$\Re(\rho) = \mathcal{N}_l J_l(\sigma\rho) \quad (76)$$

where $\sigma \equiv \sqrt{\varepsilon - k_{dB}^2}$, $\mathcal{N}_l$ is a normalization constant, and J_l is the only Bessel function of the first kind nonsingular at the origin. If we define j_{lq} to be the qth root of the Bessel function J_l, then the demand $\Re(a) = 0$ yields the transverse quantization condition

$$\sigma_{lq} a = a\sqrt{\varepsilon_{lq} - k_{lq}^2} = j_{lq} \quad (77a)$$

or

$$\varepsilon_{lq} = k_{lq}^2 + \frac{j_{lq}^2}{a^2} \quad (77b)$$

where l is the number of azimuthal nodes and q the number of radial nodes in the atom wavefunction ψ. For fixed total energy ε_{lq}, we see that the atom's longitudinal energy, $E_{lq}^{\|} = \hbar^2 k_{lq}^2/(2M)$ is also quantized, as in any waveguide. (In particular, atoms with kinetic energy $E < E_{01}^{\parallel}$ will be below the cutoff for the guide and will not propagate.)

Hence, and this is crucial, for single mode operation, atoms must be loaded into the guide with a velocity distribution that satisfies the dispersion relation, Eq. (77). In other words, if the hollow fiber is exposed to a cooled vapor of atoms, say in a magneto-optical trap (MOT), only atoms with a transverse velocity $v_{01}^{\rho} = \hbar \sigma_{01}/M$ will be guided. The longitudinal velocity v_{01}^{z} is thus not independent of v_{01}^{ρ}, but constrained by the dispersion relation Eq. (77b), and hence by the allowed range of total kinetic energy ε_{01}, as constrained by the Boltzmann velocity distribution of the atoms in the MOT. In other words, to estimate how many atoms are guided, one must integrate over a peculiarly shaped volume of phase space in the Boltzmann distribution. This will then determine the longitudinal velocity distribution of the single mode ψ_{01}-guided atoms. For the (l, q)th guided atom mode, the normalization constant $\mathcal{N}_{lq}$ is determined implicitly by

$$\frac{1}{\mathcal{N}_{lq}^{2}} = \int_{0}^{a} J_{l}^{2}\left(\frac{j_{lq}\rho}{a}\right) d\rho \tag{78}$$

which yields, for the Schrödinger modal decomposition,

$$\psi_{lq}(\rho, \phi, z) = \mathcal{N}_{lq} J_{l}\left(\frac{j_{lq}\rho}{a}\right) e^{\pm ik_{lq}z} e^{\pm il\phi} \tag{79}$$

We plot the squared moduli of ψ_{01}, ψ_{02}, ψ_{11}, and ψ_{12} in Fig. (12) at $z = 0$. These are the same as the normal elastic modes of oscillation found in a circular drumhead. To get an idea of the relationship between transverse temperature, fiber diameter, and mode of propagation, we define a transverse de Broglie wavelength

$$\lambda_{lq}^{\perp} \equiv \frac{2\pi}{\sigma_{lq}} = \frac{2\pi a}{j_{lq}} \tag{80}$$

where σ_{lq} are the transverse attenuation constants, Eq. (77a). For good, single mode, coherent atom, waveguide propagation, one would like to use the lowest order ψ_{01} mode, which would require the lowest transverse temperatures and largest transverse de Broglie wavelengths, $\lambda_{01}^{\perp} = 2\pi a/j_{01} \approx 2\pi a/2.4 \approx 2.62a$. This agrees reasonably well with the naive approximation that the lowest order guided mode would have one-half of a de Broglie wave fitting across the diameter of the hollow, or $\lambda_{01}^{\perp} \approx 4a$. The transverse temperatures $T_{lq}^{\perp}$ required to attain these wavelengths are

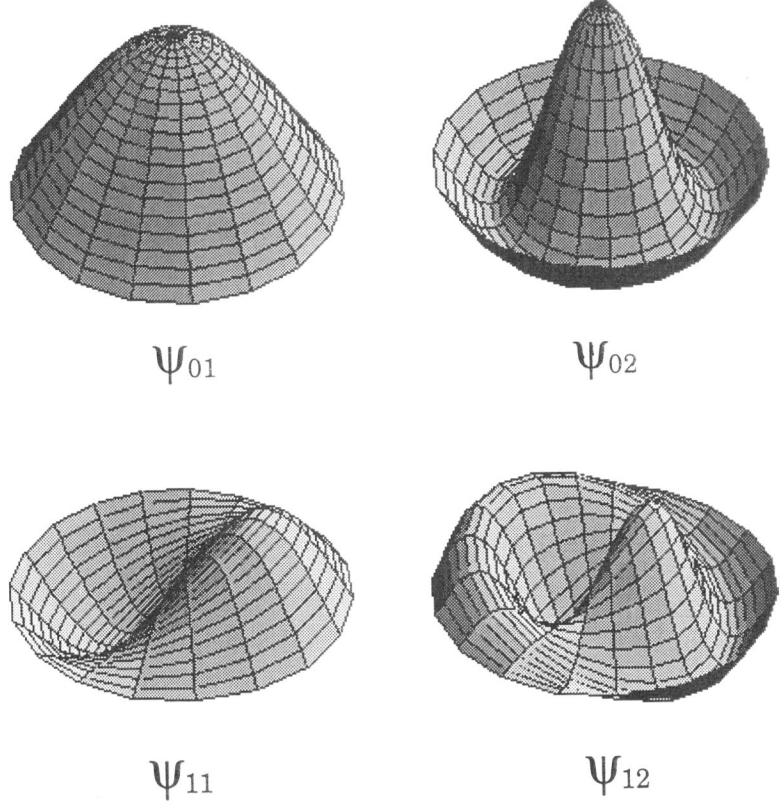

FIG. 12. First four atom wavefunctions $|\psi_{lq}|^2$ for single mode atom wave propagation in a blue-detuned evanescent-mirror, hollow fiber atom guide, as per Eq. (79). The view is taken at a cross section through the fiber at $z = 0$. These modes are the same as the elastic vibrations of an acoustic drumhead.

given by

$$T_{lq}^{\perp} = \frac{\hbar^2 \sigma_{lq}^2}{2Mk_B} = \frac{\hbar^2 j_{lq}^2}{2Mk_B a^2} \qquad (81)$$

where k_B is Boltzmann's constant.

For the lowest order (ψ_{01}) mode, we have, for M measured in atomic units (au) and a in nanometers,

$$T_{01}^{\perp} \approx \frac{1.4}{Ma^2} \qquad (82)$$

where $T_{01}^{\perp}$ is in degrees Kelvin. If we assume that the confining laser light is propagating also in a single HE_{11} mode, as discussed previously, then we also have $\lambda_{opt} \approx a$. Using rubidium atoms, as in the JILA experiment (Renn et al., 1996), we have $\lambda_{opt} = 789$ nm and $M_{Rb} = 87$ au. Equation (82) yields a transverse temperature of $T_{01}^{\perp} \approx 26$ nK. Although the recoil temperature for Rb is $T_r = 400$ nK (Table I, Section V), the cooling of atoms to below recoil temperature by as much as $T_r/70$ has been reported recently (Reichel et al., 1995), making such a quantum atom wave device not totally unfeasible. Of course, if one is willing to sacrifice lowest order single atom mode operation, then such low temperatures are not required.

As in the red-detuned scheme discussed previously, there is a limit on the transverse velocity v_ρ of atoms that may be guided, which can be estimated by setting $Mv_\rho^2/2 = V(a)$, where V is the gradient potential evaluated at the wall, $\rho = a$. Similarly, the minimum radius of curvature R of the fiber that can be used before centrifugal force drives them into the walls can be estimated from Eq. (62), replacing ΔV with $V(a)$. However, there are a few additional considerations not present in the red-detuned scheme. As discussed, in the red-detuned scheme, the atoms are localized close to the hollow axis at the bottom of the $J_0^2(\chi\rho)$ potential and far from the walls. In the blue-detuned case here, the atoms can come extremely close to the walls with maximal transverse velocity before they are reflected by the $I_0^2(\chi\rho)$ repulsive potential. One problem is that—since the evanescent potential is exponentially thin—the atoms can quantum mechanically tunnel and hit the wall, whereas they would be forbidden from doing so using simple classical estimates. This effect will tend to reduce the guided atom flux, and it has been estimated by Savage et al. (1993) and Marksteiner et al. (1994). An additional problem is that atoms can now come close enough to the walls so that van der Waals forces are important, and hence the repulsive potential $V(\rho)$ is weakened somewhat, contributing to an additional loss factor. This effect is also estimated in Marksteiner et al. (1994).

3. The Blue-Detuned Experiment at JILA

The same group at JILA that successfully demonstrated the guiding of rubidium atoms in a red-detuned OOL scheme (Renn et al., 1995) has also successfully demonstrated guidance in the blue-detuned SMZ scheme (Renn et al., 1996). We briefly summarize this result here. [In addition, independently, the Japanese–Korean collaboration of Ito et al. (1996) has also reported atom guiding in a hollow, blue-detuned, evanescent light-wave fiber guide. At the time of this writing, only the JILA paper had been accepted for publication, and so we discuss only this experiment in some detail. However, the group of Ito *et al.* have performed two-step laser photoionization of guided rubidium atoms and have also reported a quantum state selective, stable isotope separation of these atoms in the guide.]

The JILA experiment once again involved rubidium atoms using the $D2$ transition corresponding to $\lambda_{opt} = 780$ nm. The fiber was a hollow glass cylinder with dimensions inner radius $a = 5$ μm and outer radius $b = 38.5$ μm, corresponding to a shell wall thickness of $\Delta\rho = b - a = 33.5$ μm. The annular core index is $n_1 = 1.5$, and there is no cladding, corresponding to $n_2 = 1.0$, in our notation. Since $a, b, \Delta \gg \lambda_{opt}$, the theory of a planar dielectric optical waveguide in the ray approximation can be used to describe the blue-detuned evanescent guiding field that propagates, mostly in the annular core by total internal reflection, corresponding to TE_1 or TM_1 dielectric waveguide modes, as discussed previously. (Here, since $a \gg \lambda_{opt}$, there is no real HE_{11} mode; rather, it goes over into a TE_1 mode, as discussed in Savage *et al.*, 1993; and Marksteiner *et al.*, 1994.) However, there is a problem. With these parameters, if one just shoots blue-detuned light into the fiber willy-nilly, all sorts of modes are excited —not just the desired TE_1 guiding mode. In particular, our old friend the EH_{11} grazing mode with the $J_0(\chi\rho)$ field profile is excited, too—only now *blue-detuned*. Hence, while the TE_1 mode sets up an exponentially thin coating of repulsive evanescent fuzz on the hollow wall, as desired, the EH_{11} grazing mode sets up an OOL, $J_0^2(\chi\rho)$ potential along the axis as before—but since it is blue-detuned, it now acts to slam the atoms into the wall, right through the weaker evanescent field. Now the lowest order EH_{11} mode for these parameters has an attenuation length of only $1/\beta'' = 1/\text{Im}\{\beta\} \approx 4$ mm, where β satisfies the EH_{11} dispersion relations, Eq. (57). So atoms could be guided by the TE_1 evanescent mirror—*if* they ever got past the first few millimeters of fiber length. In the initial attempts at this experiment, none apparently did (Renn *et al.*, 1996). The ingenious idea of the JILA group was to then use this coupling of light into EH_{11} modes to their advantage by providing a second laser beam—an "escort" beam—that was red-detuned, coupled to the EH_{11} mode, and powerful enough to escort the atoms past the few millimeters of the "danger zone" by using the OOL scheme (Ol'Shanii *et al.*, 1993). Since this red-detuned escort also attenuates after a few millimeters, it does its job and then lets the blue-detuned TE_1 evanescent guide take over.

This double-laser approach succeeded, as can be seen in Fig. 13, where the experimental flux of guided atoms is shown as a function of the detuning of the TE_1 evanescent guiding field for two values of red-detuning for the escort EH_{11} field. In addition, Renn *et al.* measure a

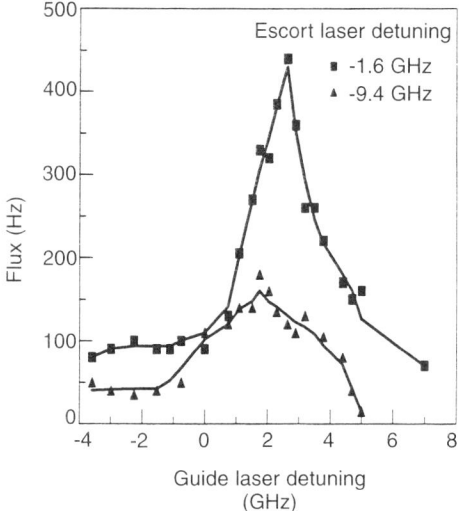

FIG. 13. Experimental plot of guided atom flux through a 6 cm length of a blue-detuned evanescent-wave, hollow fiber atom guide, as carried out by the JILA group. The atom flux is plotted as a function of evanescent-wave detuning for two values of detuning of a red-detuned "escort" laser, $\Delta = -1.6$ and -9.4 GHz. (The atom escort service is provided only in the red-detuned light district near the opening of the fiber.) The atoms being guided are thermal rubidium atoms; the fiber has a 20-μm-diameter hollow; and the optical wavelength $\lambda_{opt} \gg a = 10$ μm, the hollow radius. Hence, the optical mode is the TE_1 mode of a dielectric slab (from Renn et al., 1996, reprinted with permission).

threshold effect in the turn-on of atom flux as a function of optical intensity, indicating the presence of the van der Waals potential that must be overcome before guiding can occur. Although this and the preceding experiment are essentially classical, in that the atoms can be treated as ballistic particles, they nevertheless represent important proof-of-principle steps toward constructing quantum atom waveguides.

C. PARALLEL-MIRROR WAVEGUIDE

In Sections II.A, B, we discussed using parallel evanescent light-wave plane mirrors as atom cavities, with widely separated mirrors. In such cavities, the atom motion is primarily normal to the mirror surface, with the intent of minimizing transverse motion. In Section III.C, we discussed the planar waveguide of the pushme–pullyou scheme, but due to the red-detuning, there is little hope of using this as a coherent waveguide, because of spontaneous emission. In this section, we discuss the atom modal structure of two blue-detuned evanescent-wave light mirrors separated by only a few de Broglie wavelengths, so that the device functions as a waveguide for atoms moving primarily transversely to the mirror normal, confined to an essentially two-dimensional field-free region between the mirror surfaces (see Fig. 14). If the atoms confined to this guide are sufficiently cool, so that their de Broglie wavelength normal to the mirrors is half the intermirror spacing L, they will be guided in the fundamental mode. If in addition there are many atoms, so that their de Broglie wavelengths begin to overlap in the transverse direction, then some very interesting new physics can be done in this two-dimensional space.

One application is to the study of the anyonic behavior of a neutral vapor of atoms in a two-dimensional space. In three-dimensional space it is well known, since the covering of the rotation group is SU(2), that all quantum particles must obey Bose or Fermi statistics, depending on whether they have integer or half-integer spin, respectively. This connec-

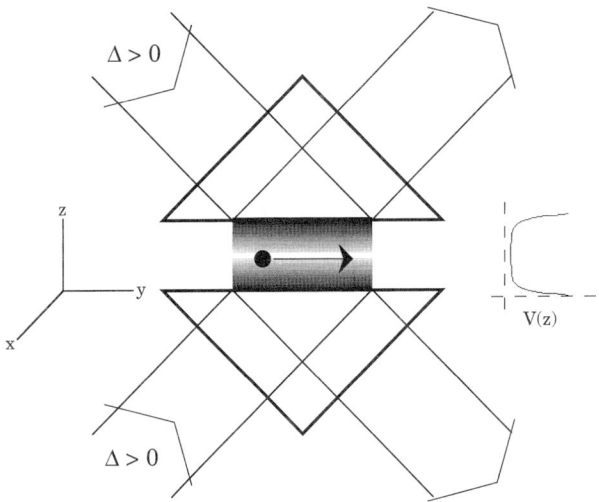

FIG. 14. Parallel-mirror atom waveguide formed by two blue-detuned, repulsive, evanescent-wave mirrors formed on the surfaces of two prisms brought close together. The atoms are confined to propagate in a plane of height $z \approx L/2$, where L is the mirror separation. The confining potential is shown to the right.

tion between spin and statistics is related to the fact that—in three dimensions—a bosonic wavefunction acquires a phase of $\phi = 2\pi$ under interchange of identical particles about the origin, while fermions acquire only $\phi = \pi$, implying an overall reversal in sign. (This is the origin of the Pauli exclusion principle; see Pauli, 1950.) However, if quantum particles are confined to a two-dimensional region of space, the rotation group is now U(1), and particles can have *any* real value of spin, not just integral or half-integral spins. Consequently, there is a generalized spin statistics theorem (Iengo and Lechner, 1992): In two dimensions, quantum particles can have arbitrary statistics, i.e., their wavefunctions can accumulate arbitrary phase ϕ under the interchange of identical particles. In particular, there can be particles with fractional statistics in two dimensions—such unusual particles that can carry *any* spin and obey arbitrary statistics are now called *anyons*. It is now widely accepted that fractional anyonic statistics are the explanation for the fractional quantum Hall effect, and a leading contender for theoretical explanation of high-T_c superconductivity also invokes anyons (Iengo and Lechner, 1992). Hitherto, only in the solid state, where electrons are confined to two-dimensional surfaces or conducting planes, has anyonic behavior been studied. Hence, our proposal to study supercooled neutral atomic vapors or Bose condensates in a parallel-evanescent-mirror waveguide could open up a new proving ground in the experimental study of anyonic behavior. In particular, what does *Bose* condensation even mean in two dimensions, where there are no true *bosons*?

Related to these considerations is the whole concept of phase transitions in a two-dimensional space. For example, there are phase transitions —such as the Kosterlitz-Thouless transition—that occur *only* in a two-dimensional space (Minnhagen, 1987). These transitions have, hitherto, been seen only in superfluid ^{4}He films and superconducting films. A two-dimensional waveguide for coherent atoms would allow for the experimental investigation of this effect in a new and unique system consisting of a two-dimensional atomic vapor.

The experimental setup we envision is to bring two prisms, with blue-detuned evanescent light-wave mirrors, close together until the system forms—nearly—a frustrated optical prism system (Fig. 14). Making the usual large-detuning approximation ($\Delta \gg \Omega_0, \gamma$), we can again approximate the evanescent potential by an infinite step function. Assuming that the mirrors are infinite in the transverse directions, x and y, the quantization takes place in the normal direction z, as shown in Fig. 14. Demanding that the atom's time independent wavefunction $\psi(x, y, z)$ vanish at $z = 0$ and $z = L$, where L is the mirror separation, gives modal atom wave solutions of the form

$$\psi_n(x, y, z) = \frac{2}{L} e^{\pm ik_x x} e^{\pm ik_y y} \sin k_n z \tag{83}$$

where $k_n = n\pi/L$, with $n \in \{0, \pm 1, \pm 2, \ldots\}$, and where the de Broglie wavenumbers k_x, k_y, and k_n must obey the dispersion relation

$$\varepsilon_0 = k_x^2 + k_y^2 + k_n^2 = \frac{Mv_0}{\hbar} \tag{84}$$

Here $\varepsilon_0 = 2ME_0/\hbar^2 = (Mv_0/\hbar)^2$ is the scaled kinetic energy of the atom, with $E_0 = Mv_0^2/2$. Hence, the fundamental mode, $n = 1$, corresponds to $k_1 = 2\pi/\lambda_1^{\text{dB}} = \pi/L$, or a fundamental de Broglie wavelength of $\lambda_1^{\text{dB}} = L/2$, as expected from the boundary conditions in the z direction. Coupling atoms into such a waveguide will present challenges similar to coupling atoms into hollow fibers, as discussed previously. Several possibilities present themselves: (1) Form a cooled ball of atoms in a magneto-optical trap (MOT) with the cooling beams propagating through the prisms, and then bring the prisms together on the ball; (2) bring the cloud over the frustrated prism turned sideways, and drop the atoms into the guide; (3) use a hollow fiber guide to couple atoms into the parallel-mirror guide. In particular, in method (3), using a fiber would allow velocity preselection in order to couple into a selected guided-atom mode. Once again, in methods (1) and (2), only those atoms in the Maxwell velocity distribution of the MOT cloud that also obey the dispersion relation, Eq. (84), will couple to the nth mode.

Choosing a particular direction of transverse atom wave propagation, $\mathbf{k}_\rho = k_x \hat{\mathbf{x}} + k_y \hat{\mathbf{y}}$, we can calculate the group velocity for an atom wave pulse of mean initial speed v_0 as it propagates in the guide. Defining a de Broglie frequency ω_0 by $\hbar \omega_0 = Mv_0^2/2$, then $\varepsilon_0 = 2M\omega_0/\hbar$ in Eq. (84), and one can immediately compute the group velocity v_g from the dispersion relation

$$v_g = \frac{d\omega_0}{dk_\rho} = \pm \sqrt{v_0^2 - \left(\frac{nh}{2ML}\right)^2} \tag{85}$$

where $h = 2\pi\hbar$, and L is the guide thickness. Hence, the group velocity in the nth mode of the guide is reduced from the free space mean speed of $v_0 = (v_x^2 + v_y^2 + v_z^2)^{1/2}$. The atom has given up some of its momentum to the mirrors. In addition, we see that no atoms with $v_0 < h/(2ML)$ can be guided. This is the cutoff velocity for the guide. Similar remarks hold for the hollow fiber guides, but the physics is more transparent here in the simple plane parallel-mirror guide.

V. Blue-Detuned Concave Atom Traps

The open geometries discussed in Section III for atom resonators have clear advantages when it comes to loading the trap and, possibly, further manipulating the atoms (e.g., cooling, mode selecting). They suffer, however, from the problem that it is relatively easy for the atoms to escape: The recoil kick produced when a spontaneous photon is emitted, for instance, is usually sufficient to remove the atom from a given mode, and possibly from the cavity altogether.

An intriguing alternative (Dowling, 1993; Dowling and Gea-Banacloche, 1994, 1995; Söding *et al.*, 1995) is to confine the atoms in transversely tighter cavities, of which we will consider three examples in this section: a parallelepiped-shaped box (Fig. 15) all whose sides are "coated" with evanescent waves from two lasers shining in opposite directions along the (111) axis; a pyramidal gravitational trap (Fig. 17a); and a conical gravitational trap (Fig. 17b). The latter two are open from above. These traps might be somewhat harder to load than those of Section III, but they would yield longer lifetimes since the only way to lose atoms would be if

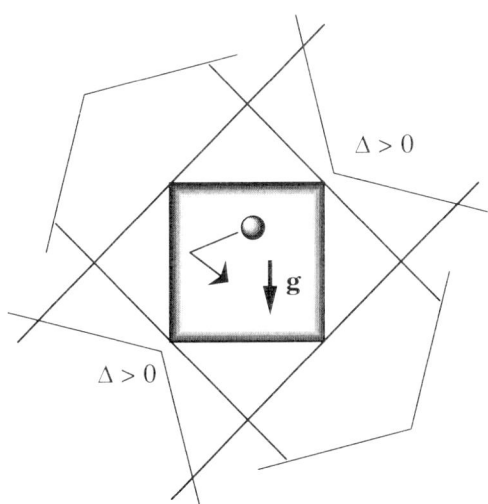

FIG. 15. Cubical quantum atom resonator in cross section. The atom is confined by the evanescent-wave mirror lining the box, which is created as two oppositely directed light beams, which are approaching the cavity from inside the dielectric along the (111) direction, are totally internally reflected. The gravitational field **g** is shown.

they made it past the evanescent-wave barrier and stuck to the wall (by e.g., emitting a spontaneous photon in the wrong direction while close to the walls, or by tunneling). If a suitable scheme can be found to cool the atoms (one will be discussed later in this section), these traps might become an effective way to store cold atoms for relatively long times.

A. Particle-in-a-Box with Gravity

The possibility of using the evanescent-wave mirror to confine an atom in a box-shaped cavity, like the one in Fig. 15, was in fact already suggested by Cook and Hill (1982) in the same paper where they proposed the evanescent-wave mirror itself. Clearly, loading such a trap would not be easy, but it might be possible to run an atom optic fiber waveguide to a hole in the trap wall, or to etch out the two halves of the trap on separate glass sheets and bring them together over the atoms (in this case, the trap starts out as a pyramid, of the sort that will be discussed in the following subsection). In any event, the system is sufficiently simple, theoretically, to allow for a detailed solution and to illustrate some of the possibilities of this kind of device.

To see clear quantum effects one should try to make the box size comparable with the thermal de Broglie wavelength λ_{dB} of the atoms. As we have seen earlier, this is typically a very small number (nanometers for Doppler-limit-cooled sodium). A micrometer-sized cavity might be about the smallest one could envision making. However, a micrometer is also the order of magnitude of the gravitational length l_g, Eq. (36), which means that at this scale gravity would be an inseparable part of the problem. This may also be understood by recalling that the particle-in-a-box (PIB) ground state energy, in the absence of gravity, is given by

$$E_1 = \frac{\hbar^2 \pi^2}{2ML^2} \tag{86}$$

for a box of side L. The PIB level spacings—of the order of Eq. (86) for the lowest lying states—become comparable with the gravitational potential energy difference MgL between the bottom and the top of the box, when the box dimension L is of the order of the gravitational length l_g, Eq. (36). If L were much smaller, gravity would play a less important role; if L were much larger, as in Section III, the quantum structure would be very difficult to resolve.

Table I (adapted from Hall et al., 1989) looks at some of the aspects involved in the design of a quantum resonator of this kind, from the perspective of making it maximally coupled to gravity. By this, we mean that we have calculated the gravitational temperature T_g necessary to

TABLE I

	^{7}Li	^{23}Na	^{24}Mg	^{39}K	^{87}Rb	^{133}Cs
λ (nm)	671	589	285	766	780	852
δ (nm)	108	95	46	124	126	138
l_g (nm)	1610	727	706	511	299	226
T_g (nK)	9	13	14	16	21	24
T_r (μK)	6.0	2.4	9.8	0.83	0.40	0.20
T_g/T_r ($\times 10^{-3}$)	1.50	5.42	1.43	19.2	52.5	120

Notes. Cooling laser wavelength, λ; corresponding evanescent-wave penetration depth, δ (assuming $n = 2.42$), characteristic gravitational length l_g, temperature T_g corresponding to a de Broglie wavelength $\lambda_{dB} = l_g$, and recoil temperature T_r, for several atomic species used in cooling and atom fountain experiments. The last row shows the important ratio of T_g to T_r.

make $\lambda_{dB} \approx l_g$. We see that these are extremely low (nano-Kelvin) temperatures, especially when compared with the recoil-cooled temperatures T_r for the same atomic species (also shown in the table). However, Reichel *et al.* (1995) have recently demonstrated Raman cooling of cesium to below 3 nK, utilizing Lévy flight statistics, which is a factor of 70 below the recoil temperature. There are also some problems in making the cavity length L comparable with l_g: Table I lists the evanescent field penetration depth $\delta = 1/\alpha$ into the cavity, assuming a large refractive index n in Eq. (14) (we have used $n = 2.42$, characteristic of diamond), and using the cooling laser wavelength appropriate to each species. We see that for the heavier atoms, δ actually becomes comparable with the gravitational length itself, so that the approximation of a sharp, "infinite" potential barrier at the walls would not hold for those atoms, and indeed, in a cavity of those dimensions, the atoms would never be in a truly field-free region. This is not a problem with the lighter atomic species, but their lighter masses yield larger values for the recoil temperature:

$$T_r = \frac{\hbar^2}{k_B M \lambda_{opt}^2} \quad (87)$$

where k_B is Boltzmann's constant. For these atoms, therefore, even if one could make the cavity of the order of the gravitational length, the de Broglie wavelength (when cooled to the recoil limit) would be several orders of magnitude smaller, so that the cavity would still be excited in some relatively high mode. It is conceivable that sub-recoil cooling, as in Reichel *et al.* (1995), might be able to overcome this problem and achieve ground state mode operation. If the penetration depth δ is reasonably small compared with the box dimension L and the gravitational length l_g, then the approximation of replacing the exponential potential rise with an infinite one is actually very good, as has been recently demonstrated in careful theoretical detail by Chen and Milburn (1995).

The more massive atoms, such as cesium, also appear to be preferable for cooling in the pyramidal and conical traps, as simulated numerically by Ovchinnikov et al. (1995a). As Table I shows, these species would indeed be ideal if the problem of the large penetration depths could be circumvented by using higher frequency transitions (sub-recoil cooling might also help here). Equation (14) suggests that this would be possible if a shorter wavelength laser were used for the evanescent wave (by coupling to a higher excited state); obviously, such an approach has limitations of its own.

From the point of view of the theory, the problem of an atom in a box with an infinite potential at the walls, and in the presence of gravity, is fairly straightforward. Consider the one-dimensional situation first, with the bottom of the box at $z = 0$ and the top at $z = L$. The Schrödinger equation reads

$$-\frac{\hbar^2}{2M}\frac{d^2\Psi}{d^2z} + Mgz\Psi = E\Psi \tag{88}$$

Introducing the gravitational length l_g defined by Eq. (36), we can scale the z variable to $\zeta = z/l_g$ and the energy to a dimensionless

$$\varepsilon = \frac{E}{Mgl_g} = E\left(\frac{2}{\hbar^2 Mg^2}\right)^{1/3} \tag{89}$$

to obtain the equation

$$\frac{d^2\Psi}{d^2\zeta} - (\zeta - \varepsilon)\Psi = 0 \tag{90}$$

With a further change of variable $\zeta' = \zeta - \varepsilon$, Eq. (90) becomes the equation for the Airy function; the solutions, therefore, are of the form

$$\Psi(\zeta) = C_1 \operatorname{Ai}(\zeta - \varepsilon) + C_2 \operatorname{Bi}(\zeta - \varepsilon) \tag{91}$$

The constants C_1 and C_2 are to be determined from the condition that the wavefunction vanish at $\zeta = 0$ and $\zeta = L/l_g \equiv \gamma$. The resulting two homogeneous equations can be solved only if the determinant is equal to zero, which results in the eigenvalue equation

$$D_\gamma(\varepsilon) = \operatorname{Ai}(-\varepsilon)\operatorname{Bi}(\gamma - \varepsilon) - \operatorname{Ai}(\gamma - \varepsilon)\operatorname{Bi}(-\varepsilon) = 0 \tag{92}$$

which determines the discrete dimensionless energy eigenvalues $\varepsilon = \varepsilon_n$ as its roots, $D_\gamma(\varepsilon_n) = 0$, where $n \in \{1, 2, 3, \ldots\}$. (Here, the dimensionless parameter γ is the ratio of the two length scales L/l_g, and not a decay rate.)

The solutions to Eq. (92) for specific values of γ are easily calculated numerically. As an example, for a cavity of length $L = 4l_g$ we obtain for the lowest three dimensionless energies $\varepsilon_1^g = 2.36$, $\varepsilon_2^g = 4.53$, and $\varepsilon_3^g = 7.60$. These can be compared with the corresponding energies for the PIB without gravity. The PIB energies (in one dimension) are given by $E_n^{PIB} = n^2 \hbar^2 \pi^2 / (2ML^2)$, and the corresponding dimensionless energies can be written as

$$\varepsilon_n^{PIB} = \frac{E_n^{PIB}}{Mgl_g} = \left(\frac{n \pi l_g}{L} \right)^2 \tag{93}$$

which for $l_g/L = 1/4$ yields $\varepsilon_1^{PIB} = 0.62$, $\varepsilon_2^{PIB} = 2.47$, and $\varepsilon_3^{PIB} = 5.55$. The level differences are $\varepsilon_2 - \varepsilon_1 = 2.17$ with gravity, versus 1.85 without (a change of about 15%); $\varepsilon_3 - \varepsilon_1 = 5.24$ with gravity, versus 4.93 without (a change of about 6%); and $\varepsilon_3 - \varepsilon_2 = 3.07$ with gravity, versus 3.08 without (a change of about 0.3%). Thus, the changes in the particle's transition frequencies due to gravity are most pronounced in the transitions to the ground state.

Using the eigenvalues ε_n, we can compute the eigenfunctions of Eq. (91) and compare them with the ordinary sinusoidal PIB wavefunctions that one obtains in the absence of gravity. Figure 16 shows $\Psi_n(\zeta)$ for $n = 1, 2, 3$, with $L/l_g = 4$. The vertical origin of each Ψ_n is set at ε_n, for clarity. The floor and ceiling of the box are at $z = 0$ and L, corresponding to $\zeta/4 = 0$ and 1, respectively. The dimensionless gravitational potential $Mgz/(Mgl_g) = \zeta$ is plotted as a dashed line. We see that Ψ_1 is most affected by g, whereas Ψ_2 and Ψ_3 feel the potential shift less and look more nearly sinusoidal. In general, we would expect the levels with energies $E \gg MgL$ ($\varepsilon \gg L/l_g$) not to be very much affected by the gravitational potential, since for such states most of the energy would be kinetic.

The dependence of the energy differences on gravity opens up the intriguing possibility of using an atom, or a cloud of atoms, in the low lying states of a trap of this type as a gravity or acceleration sensor. Since the energy levels depend on g basically as $g^{2/3}$, Eq. (89), for a single atom we expect a sensitivity to changes in g of the order of $\Delta \omega / \omega = (2/3) \Delta g / g$ (this is not entirely correct, since the dimensionless energy eigenvalues ε also depend on g, for a fixed cavity length L, through the ratio L/l_g; it is a more exact estimate for the open gravitational cavities of the pyramid

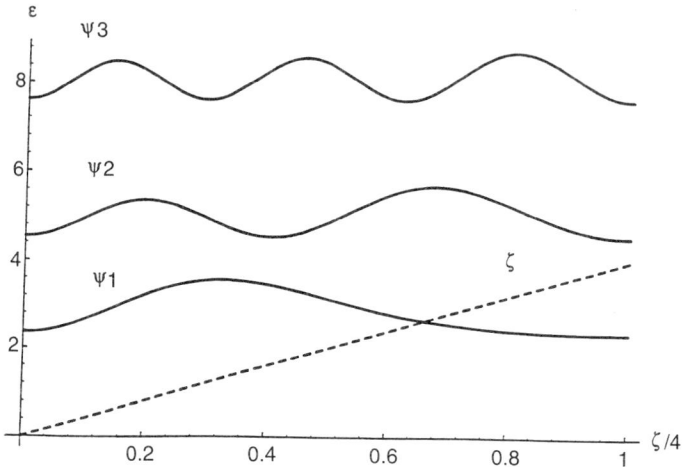

FIG. 16. The norm square of first three normalized eigenfunctions for a particle in a one-dimensional cavity in the presence of gravity, for the case $L/l_g = 4$, where L is the box height and l_g the gravitational length. Each function is shown shifted vertically by its corresponding (dimensionless) energy, and the dimensionless gravitational potential is also plotted (dashed line).

and cone type, to be discussed in the next subsections). For a cloud of N atoms, an enhancement of sensitivity by a factor of $\sqrt{N}$ would be expected, but for a Bose condensate, an enhancement factor of N could be heuristically argued (Scully and Dowling, 1993).

The $|1\rangle \rightarrow |2\rangle$ energy level difference in the resonator with gravity corresponds to a frequency of the order of $\omega_{12}/(2\pi) \approx 1500$ Hz for our numerical example. Shift measurements at such low frequencies would be difficult or impossible to obtain by such direct techniques as the Ramsey method. It is interesting to note, however, that Cirac *et al.* (1993a, b) have recently suggested an intriguing scheme for optically accessing such low frequency trap transitions. The method involves the placement of the trap —or, in this case, the resonator—at a mutual node of two detuned, monochromatic, standing wave light fields. When the detuning of the two light fields is close to the trap frequency, information can be obtained about the quantum state of the atom (or ion) in the trap. In this fashion, it might even be possible to read off the resonator level shifts and determine g.

The extension of this particle-in-a-box analysis to three dimensions presents no particular problem. It is interesting to note, however, that with the three-dimensional resonator both the magnitude and the direction of **g** may be determined. The Schrödinger equation for this case reads

$$\left(-\frac{\hbar^2}{2M}\nabla^2 + M\mathbf{g}\cdot\mathbf{r}\right)\Psi = E\Psi \tag{94}$$

where $\mathbf{g}$ is the gravity (acceleration) vector. The equation is separable in a right parallelepiped, and hence $\Psi = \mathcal{N}X(x)Y(y)Z(z)$, the solution, will be the product of three Airy solutions of the form of Eq. (91), up to the normalization constant $\mathcal{N}$. The boundary conditions in x, y, and z will result in three eigenvalue equations of the form of Eq. (92), each with its own countable set of eigenvalues E_l, E_m, and E_n, in x, y, and z, respectively. By making the three dimensions of the box different, one could lift the degeneracies and distinguish level shifts due to the components of $\mathbf{g}$ in the x, y, and z directions.

B. Pyramidal Gravitational Trap

It is also possible to let gravity confine the particle in the vertical direction, that is, to remove the top surface of the box. With one vertex pointing down, the result would be a pyramidal gravitational trap, formed by three planes intersecting at right angles, a trap first suggested by us (Dowling, 1993; Dowling and Gea-Banacloche, 1994, 1995). Of course, other angles can be envisioned; alternatively, one may consider a conical gravitational trap along the same lines, and this will be discussed in the following subsection.

The pyramidal and conical traps are reminiscent of the parabolic trap of Section III.B, but they differ in important respects. Because the opening angle is constant, a single laser beam aimed at the vertex along the axis of the cone or pyramid can meet the condition for total internal reflection at any height above the vertex; hence, the walls of the trap can be "coated" with an evanescent field up to any height (limited only by the laser beam spot size), which should result in a much narrower transverse confinement and a much longer storage time for the atoms (Fig. 17).

In contrast to the parabolic trap, there are no paraxial trajectories in the pyramidal or conical traps; because of the sharp corner at the vertex, an atom falling vertically can be reflected at a rather large angle to the vertical. Nonetheless, a sufficiently narrow opening will provide confinement in the transverse direction: The atoms cannot bounce much higher than the point where they were dropped, if they are dropped with small kinetic energy, unless they absorb many photons from the evanescent-wave field.

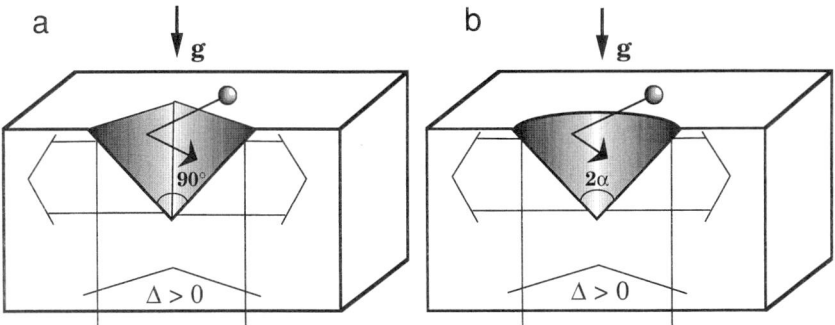

FIG. 17. Pyramidal, evanescent light-wave, gravitational atom trap (a) and conical trap (b). A blue-detuned laser (big arrow) is incident from below and totally internally reflects off the sides of the traps, producing a repulsive evanescent field, lining the inside. The minimum indices of refraction required for total internal reflection are $n = \sqrt{3/2}$ for (a), and $n = 1/\cos\alpha$ for (b).

Figure 18 shows some of the trajectories of an atom bouncing in a two-dimensional wedge under the influence of gravity. Interestingly, this classical two-dimensional motion, called the gravitational billiard, is known to be chaotic, except when the half-opening angle of the wedge is precisely 45° (as in Fig. 18a). In this case, the two planes meet at right angles and the equations of motion are separable (Lehtihet and Miller, 1986).

The quantum equations of motion for the corresponding three-dimensional right pyramid are also separable in Cartesian coordinates, and this is the only case we shall study in this section. We have to solve the Schrödinger equation (94) with the condition that the wavefunction Ψ be zero at the walls formed by the $x = 0$, $y = 0$, and $z = 0$ planes; we assume that the walls of the trap continue indefinitely, although in practice this simply means that they should extend over a length substantially greater than l_g (how much greater depends on how many of the lower lying eigenstates we are interested in). Under these circumstances, the second Airy function, $\text{Bi}(\zeta - \varepsilon)$ in Eq. (91), is not acceptable since it diverges at infinity. The solutions are then of the form

$$\Psi_{ijk}(x,y,z) = \mathcal{N}_{ijk} \,\text{Ai}(\xi + r_i)\, \text{Ai}(\eta + r_j)\, \text{Ai}(\zeta + r_k) \qquad (95)$$

where $\mathcal{N}_{ijk}$ is a normalization constant, r_i are the roots of the Airy function Ai, given by $-2.338, -4.088, -5.521, \ldots$, and the dimensionless variables ξ, η, ζ are x, y, and z scaled by the corresponding gravitational lengths,

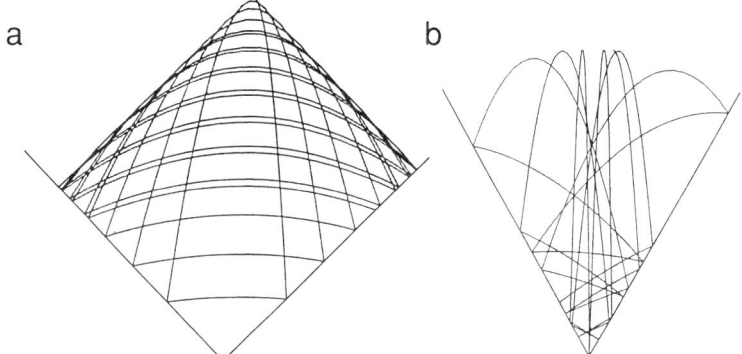

FIG. 18. The first 20 bounces of a particle in a wedge under the influence of gravity for a wedge of half-opening angle 45° (a) and of 30° (b).

i.e.,

$$\xi = x\left(\frac{2g_x M^2}{\hbar^2}\right)^{1/3}, \quad \eta = y\left(\frac{2g_y M^2}{\hbar^2}\right)^{1/3}, \quad \zeta = z\left(\frac{2g_z M^2}{\hbar^2}\right)^{1/3} \quad (96)$$

Here g_i are the (absolute values of the) components of the gravity vector **g**. Since the Airy function Ai drops off exponentially for positive values of its argument, the wavefunction of Eq. (95) is confined to a neighborhood of the vertex of the pyramid. The probability amplitude of the trapped atom does not extend much farther (in the scaled coordinates) than the largest of the values of r_i, r_j, r_k appearing in Eq. (95), corresponding to the classical turning points.

It is well known (see, e.g., Sakurai, 1985) that the nth zero, r_n, of the Airy Function Ai is given to an excellent approximation by the asymptotic formula

$$r_n \underset{n \gg 1}{\sim} \left[\frac{3\pi}{2}\left(n - \frac{1}{4}\right)\right]^{2/3} \quad (97)$$

even for relatively small integers n. [This result can be obtained by using the WKB approximation to solve (94).] The energy eigenvalue associated with the wavefunction (95) is

$$E_{ijk} = -\left(r_i g_x^{2/3} + r_j g_y^{2/3} + r_k g_z^{2/3}\right)\left(\frac{M\hbar^2}{2}\right)^{1/3} \quad (98)$$

Clearly, there are multiple degeneracies if the gravity vector is pointing along the axis of the pyramid [the (111) direction], i.e., if $g_x = g_y = g_z$. On the other hand, these degeneracies may be lifted easily by just tilting the trap slightly. In any event, it is clear that the energy level differences contain information not only on the magnitude of **g** but on the direction as well. The shifts in the energy level differences that would have to be resolved are of the same order of magnitude as those considered in the previous subsection for the PIB.

If the angular opening of the pyramid is not 90°, the differential equation (94) is still separable in Cartesian coordinates, but the boundary conditions are not, which makes the problem much more difficult to solve (one may introduce nonorthogonal axes to deal with the boundary conditions, but then the Schrödinger equation is no longer separable). We shall use, in the following subsection, the relatively simpler situation of a conical trap to illustrate what one may expect to find as the opening angle of the trap is varied. Conical traps may also be easier to make than pyramidal ones (see, e.g., Pangaribuan *et al.*, 1992), although Ovchinnikov *et al.* (1995a) have shown that higher phase-space densities might be more easily obtainable in the pyramid (see Section IV.D).

C. CONICAL GRAVITATIONAL TRAP

The Schrödinger equation (94) in spherical coordinates reads

$$-\frac{\hbar^2}{2M}\left[\frac{1}{r}\frac{\partial^2}{\partial r^2}(r\Psi) + \frac{1}{r^2 \sin\theta}\frac{\partial}{\partial\theta}\left(\sin\theta\frac{\partial\Psi}{\partial\theta}\right) + \frac{1}{r^2 \sin^2\phi}\frac{\partial^2\Psi}{\partial\phi^2}\right]$$
$$+ Mgr\cos\theta\,\Psi = E\Psi \qquad (99)$$

assuming that the gravity vector **g** points straight down along the axis of the cone (z axis). The boundary condition on the cone surface is simply that $\Psi = 0$ at $\theta = \alpha$, where α is the cone's half-apex angle (Fig. 17b).

The ϕ part of the equation is separable because of the conservation of the L_z component of angular momentum: The ϕ dependence is then of the simple form $e^{\pm im\phi}$. The remaining equation for the r and θ dependence must be solved numerically. We have reported on solutions for the first few $m = 0$ azimuthal eigenfunctions in Dowling and Gea-Banacloche (1994, 1995); we used an expansion in Legendre functions for the θ dependence and solved the coupled radial equations by a finite element method. Plots of the first few eigenvalues as functions of the cone's angle, and of the first few eigenfunctions for a few special angles, are reproduced here (Figs. 19 and 20).

MIRRORS, RESONATORS, WAVEGUIDES, AND TRAPS

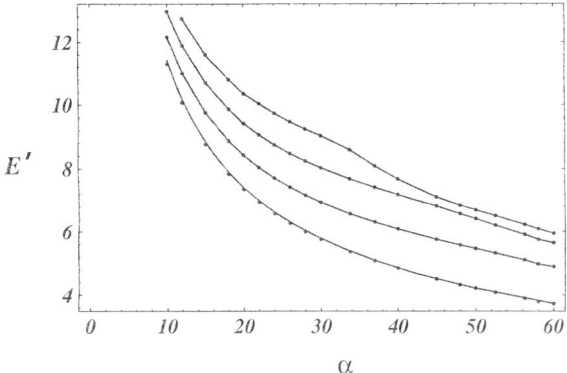

FIG. 19. The four lowest (numerically calculated) eigenvalues for the azimuthally symmetric wavefunctions inside a conical gravitational trap, as a function of the half-opening angle α. The energy is in units of Mgl_g. The lowest solid curve is the energy of the ground state, calculated by the variational method.

To calculate the lower eigenfunctions, we have found that $n \leq 5$ Legendre functions are usually sufficient to approximate the θ dependence to good accuracy; in fact, for the narrower cones the Legendre functions P_{ν_j} with $j > 1$ hardly contribute at all to the three lowest eigenfunctions. This is probably because the higher order Legendre functions have more nodes in the transverse direction; for a narrow cone, that means a component with a very small transverse wavelength and hence a very large transverse kinetic energy. For the narrow cones, it is therefore favorable, energetically, to place the nodes in the radial direction at first, as Fig. 20 shows. This means, actually, that for very narrow cones ($\alpha \leq 20°$, say) the first few wavefunctions ought to be well approximated by $U(r)P_{\nu_1}(\cos\theta)$, where ν_1 is the first solution to $P_{\nu_1}(\cos\alpha) = 0$ and $U(r)$ is the solution of the radial equation with only the first Legendre function:

$$\frac{1}{\rho}\frac{d^2}{d\rho^2}(\rho U) - \frac{1}{\rho^2}\nu_1(\nu_1 + 1)U + \rho U = \varepsilon U \qquad (100)$$

Despite its apparent simplicity, however, we have not been able to solve Eq. (100) in closed form. In Eq. (100), as in the figures, ρ and ε are dimensionless length and energy variables, scaled by the characteristic gravitational length and energy, respectively: $\rho = r/l_g$, $\varepsilon = E/(Mgl_g)$.

We have been able to obtain a very good approximation to the energy of the ground state using a variational trial wavefunction

$$\psi(\rho,\theta) = \rho^a e^{-b\rho}(\cos\alpha - \cos\theta) \qquad (101)$$

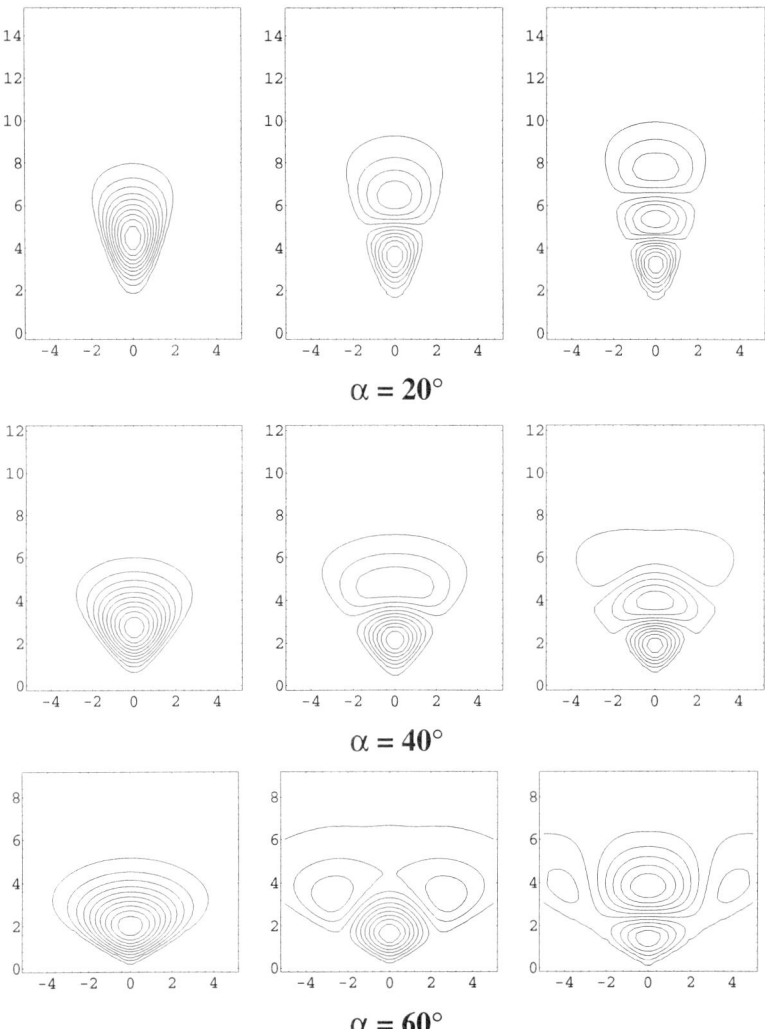

FIG. 20. Contour plots of the three lowest energy wavefunctions in a conical trap for different values of the conical half-angle α. Vertical and horizontal coordinates are in units of the gravitational length l_g.

The result is, in parametric form,
$$\cos \alpha = \frac{2a^3 + 3a^2 - 12a - 11}{2a^3 + 3a^2 + 6a + 4} \tag{102}$$
where
$$b = \left[\frac{1}{16} \frac{(a+1)(2a+1)(2a+3)(3+\cos\alpha)(1-\cos\alpha)}{a+5-(a-1)\cos\alpha} \right]^{1/3} \tag{103}$$
and
$$\varepsilon = \frac{3}{2^{8/3}} \frac{(1+2a)^2(3+2a)}{(5+6a)^{1/3}(4+6a+3a^2+2a^3)^{2/3}} \tag{104}$$

From Eqs. (102) and (103), it can be seen that for narrow cones the exponents a and b scale as
$$a \approx \frac{3\sqrt{2}}{\alpha}, \qquad b \approx \left(\frac{9\alpha}{\sqrt{2}} \right)^{1/3} \tag{105}$$
and the dimensionless energy goes as
$$\varepsilon = \frac{E}{Mgl_g} \approx 3 \left(\frac{3}{2} \right)^{1/3} \frac{1}{\alpha^{2/3}} \tag{106}$$

The dependence of the energy on the angle α, for small α, can be qualitatively understood from the following argument. Suppose the ground state corresponds to the particle being roughly at a height r above the vertex; then the potential energy is Mgr, the uncertainty in the transverse position is $\Delta x = \Delta y \approx r \tan \alpha$, and the corresponding transverse momentum uncertainty is $\Delta p_x = \Delta p_y \approx \hbar/(2r \tan \alpha)$. There is a kinetic energy $(\Delta p)^2/(2M)$ associated with this, and so we may estimate the total energy as
$$E \approx Mgr + \frac{\hbar^2}{4r^2 \tan^2 \alpha} \tag{107}$$
Minimizing E with respect to r shows that $r \approx (\tan \alpha)^{-2/3}$; more precisely, $r = l_g/(\tan \alpha)^{2/3}$. Substituting this back in (107), we obtain, for the ground state energy,
$$E \approx \frac{3}{2} \frac{Mgl_g}{(\tan \alpha)^{2/3}} \tag{108}$$

that has the correct α dependence, showing that the ground state energy diverges as $\alpha^{2/3}$ as the apex angle gets narrower and the particle is pushed higher up the cone. We expect this argument to hold equally well for the pyramidal gravitational trap.

D. Evanescent-Wave Cooling in Gravitational Traps

In several publications (Söding *et al.*, 1995; Ovchinnikov *et al.*, 1995a, b), Ovchinnikov, Söding, and Grimm (OSG) have proposed a way to cool the atoms in the pyramidal or conical gravitational traps by using the evanescent-wave field itself, coupled with a repumping laser beam that would be shined on the atoms from above. The method makes use of the hyperfine splitting of the ground-state level, characteristic of alkali atoms.

The level scheme considered is as shown in Fig. 21. The evanescent-wave laser is detuned by an amount Δ from the lowest ground state $|g_1\rangle$, which is where one expects to find most of the atoms. As a result of the hyperfine splitting Δ_{hfs}, this field is detuned by an amount $\Delta + \Delta_{hfs}$ from the upper ground state $|g_1\rangle$. Since this is greater than the detuning for $|g_1\rangle$, one expects that an atom in the upper ground state will be more weakly

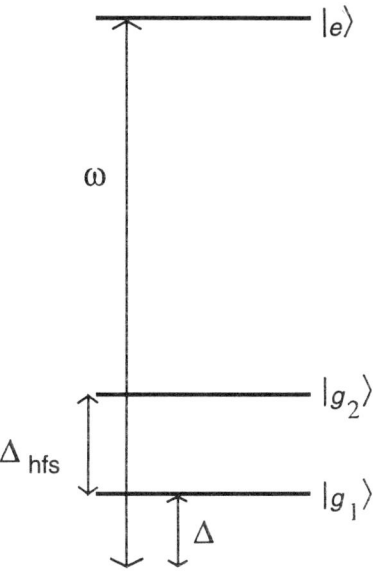

Fig. 21. Atomic level scheme, with ground state hyperfine splitting, for use in evanescent-wave cooling in a pyramidal or conical evanescent light-wave, gravitational trap.

repelled at the mirror [since, for large Δ, the dipole force is inversely proportional to Δ; see Eq. (18c)]. Söding *et al.* (1995) give, for the effective repulsive potential seen by atoms in either of the two states,

$$U_1(\xi) = \frac{2}{3}\frac{\hbar\Omega^2(\xi)}{4\Delta} \tag{109}$$

for the lower ground state, and

$$U_2(\xi) = \frac{2}{3}\frac{\hbar\Omega^2(\xi)}{4(\Delta + \Delta_{\text{hfs}})} \tag{110}$$

for the upper ground state (compare with Eq. (18c); the extra factor of 2/3 is due to a sum over the transition strengths of the excited state hyperfine levels). Here, ξ is the scaled coordinate in the direction perpendicular to the wall.

Imagine an atom that approaches the wall initially in the lower ground state. As it begins to feel the evanescent wave, there is a small probability that it might make a transition (by scattering a photon from the evanescent-wave field) to the upper ground state. If it does this while it is approaching the wall, there is a chance that it might be lost, if its kinetic energy is greater than $U_2(0)$ (this may happen to a few atoms in the early stages of the cooling process). If, however, it is reflected, it is pushed away from the wall by the weaker potential, Eq. (110), and therefore it does not recover all the kinetic energy that it lost as it approached the wall under the influence of the steeper potential, Eq. (109). Specifically, if the jump from the lower to the upper ground state occurs at a point ξ_0, the total change in the atom's kinetic energy K will be

$$\delta K = \frac{2}{3}\frac{\hbar\Omega^2(\xi_0)}{4}\left(\frac{1}{\Delta + \Delta_{\text{hfs}}} - \frac{1}{\Delta}\right) \tag{111}$$

that is, the sum of the (positive) work done by the potential U_2 as the atom moves away from the wall, plus the (negative) work done by the potential U_1 as the atom approached the wall (both effective potentials go to zero at a sufficiently large distance). This is reminiscent of the Sisyphus cooling effect (Dalibard and Cohen-Tannoudji, 1989), in that the atom is made to climb a steeper hill on the way up than on the way down.

To return the atom to the lower ground state, Söding *et al.* propose to have a weak "repumping" laser shining downward along the pyramid's (or cone's) axis, tuned to the transition from the upper ground state $|g_2\rangle$ to the excited state $|e\rangle$. The atom will eventually decay spontaneously to the lower ground state $|g_1\rangle$ and stay there, since the repumping laser is weak and not resonant with the $|g_1\rangle \rightarrow |e\rangle$ transition. This process, which is most likely to take place outside the evanescent field region (because the atom spends relatively little time there), may lead to a small heating due to the random spontaneously emitted photons, but this is more than compensated for by the loss of kinetic energy due to the downward-directed recoil as the atom absorbs photons from the repumping field (the atom is most likely to absorb photons earlier on, as it is moving away from the wall and hence upward; presumably, one could also detune the repumping field slightly toward the red to make sure more absorptions take place with the atom moving against the field). This geometric cooling is an important ingredient in determining the final equilibrium temperature; it becomes more important as the atom slows down, whereas the Sisyphus cooling is mainly responsible for the fast initial cooling rate.

The decay to the upper ground state is not the only effect possible as the atom moves into the evanescent-wave field. A transition to the excited state is also possible, in principle, but this is unlikely because of the large detuning. The atom does not live long in the attractive state; it decays in about one spontaneous lifetime. There is a small amount of heating associated with this decay, but it turns out to be negligible compared with the heating due to the scattering of evanescent-wave photons by ground state atoms.

Ovchinnikov *et al.* and Söding *et al.* have carried out Monte Carlo simulations of the evanescent-wave and geometric cooling for both pyramidal and conical gravitational traps. (Their pyramid is four-sided, with an apex angle of 90°, unlike the three-sided pyramid we considered in Section IV.B.) They find equilibrium temperatures near the recoil limit for potassium, rubidium, and cesium in the pyramidal trap, but generally larger (by about a factor of 20) for the cones. The reason for this is that, in the conical geometry, the z component of angular momentum, L_z, is preserved due to azimuthal symmetry, and there is some amount of kinetic energy that is tied up in this. Specifically, even if one could cool down to zero all the components of the velocity but the azimuthal one, a particle at height r above the apex of the cone, moving in an orbit of radius $r \tan \alpha$, would still have a total energy

$$E = Mgr + \frac{L_z^2}{2Mr^2 \tan^2 \alpha} \tag{112}$$

which, when minimized with respect to r, yields a kinetic energy

$$K_{\min} = \tfrac{1}{2}\left(Mg^2 L_z^2\right)^{1/3} \tag{113}$$

below which the cooling methods are ineffective. To overcome the limit (113) the rotational symmetry has to be broken, e.g., by tilting the cone.

The equilibrium rms momentum, $\mathscr{P}_{rms}$, for the atoms in a pyramidal trap is given, according to Söding *et al.*, by the equation

$$-\frac{2}{3}\frac{\Delta_{hfs}}{\Delta + \Delta_{hfs}}\left(\frac{\mathscr{P}_{rms}}{\hbar k}\right)^2 - \frac{\sin \alpha}{q_r}\frac{\mathscr{P}_{rms}}{\hbar k} + \frac{2}{q_r^2} + \frac{2}{1 - q_{ev}} = 0 \quad (114)$$

where q_{ev} is the mean branching ratio to the lower hyperfine ground state for the scattering of an evanescent-wave photon, and q_r the same but calculated for the excitation from the upper hyperfine to the lower hyperfine ground state by the repumping laser. The first term in this equation represents the evanescent-wave cooling, the second one the "geometric cooling" by the repumping laser, and the last two the heating effects associated with scattering and spontaneous emission. For cesium, in the simulations described in Söding *et al.* (1995), Eq. (114) predicts $\mathscr{P}_{rms}$ = $3.2\hbar k_{opt}$, in good agreement with the $3.4\hbar k_{opt}$ obtained from the Monte Carlo simulation.

The energy corresponding to this rms momentum for cesium, with a cooling wavelength of 852 nm (Table I), is about $84 Mgl_g$. Comparing this with the energy levels of the three-sided pyramid in Section VI.B, we see that it corresponds to about the tenth zero of the Airy function in every one of the three orthogonal directions, which is about the $n = 10^3$ energy level (including degeneracies). Some experimental progress in making pyramidal and conical traps has been reported by Lee *et al.* (1996).

VI. Red-Detuned Convex Atom Traps and Guides

In the previous two sections we discussed atom resonators and hollow fiber guides that were concave, and for which the restoring confining force was provided by repulsive, blue-detuned, evanescent light-wave mirrors and possibly gravity. In this section, we discuss resonators in which a convex dielectric surface presents an attractive, red-detuned evanescent field to the vacuum. In these traps, a second effective repulsive force is required to balance the attractive evanescent field so that the trapped atoms may orbit the convex glass body in a stable configuration. Prototypical of these traps is the microsphere "whispering-gallery" trap of Mabuchi and Kimble (1994), discussed next.

A. MICROSPHERE WHISPERING-GALLERY TRAP

An acoustical *whispering gallery* is a gallery or dome of circular or elliptical horizontal cross section in which faint sounds propagate with startling clarity around the circumference of the inner wall with little attenuation. Hence, a whisper produced near the wall of the gallery can be heard distinctly by a listener at the wall on the opposite side, but heard not at all by an unwary subject who is standing out away from the wall in the middle of the gallery where little sound reaches. One of the most famous of whispering galleries is St. Paul's Cathedral in London, where legend has it that misplacement of a confessional too close to the wall once allowed gossipy parishioners to inconspicuously listen in on privileged conversations between sinner and priest by innocently loitering close to the cathedral wall at another point far around the circumference. Similarly, optical whispering-gallery modes (WGMs) occur in dielectric resonators of circular cross section—such as spheres or disks—where light propagates by total internal reflection around the circumference. If the circle has radius R, then (in the limit $R \gg \lambda_{opt}$) resonance between the geometry and the light occurs whenever an integral number of wavelengths λ_{opt} fit around the circumference, $l\lambda_{opt} = 2\pi R$, $l \in \{0, 1, 2, \ldots\}$. These resonances correspond to Mie resonances of the dielectric resonator (van der Hulst, 1981). The fact that these optical modes propagate by total internal reflection, with extremely low loss, allows optical fields to build to very high values in these modes, forming an extremely high-Q cavity. This principle is behind lasing and nonlinear optical effects in dye droplets (Campillo *et al.* 1991), microdisk laser diodes (McCall *et al.*, 1992), and solid dielectric microspheres (Treussart *et al.*, 1994). In particular, as emphasized by Mabuchi and Kimble (1994) and Dutta and Agarwal (1995), if atoms could be made to interact with these optical whispering-gallery modes, cavity QED couplings to the atoms could be increased by orders of magnitude over current schemes. Large atom–cavity couplings are crucial for making, e.g., the types of controlled-NOT gates required for quantum computers (Feynman, 1982, 1985, 1986; Deutsch and Joza, 1992; Ekert, 1995; Barenco *et al.*, 1995; Monroe *et al.*, 1995; Pellizzari *et al.*, 1995). Since the atom–cavity coupling is enhanced by long interaction times, Mabuchi and Kimble suggested that to maximize this time one could bind the atom to the WGM by the evanescent vacuum field, with the atom orbiting, say, a dielectric microsphere. We first consider a simpler trap model than that of Mabuchi and Kimble; in particular, we assume that the only two forces on the atom are an attractive red-detuned evanescent field from a WGM and a repulsive centrifugal force.

For a dielectric microsphere of radius R and index of refraction n, a TM-polarized WGM propagating around the inside walls of the sphere at the equator produces an evanescent wave $\mathbf{E}(r, \theta, \phi)$ extending into the vacuum (Fig. 22a), whose radial component $E_r(r, \theta, \phi)$ is given by (Mabuchi and Kimble, 1994; van der Hulst, 1981, Section 9.2)

$$E_r = \mathcal{N}_{lmp} E_0 P_l^m (\cos \theta) \frac{h_l^{(1)}(k_{lp} r)}{k_{ln} r} e^{i(m\phi - \omega_{lp} t)} \qquad (115)$$

where $k_{lp} = \omega_{lp}/c$, E_0 is the field in the dielectric, $\mathcal{N}_{lmp}$ is a normalization constant, P_l^m is the associated Legendre function, and $h_l^{(1)}$ is the spherical Hankel function (Abramowitz and Stegun, 1965). Here, l and m are the angular and azimuthal mode numbers, respectively, for the optical whispering-gallery field, and p is a radial mode number that—for each l—labels one of a discrete set of complex resonant wavenumbers $k_{lp} = k'_{lp} + ik''_{lp}$, which are determined from the characteristic equation that arises when $\mathbf{B}$ and $\mathbf{E}$ are made to satisfy the appropriate boundary conditions at the sphere's surface (van der Hulst, 1981). (The indices p, l,

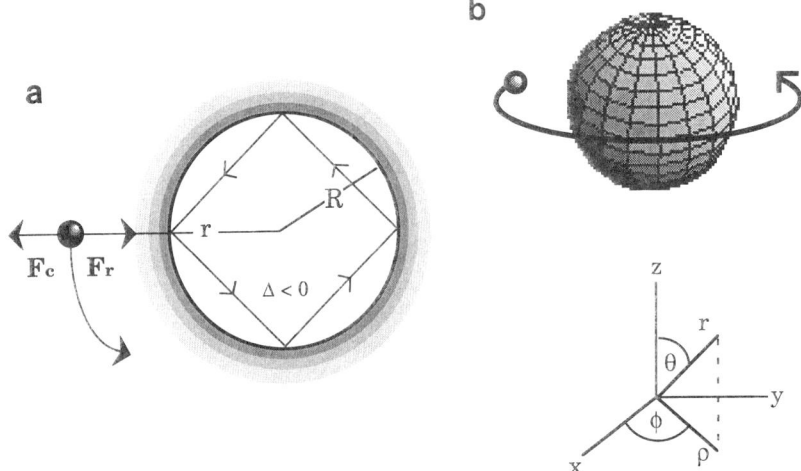

FIG. 22. Atom orbiting microsphere at equator, trapped by centrifugal force $\mathbf{F}_c$ and the attractive evanescent field $\mathbf{F}_r$ of a red-detuned ($\Delta < 0$) whispering-gallery mode propagating around the inside wall of the sphere by total internal reflection. In (a), we show the cross section in the orbital plane at the equator ($\theta = \pi/2$). The gray-level gradation indicates the evanescent field. In (b), we show a schematic of the spherical microsphere trap and the coordinate system.

and m count the number of nodes in the electric field in the r, θ, and ϕ directions, respectively.) The other components, E_θ and E_ϕ, can be obtained from E_r, Eq. (115), by appropriate partial differentiation with respect to r, θ, and ϕ (van der Hulst, 1981).

For most of the experimental considerations for dielectric microspheres (Treussart et al., 1994; Mabuchi and Kimble, 1994), the radius of the sphere is much larger than the free space optical wavelength, $R \gg \lambda_{opt}$ or $k_{opt} R \gg 1$, where $k_{opt} \equiv 2\pi/\lambda_{opt}$. (For example, $R = 50$ μm and $\lambda_{opt} \lesssim 1$ μm.) In this limit, we may approximate the spherical Hankel function $h_p^{(1)}$ by its asymptotic value (van der Hulst, 1981),

$$h_p^{(1)}(kr) \underset{kr \gg 1}{\sim} (-i)^{p+1} \frac{e^{ikr}}{kr} \tag{116}$$

Making the assumption that we are on the equator, $\theta = \pi/2$, allows us to define a radial, position dependent Rabi frequency $\Omega_{lmp}(r)$ by

$$\Omega_{lmp}(r) \equiv \mathcal{N}_{lmp} |P_l^m(0)| \frac{e^{-k_{lp}'' r}}{|k_{lp}|^2 r^2} \Omega_0 \tag{117}$$

where we have combined Eqs. (115) and (116). Here, Ω_0 is on the order of $n\Omega_i$, where n is the index of refraction of the sphere and Ω_i is the Rabi frequency associated with the incident free space laser that is pumping the WGM, as per Eq. (12). The factor $|P_l^m(0)|$ can be computed from (Gradshteyn and Ryzhik, 1965)

$$P_l^m(0) = \frac{2^m \sqrt{\pi}}{\Gamma\left(\frac{1+l-m}{2} + \frac{1}{2}\right) \Gamma\left(\frac{1-l-m}{2} + \frac{1}{2}\right)} \tag{118}$$

where Γ is Euler's gamma function (Abramowitz and Stegun, 1965). This factor can be very large, especially when $m \approx l$, as can be seen from the special case of Eq. (118), namely,

$$P_l^l(0) = (-1)^l (2l-1)!! \tag{119}$$

where the "!!" are not for emphasis, but rather represent the odd factorial, namely, $(2l-1)!! \equiv 1 \cdot 3 \cdot 5 \cdots (2l-1)$. However, the normalization factor N_{lmp} can be very small at these points. If we write $\mathcal{N}_{lmp} = \mathcal{N}_{lm} \mathcal{N}_p$, as the product of angular and radial normalization factors, then (Abramowitz and

Stegun, 1965)

$$\mathcal{N}_{lm}^2 = \frac{2l+1}{2} \frac{(l-|m|)!}{(l+|m|)!} \quad (120)$$

with $0! \equiv 1$. Hence, the product $|P_l^m(0)|^2 \mathcal{N}_{lm}^2$, which corresponds to a numerical evanescent-wave enhancement factor, is not necessarily all that large. However, for certain large values of l and m, this factor—in conjunction with the Mie resonance, whispering-gallery, wavenumber solutions k_{lp}—can give a very large enhancement to the Rabi frequency $\Omega_{lmp}(r)$ in Eq. (117). This WGM enhancement of the evanescent wave can be of several orders of magnitude, due to the high Q value of these modes, and hence it offers an alternative scheme to that of the surface-plasmon or waveguide enhancement mechanisms discussed in Section II.

1. Red-Detuned Centrifugal Trap

In the large detuning limit, $\Delta \gg \Omega_{lmp}$, we can use the gradient potential $V(r)$, in the form of Eq. (18c), namely,

$$V_{lmp}(r) \approx \frac{\hbar \Omega_{lmp}^2(r)}{4\Delta}$$

$$= \mathcal{N}_{lmp}^2 |P_l^m(0)|^2 \frac{\hbar \Omega_0^2}{4\Delta} \frac{e^{-2k_{lp}''r}}{|k_{lp}|^4 r^4} \quad (121)$$

which has the form of a Yukawa-type potential. For $\Delta < 0$, the potential is attractive, with an approximate gradient force

$$F_{lmp}(r) = -\frac{dV_{lmp}}{dr}$$

$$\approx 2k_{lp}'' V_{lmp}(r) \quad (122)$$

where we have neglected a term of order $\exp(-2k_{\text{opt}}r)/|k_{\text{opt}}r|^5$, assuming $k_{\text{opt}}R \gg 1$. Here, $k_{lp}'' = \text{Im}\{k_{lp}\}$, with k_{lp} the complex propagation constant associated with the evanescent field. We can set this equal to the centrifugal force $F_c = Mv_\phi^2/R$, to estimate the transverse orbital velocity v_ϕ of an atom circling just above the sphere. Taking from Mabuchi and Kimble: $R = 50\ \mu$m, $\lambda = 894$ nm, $l = 492$, $m = 488$, and $p = 1$, gives a value of $\Omega_{lmp}(R)/(2\pi) \approx 3.45 \times 10^8$ Hz. Taking a detuning of $\Delta/(2\pi) = -2.2 \times 10^{12}$ Hz from that paper also, we see $|\Delta| \gg \Omega_{lmp}(R)$ is satisfied.

This gives an orbital velocity

$$v_\phi = \sqrt{\frac{2k''_{lp}\hbar R}{M|\Delta|}}\ \Omega_{lmp}(R) \qquad (123)$$

Approximating $k''_{lp} \approx 2\pi/\lambda_{\text{opt}}$, and using the mass of Cs^{133}, gives a velocity of $v_\phi \approx 34$ cm/sec, corresponding to an orbital kinetic energy temperature of $T_\phi \approx 1.22$ mK. From Eq. (127), following, this corresponds to a radial temperature of $T_r = 2\pi R T_\phi/\lambda_{\text{opt}} \approx 3.5$ μK, in qualitatively good agreement with the full numerical result of Mabuchi and Kimble, i.e., about 10% of 40 μK.

2. Red–Blue Pushme–Pullyou Trap

In the actual resonator scheme of Mabuchi and Kimble, however, there is not just one optical field, but two (Fig. 23). There is the attractive red-detuned field at $\lambda_r = 894$ nm propagating in whispering-gallery mode $(l, m, p) = (492, 488, 1)$ and also a repulsive *blue*-detuned field propagating in the WGM $(971, 971, 1)$ with $\lambda_b = 456$ nm. Since many more of the λ_b fit around the inside of the spherical shell wall, they have a much shallower angle of incidence than red-detuned waves. In the $R \gg \lambda_b, \lambda_r$ limit, this is precisely the same sort of pushme–pullyou trap first suggested by Ovchinnikov *et al.* (1991), as discussed in Section III.C. Here, too, the angle of incidence of the blue-detuned ray is much shallower than that of the red-detuned ray, $\theta_b \gg \theta_r \gtrsim \theta_c$, where θ_c is the critical angle for total internal reflection. Hence, by Eq. (14), the blue-detuned ray has a much smaller evanescent penetration depth $\delta_b \ll \delta_r$, as discussed in Section II.C. Therefore, the red-detuned evanescent field provides a long range attractive force far out from the surface, while the blue-detuned evanescent field provides a short range repulsive force to keep the atom away from the surface (see Fig. 23). In the Ovchinnikov *et al.* pushme–pullyou scheme, a tightly focused laser beam, with a Gaussian profile, striking a planar interface can also provide confinement transverse to the mirror normal, making a real three-dimensional trap. In the Mabuchi–Kimble scenario, the spherical geometry elongates this trap into a toroid that encircles the sphere. If the red- and blue-detuned Rabi frequencies $\Omega_{lmp}^{(r,b)}(R)$ are strong enough, then the centrifugal force is only a small (10%) perturbation of this pushme–pullyou potential, as just shown. There are several advantages to the pushme–pullyou scheme, over the simple centrifugal orbits: (1) One can tune both the repulsive and attractive potentials independently; (2) the atoms can be held in regions of high field

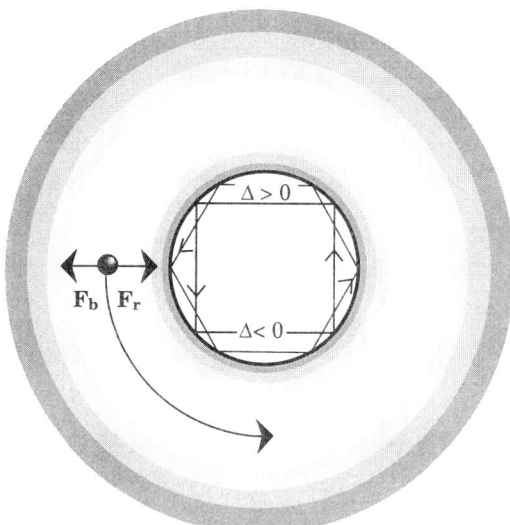

FIG. 23. "Red–blue pushme–pullyou," whispering-gallery mode (WGM) microsphere trap of Mabuchi and Kimble (1994). Two lasers, one red-detuned ($\Delta < 0$) and one blue-detuned ($\Delta > 0$), propagate by total internal reflection around the wall of the dielectric sphere. At each reflection, the blue-detuned rays have a much larger angle of incidence than the red-detuned ones, since many more of these wavelengths, λ_b, fit around the wall. Hence, the red-detuned evanescent field extends farther into space than the blue-detuned one, together forming a potential well a short distance from the sphere surface, as per Section III.C. The atom orbits the sphere in this potential, as shown.

strength at large detunings, giving huge atom–cavity coupling factors; (3) the atoms are stable against precession of the orbital plane.

B. External Solid Fiber Guide

In the previous discussion of microspherical, WGM, evanescent light-wave traps, we bound atoms in orbits around dielectric microspheres. In that scenario, a red-detuned evanescent wave provided a radial attractive force, while either centrifugal force or a blue-detuned evanescent wave provided a radial repulsive force, confining the atoms in stable equatorial orbits. Similar configurations can be envisioned in a cylindrical—rather than spherical—geometry, with light propagating in solid fibers by total internal reflection, providing the requisite evanescent waves (see Figs. 24 and 25). We present two such schemes for the first time here.

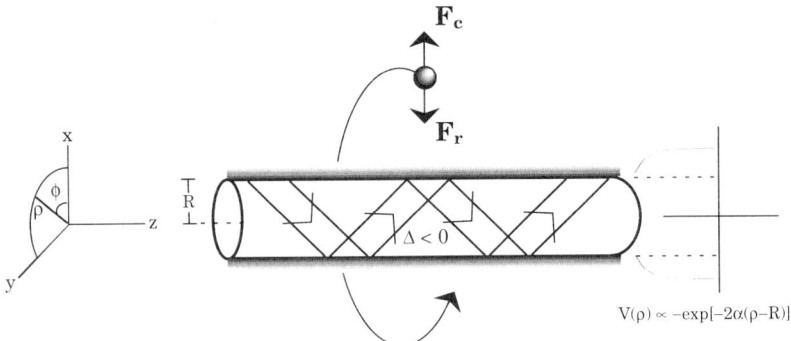

FIG. 24. Red-detuned solid fiber centrifugal waveguide. Red-detuned ($\Delta < 0$) laser light propagates down large ($R \gg \lambda_{\text{opt}}$) solid fiber by total internal reflection in TE or TM modes. This produces an attractive evanescent field (gradation) that "coats" the outside of the fiber. We propose to guide the atoms in helical orbits moving in the z direction and confined radially by a balance of centrifugal, $\mathbf{F}_c$, and red-detuned gradient, $\mathbf{F}_r$, forces.

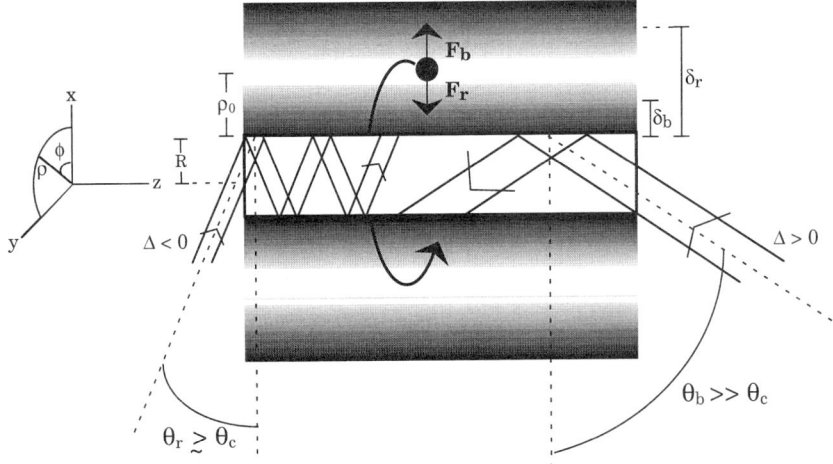

FIG. 25. Red–blue detuned "pushme–pullyou," solid fiber orbital guide. Here, we show a solid fiber in which a red-detuned ($\Delta < 0$) and a blue-detuned ($\Delta > 0$) laser counterpropagate in the fiber in TE or TM modes, in the slab waveguide approximation, good for $R \gg \lambda_{\text{opt}}$. The red-detuned and blue-detuned beams are arranged to totally internally reflect down the fiber with two different angles of incidence $\theta_r \gtrsim \theta_c$ and $\theta_b \gg \theta_c$, respectively, where θ_c is the total internal reflection critical angle. The red- and blue-detuned beams have evanescent skin depths of $\delta_r \gg \delta_b$, and the atom is attracted to an annular sheath at a radial distance ρ_0 where the radial potential $V(\rho_0)$ is a minimum. The atoms are confined to circular radial orbits in this potential—but may translate freely in the z direction.

1. Red-Detuned, Centrifugal, Solid Fiber Guide

Let us first consider the case of a centrifugal restoring force, as shown in Fig. 24. We consider here only a ballistic classical trajectory for the atom and a solid fiber of large radius $R \gg \lambda_{\text{opt}}$, where λ_{opt} is the free space wavelength of the light. Light rays that enter the fiber within the aperture angle, $\theta \lesssim \theta_{\text{ap}} \equiv \pi - \theta_c$, where θ_c is the critical angle of total internal reflection, will propagate down the fiber. In the $R \gg \lambda_{\text{opt}}$ limit, we may use the slab approximation (Savage *et al.*, 1993; Solimeno *et al.*, 1986) and treat the fiber as a dielectric slab of index n and thickness $2R$. (Since these red-detuned schemes involve a lot of spontaneous emission diffusion, there is no sense in looking at small fibers with $R \approx \lambda_{\text{opt}}$, since coherent atom wave propagation is not feasible anyway. The current scheme could, however, be used for coherent ballistic guiding of atoms.) From Eq. (10), we see that the evanescent electric field just outside the solid fiber has the form

$$\mathbf{E}(\rho, \phi, z) = \hat{\mathbf{e}} E_0 e^{-\alpha(\rho - R)} e^{-i(k_{\text{opt}} z - \omega t)} \tag{124}$$

where we have switched to cylindrical coordinates appropriate for light propagating in the z direction (see Fig. 24). The evanescent field strength E_0 is computed from the propagating incident field E_i as per Eq. (12), and the inverse skin depth $\alpha = 1/\delta \approx 2\pi/\lambda_{\text{opt}}$ is given by Eq. (14). Taking the position dependent Rabi frequency as $\Omega(\rho) \propto |\mathbf{E}(\rho, \phi, z)|$, in the large detuning limit we have, from Eq. (18c), a radial potential

$$V(\rho) \underset{\Delta \gg \Omega_0, \gamma}{=} \frac{\hbar \Omega_0^2}{4\Delta} e^{-2\alpha(\rho - R)} \tag{125}$$

which is clearly attractive for red detunings ($\Delta < 0$). Setting the resultant gradient force, $F_r = -dV/d\rho$, equal to the centrifugal force, $F_c = Mv_\phi^2/\rho$, yields for the kinetic energy of orbital motion a temperature equivalent, $T_\phi \equiv Mv_\phi^2/k_B$, where k_B is Boltzmann's constant and v_ϕ is the azimuthal orbital velocity. This can be written

$$T_\phi = \frac{\hbar \Omega_0^2}{2|\Delta| k_B} \alpha \rho e^{-2\alpha(\rho - R)} \tag{126}$$

To be trapped in such orbits, the atoms must have a mean radial velocity less than v_ρ, where $Mv_\rho^2/2 \approx V(\rho) \equiv k_B T_\rho/2$, where T_ρ is an associated temperature, which can be written

$$T_\rho = \frac{\hbar \Omega_0^2}{2|\Delta| k_B} e^{-2\alpha(\rho - R)} \tag{127}$$

and hence $T_\phi = \alpha \rho T_\rho$.

Let us take an experimental setup similar to the red-detuned hollow fiber guide experiments of Renn et al. (1995), but with the hollow filled in and the atoms to be guided in helical orbits along the outside of a solid fiber of radius $R = 77$ μm. Let us assume similar experimental parameters, such as ^{87}Rb atoms to be guided by 500 mW of laser power with an incident angle of $\theta_i \approx 0.26$ rad. From Table I, we see that the cooling transition of ^{87}Rb corresponds to $\lambda_{opt} = 780$ nm. With these numbers, Eq. (127) gives $T_\rho \approx 22$ mK and hence $T_\phi \approx 14$ K, since $\alpha\rho \approx 2\pi R/\lambda_{opt} \approx 620$. These values correspond to radial and azimuthal velocities of $v_\rho \approx 144$ cm/sec and $v_\phi \approx 3600$ cm/sec, respectively. (The azimuthal velocity v_ϕ can be large because the centrifugal force, $F_c \approx Mv_\phi^2/R$, scales with $1/R$ and R is large, so that $F_c \approx F_r$, the relatively small gradient force.) Hence, if the end of this solid fiber is stuck in a thermal atomic vapor, atoms with these values of v_ϕ and v_ρ will be guided in helical orbits down the fiber at net forward velocity v_z. (In this classical ballistic regime, the atom motion is not quantized and no dispersion relation or quantization condition relates v_z to v_ρ and v_ϕ, as in the blue-detuned quantum atom waveguides discussed previously.) When the guided atoms reach the other end of the fiber, they will fan out into a cone of angular opening $\vartheta = 2\arctan(v_\rho/v_z)$. Since the range of v_ρ is known, atoms at a particular angle ϑ will have a velocity $v_z = v_\rho \tan 2\vartheta_0$, and so the device can be used as a velocity selector.

2. Red–Blue Pushme–Pullyou, Solid Fiber Atom Guide

We now consider the case where the dominant outward radial restoring force is not a centrifugal force, but a blue-detuned evanescent wave in the pushme–pullyou configuration of Ovchinnikov et al. (1991), discussed in Section III.C. (This is the same idea behind the Mabuchi–Kimble WGM trap, Section VI.A.) In the planar slab dielectric guide approximation ($R \gg \lambda_{opt}$), we may transform the pushme–pullyou gradient potential, Eq. (52), as

$$V(\rho) \underset{\Delta \gg \Omega_0, \gamma}{=} \frac{\hbar}{4}\left[\frac{\Omega_b^2}{|\Delta_b|}e^{-2\alpha_b(\rho-R)} - \frac{\Omega_r^2}{|\Delta_r|}e^{-2\alpha_r(\rho-R)}\right] \quad (128)$$

where Ω_b and Ω_r are the blue- and red-detuned Rabi frequencies, with detunings Δ_b and Δ_r, respectively (see Fig. 25).

As in the Mabuchi–Kimble setup (1994), we need two transitions near λ_r and λ_b. Using cesium, as in the WGM trap, we take $\lambda_r = 894$ nm and

$\lambda_b = 456$ nm. Now, to implement the pushme–pullyou configuration, we need to have the red- and blue-detuned rays with angles of incidence $\theta_r \gtrsim \theta_c$ and $\theta_b \gg \theta_c$, respectively, where θ_c is the critical angle for total reflection. Let us assume a fiber of index $n = 1.45$, in which case $\theta_c \approx 0.76$ rad. Hence, a choice of $\theta_r = \pi/4 \approx 0.79$ rad and $\theta_b = \pi/3 \approx 1.1$ rad will do. With these parameters, we can calculate vacuum penetration of "skin" depths from Eq. (14), yielding $\delta_r \approx 634$ nm and $\delta_b \approx 102$ nm. Hence, the red-detuned light has a larger penetration depth than the blue-detuned, as required by the pushme–pullyou scheme (Section III.C). Atoms will be localized at a potential minimum a radial distance $\rho_0 - R = 222$ nm above the fiber surface, as computed from Eq. (53), assuming $\Omega_r = \Omega_b \approx 8.9 \times 10^8$ rad/sec, and $\Delta_b = -\Delta_r = 1.4 \times 10^{10}$ rad/sec, which are comparable with the numbers of Renn et al. (1995). From Eq. (12), these Rabi rates correspond to a laser power of about $P = 500$ mW. The potential well of Eq. (128) has walls corresponding to transverse temperatures of confinement, T_ρ, given by Eq. (127), namely $T_\rho \approx 10$ mK, comparable with the transverse well depth of the blue-detuned hollow fiber guide experiment of Renn et al. (1996). The red-detuned potential must be large enough to overcome centrifugal force. The corresponding orbital kinetic temperature for this is given from Eq. (126) by $T_\phi = \alpha_r \rho T_\rho \approx 2\pi R T_\rho / \lambda_r \approx 88.5$ mK, for $R = 50$ μm. Once again, atoms in these orbits will be guided along the fiber in an annular sheath at a distance $\rho - R = 222$ nm radially out from the fiber surface, within a potential well about 266 nm wide. The tricky part in experimentally setting this up would be arranging the two different angles of incidence, θ_r and θ_b. One idea is to send the red-detuned light in one end of the fiber and the blue-detuned light in the other, using coupling lenses with different numerical apertures, as shown in Fig. 25.

We note that in the large radius limit discussed here, $R \gg \lambda_{opt}$, the physics of TM_1 or TE_1 mode optical propagation in the cylindrical solid fiber dielectric is the same as for light propagating in an infinite, planar, right-rectangular dielectric slab of index n and thickness R, bounded on either side by vacuum. Although centrifugal force cannot now be used as the restoring force to make a guide, the pushme–pullyou scheme will still work. Considering Fig. 25 to represent a cross section through an infinite dielectric slab—instead of a fiber—we can see that atoms will now be guided on either side of the slab in either of two planar regions at heights $y = \pm y_0$ (where $\rho \to y$ in this coordinate system). This is essentially the original Ovchinnikov et al. (1991) pushme–pullyou two-dimensional atom waveguide scheme (Fig. 7), as discussed in Section II.C, combined with the dielectric waveguide, evanescent light-wave, mirror enhancement proposal of Kaiser et al. (1994) and Seifert et al. (1994b), as can be seen in Fig. 2.

C. INVERTED CONE YUKAWA-POTENTIAL TRAP

Recent experimental work by Pangaribuan *et al.* (1992) at the Tokyo Institute of Technology has concentrated on selective etching techniques with optical fibers. In particular, the core and the cladding of the fiber respond differently to different etching solutions. If the core etches faster than the cladding, a concave hollow cone can be microfabricated at the end of the fiber, with the cone and fiber axis the same. Light propagating down the fiber would totally internally reflect off the sides of this conical fissure, providing the necessary ingredients for the conical gravitational trap discussed in Section V.C. However, if the cladding etches faster than the core, the result this time is a protruding conical peak at the end of the fiber. These cones can have a base with a diameter on the order of 2 μm. The tip of the cone is very small, approximated by a sphere with a radius of $R \approx 15$ nm, much smaller than an optical wavelength (see Fig. 26). In this case, light propagating up the fiber produces a very strong evanescent field around the fiber tip that extends a penetration depth $\delta \approx R$ around the tip. Hence, the tip produces an approximately spherical evanescent field intensity that falls off exponentially with radial distance. Such an intensity profile provides a Yukawa-type potential that is attractive if the light is red-detuned with respect to an atomic resonance (Hori *et al.*, 1992).

To make a stable atom trap from this structure, we must balance centrifugal, gradient, and spontaneous forces. We model the fiber cone dip as a sphere of radius $R \approx 15$ nm, as shown in Fig. 26. The penetration depth $\delta = 1/\alpha$ is no longer determined in a simple fashion, as in Eq. (14). For the ^{87}Rb transition of D2, with $\lambda_{\text{opt}} \approx 780$ nm, we have $R \ll \lambda_{\text{opt}}$. In this regime, $\delta = 1/\alpha \approx R$ (Hori *et al.*, 1992). With this proviso, the gradient force, $\mathbf{F}_r$, of Eq. (11), and the spontaneous force, $\mathbf{F}_s$, Eq. (13), have the form

$$\mathbf{F}_r \underset{\Delta \gg \Omega_0}{\approx} \frac{2\hbar\alpha\Delta\Omega_0^2}{\gamma^2 + 4\Delta^2} e^{-2\alpha(r-R)}\hat{\mathbf{r}} \tag{129a}$$

and

$$\mathbf{F}_s \underset{\Delta \gg \Omega_0}{\approx} -\frac{\hbar k \gamma \Omega_0^2}{\gamma^2 + 4\Delta^2} e^{-2\alpha(r-R)}\hat{\boldsymbol{\theta}} \tag{129b}$$

where $\hat{\mathbf{r}}$ and $\hat{\boldsymbol{\theta}}$ are unit vectors in the coordinate system of Fig. 26. These two forces must be balanced by a centrifugal force $\mathbf{F}_c$ of the form

$$\mathbf{F}_c = \frac{Mv^2}{r \sin \theta}(\hat{\mathbf{r}} \sin \theta + \hat{\boldsymbol{\theta}} \cos \theta) \tag{130}$$

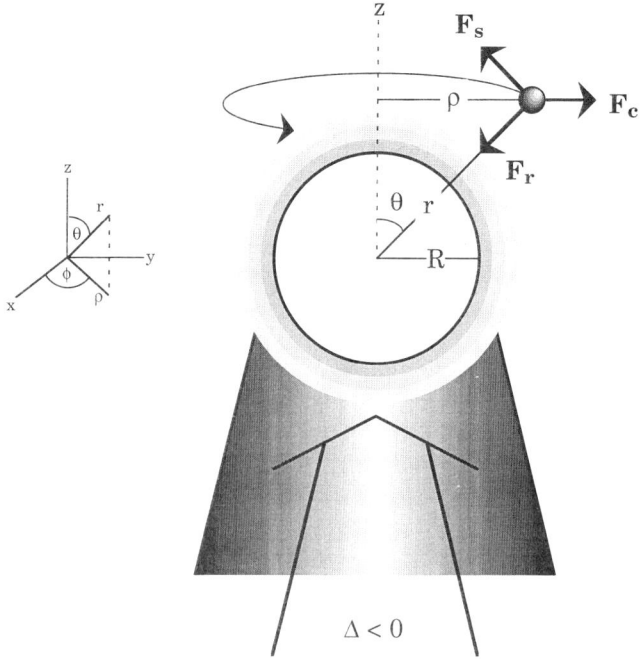

FIG. 26. Red-detuned Yukawa-potential trap at the tip of an inverted cone. Selective etching techniques (Pangaribuan *et al.*, 1992) produce a sharp-tipped cone that protrudes from the end of a solid glass fiber. The cone has a base diameter of about $2\,\mu$m and a tip that has a radius of curvature of $R \approx 15$ nm. Hence, light flowing up the fiber is concentrated at the tip by an intensity enhancement factor of (area of base/area of tip) $\approx 10{,}000$. This produces a large evanescent field at the tip, shown here red-detuned and hence attractive. The balancing of red-detuned gradient, spontaneous, and centrifugal forces, $\mathbf{F}_r$, $\mathbf{F}_s$, and $\mathbf{F}_c$, respectively, trap atoms in orbits that circle the z axis (Hori *et al.*, 1992).

where we assume the atom orbits the z axis in a circular orbit with a tangential azimuthal orbital velocity v_ϕ. Then, Newton's second law, $\mathbf{F}_r + \mathbf{F}_s + \mathbf{F}_c = 0$, implies a general consistency relationship

$$\tan\theta = \frac{2\alpha\,|\Delta|}{k_{\text{opt}}\gamma} \tag{131}$$

where θ is the spherical polar angle of the orbit and $k_{\text{opt}} = 2\pi/\lambda_{\text{opt}}$. For $\lambda \approx 780$ nm and $R \approx 15$ nm, we have $\alpha/k_{\text{opt}} \approx \lambda_{\text{opt}}/R \approx 52$. Hence, if the detuning is comparable with the linewidth, $|\Delta| \approx \gamma$, then $\tan\theta \approx 104$, or

$\theta \approx \pi/2$, and therefore the atom orbits nearly in the equatorial plane of the inscribed sphere of the cone's tip. The fact that light is originally propagating in a fiber core whose radius drops from $R \approx 10$ μm to $R \approx 15$ nm gives a large enhancement of the evanescent field. However, the base of the cone is only around 2 μm in diameter, so that not all of the light in the fiber couples in. Nevertheless, the enhancement can be still quite substantial. For a rough estimate, 500 mW of laser power will trap atoms with a mean radial kinetic energy equivalent to a temperature of $T_\rho \approx 0.26$ K or less, corresponding to radial velocities $v_\rho \approx 500$ cm/sec. This is much larger than the radial temperatures allowed in the WGM trap, $T_r \approx 3.5$ μK, or the external solid fiber guide, $T_\rho \approx 22$ mK, as previously discussed for similar laser powers. Hence, faster and hotter atoms can be confined. (We have taken the intensity enhancement factor to be $A_{\text{cone}}/A_{\text{tip}}$, where A_{cone} is the area of the cone base, and $A_{\text{tip}} = \pi R^2$, the cross-sectional area of the tip.)

Notice that the spontaneous force, Eq. (129b), is required to stabilize the atom in the trap. Without it, the orbital plane would migrate in the negative z direction until the atom hit the cone. Hence, spontaneous emission heating will always be a problem in this trap.

VII. Conclusions and Summary

We have presented a review of atom mirrors, resonators, waveguides, and traps that operate for the most part on the evanescent light-wave mechanism for atom manipulation. When we first set about this project, we contacted numerous authors who had done work in this field for advice and copies of reprints and preprints. Several researchers remarked that they thought the scope of this chapter might be too narrow, and that other atom optical trapping and manipulation schemes, such as optical lattices, might have to be included to make a substantial enough body of work for this chapter. However, as the preprints and reprints started coming in, and as we searched various bibliographic databases, it became clear that the research area of evanescent light-wave mirrors and traps was a quite large and rapidly growing field, as indicated by the reference list following.

In particular, after a slow start, this field has progressed by leaps and bounds with a close competition between theory and experiment. The experimental efforts of the *Ecole Normale* group in producing 10 bounces in a stable parabolic gravitational cavity (Aminoff *et al.*, 1993b), as well as the heroic efforts of the JILA group in fabricating hollow fiber atom guides (Renn *et al.*, 1995, 1996), have been truly remarkable. The complementary theoretical effort has been no less intrepid, with some of the best theory groups devoting significant resources to the task.

The significance and potential of evanescent light-wave trapping and guiding schemes has just begun to be realized. Of particular importance are the theoretical demonstrations by Ovchinnikov *et al.* (1995a, b) that atoms in such traps may be cooled there as well by Sisyphus and geometric techniques that actually reduce the volume of the atom phase space. This discovery has overcome the objections of Ketterle and Pritchard (1992) to an early cooling proposal of Shimoda (1989), and predictions of Bose condensation in pyramidal gravitational traps are now being made (Ovchinnikov *et al.*, 1995a). At such low temperatures, these evanescent traps begin to function as true quantum resonators, with at least one confinement dimension on the order of a de Broglie wavelength. It is the blue-detuned scheme for such resonators that is most appealing since the atoms can maintain their coherence over long times or over long propagation distances, provided the light wave is detuned sufficiently to minimize decohering spontaneous emission events. In this regime, these devices begin to function as true quantum, atom optical elements. There is much interest in using blue-detuned hollow fiber waveguides to make atom interferometers, since the Sagnac effect is so much more pronounced with atoms than with photons (Scully and Dowling, 1993). Due to their large mass, atom waves confined in hollow fibers or particle-in-a-box resonators can be used as gravimeters or accelerometers, as suggested by Harris and Savage (1995), as well as by Dowling and Gea-Banacloche (1995). In addition, the fact that atoms confined in blue-detuned, boxlike cavities have well-defined energy levels that decohere slowly could be used to fabricate a quantum bit or qubit for use in a quantum computer (Feynman, 1982, 1985, 1986; Deutsch and Joza, 1992; Ekert, 1995). As another application to quantum computing, Mabuchi and Kimble (1994), as well as Dutta and Agarwal (1995), have pointed out that atoms bound to microspheres by whispering-gallery mode evanescent waves have enormously large atom–cavity coupling constants—several orders of magnitude larger than found in current optical atom–cavity schemes. This large coupling constant would allow for the making of the controlled-NOT gate, which is the rudimentary element of any quantum computer (Pellizzari *et al.*, 1995; Monroe *et al.*, 1995; Barenco *et al.*, 1995). Even without the maintenance of coherence, evanescent light-wave traps and guides have many applications, such as to atomic clocks, atom delivery systems, and atom velocity selectors.

The field of evanescent light-wave trapping and guiding is new and rapidly growing. We hope this review chapter will serve as a primer for the novice, and an omnibus for the *Meister*.

Acknowledgments

J. P. D. would like to thank O. S. Mitchell for patience and encouragement during the preparation of this work. J. G.-B. would like to acknowledge the NSF for support. Both authors would like to thank the following people for interesting discussions and comments, and for kindly providing us with reprints and preprints of their work: D. Z. Anderson, V. I. Balykin, O. Carnal, R. J. Cook, S. Chu, R. Y. Chiao, J. I. Cirac, C. Cohen-Tannoudji, E. A. Cornell, J. Dalibard, W. Ertmer, R. Grimm, T. W. Hänsch, S. Haroche, K. Helmerson, W. Jhe, M. A. Kasevitch, W. Ketterle, H. J. Kimble, P. Meystre, G. J. Milburn, J. Mlynek, C. Monroe, Yu. B. Ovchinnikov, W. D. Phillips, D. E. Pritchard, M. J. Renn, C. M. Savage, W. Seifert, T. Sleator, D. F. Walls, H. Wallis, C. E. Wieman, M. Wilkens, D. Wineland, and P. Zoller. Finally, both authors would like to thank S. Troglen and G. McCrary for their unflagging assistance in the preparation of this work.

References

Abramowitz, M., and Stegun, I. A., eds. (1965). "Handbook of Mathematical Functions." Dover, New York.
Adams, C. S. (1994). *Contemp. Phys.* **35**, 1–19.
Adams, C. S., Carnal, O., and Mlynek, J. (1994a). *Adv. At. Mol. Opt. Phys.* **34**, 1–33.
Adams, C. S., Sigel, M., and Mlynek, J. (1994b). *Phys. Rep.* **240**, 143–210.
Almaas, E., and Brevnik, I. (1995). *J. Opt. Soc. Am. B* **12**, 2429–2438.
Aminoff, C. G., Bouyer, P., and Desboilles, P. (1993a). *C.R. Hebd. Seances Acad. Sci., Ser. 2* **316**, 1535–1541.
Aminoff, C. G., Steane, A., Bouyer, M. P., Desbiolles, P., Dalibard, J., and Cohen-Tannoudji, C. (1993b). *Phys. Rev. Lett.* **71**, 3083–3086.
Anderson, A., Haroche, S., Hinds, E. A., Jhe, W., Meschede, D., and Moi, L. (1986). *Phys. Rev. A* **34**, 3513–3516.
Ashkin, A. (1970). *Phys. Rev. Lett.* **24**, 156–159.
Ashkin, A. (1978). *Phys. Rev. Lett.* **40**, 729–732.
Aspect, A., Kaiser, R., Vansteenkiste, N., Vignolo, P., and Westbrook, C. I. (1995). *Phys. Rev. A* **52**, 4704–4708.
Averbukh, I. Sh., Akulin, V. M., and Schleich, W. P. (1994). *Phys. Rev. Lett.* **72**, 437–441.
Baldwin, K. G. H., Hajnal, J. V., and Fisk, P. T. H. (1990). *J. Mod. Opt.* **37**, 1839–1848.
Balykin, V. I. (1989). *Appl. Phys. [Part] B* **B49**, 383–388.
Balykin, V. I., and Letokhov, V. S. (1989a). *Appl. Phys. [Part] B* **B48**, 517–523.
Balykin, V. I., and Letokhov, V. S. (1989b). *Phys. Today* **48** (April), 23–28.
Balykin, V. I., and Letokhov, V. S. (1990). *Sov. Phys.-Usp.* (*Engl. Transl.*) **33**, 79–85.
Balykin, V. I., Letokhov, V. S., Ovchinnikov, Yu. B., and Sidorov, A. I. (1987). *JETP Lett.* (*Engl. Transl.*) **45**, 353–356; *Pis'ma Zh. Eksp. Teor. Fiz.* **45**, 282–285 (1987).
Balykin, V. I., Letokhov, V. S., Ovchinnikov, Yu. B., and Sidorov, A. I. (1988). *Phys. Rev. Lett.* **60**, 2137–2140.
Barenco, A., Bennett, C. H., Cleve, R., Divicenzo, D. P., Margolus, N., Shor, P., Sleator, T., Smolin, J. A., and Weinfurter, H. (1995). *Phys. Rev. A* **52**, 3457–3467.
Berkhout, J. J., and Walraven, J. T. M. (1993). *Phys. Rev. B: Condens. Matter* **47**, 8886–8904.
Berkhout, J. J., Luiten, O. J., Setija, I. D., Hijmans, T. W., Mizusaki, T., and Walraven, J. T. M. (1989). *Phys. Rev. Lett.* **63**, 1689–1692.

Born, M., and Wolf, E. (1985). "Principles of Optics," 6th ed. Pergamon, Oxford.
Campillo, A. J., Eversole, J. D., and Lin, H.-B. (1991). *Phys. Rev. Lett.* **67**, 437–440.
Carnal, O., Sigel, M., Sleator, T., Takuma, H., and Mlynek, J. (1991). *Phys. Rev. Lett.* **67**, 3231–3234.
Chang, S., Jo, J. H., and Lee, S. S. (1994). *Opt. Commun.* **108**, 133–143.
Chen, W.-Y., and Milburn, G. J. (1995). *Phys. Rev. A* **51**, 2327–2333.
Christ, M., Scholz, A., Schiffer, M., Deutschmann, R., and Ertmer, W. (1994). *Opt. Commun.* **107**, 211–217.
Chu, S., and Wieman, C., eds. (1989). *J. Opt. Soc. Am. B*, **6**, Spec. Issue.
Cirac, J. I., Blatt, R., Parkins, A. S., and Zoller, P. (1993a). *Phys. Rev. A* **48**, 2169–2181.
Cirac, J. I., Parkins, A. S., Blatt, R., and Zoller, P. (1993b). *Phys. Rev. Lett.* **70**, 556–559.
Cook, R. J. (1979). *Phys. Rev. A* **20**, 224–228.
Cook, R. J., and Hill, R. K. (1982). *Opt. Commun.* **43**, 258–260.
Dalibard, J., and Cohen-Tannoudji, C. (1985). *J. Opt. Soc. Am. B* **2**, 1707–1720.
Dalibard, J., and Cohen-Tannoudji, C. (1989). *J. Opt. Soc. Am. B* **6**, 2023–2045.
Davidson, N., Lee, H. J., Adams, C., Kasevich, M., and Chu, S. (1995). *Phys. Rev. Lett.* **74**, 1311–1314.
Deutsch, D., and Joza, R. (1992). *Proc. R. Soc. London, Ser. A* **439**, 553–558.
Deutschmann, R., Ertmer, W., and Wallis, H. (1993a). *Phys. Rev. A* **47**, 2169–2185.
Deutschmann, R., Ertmer, W., and Wallis, H. (1993b). *Phys. Rev. A* **48**, R4023–4026.
Dowling, J. P. (1993). *Opt. Soc. Am. Tech. Dig.* **16**, 222.
Dowling, J. P., and Gea-Banacloche, J. (1994). *Quantum Electron. Laser Sci. Conf. Tech. Dig.* **9**, 185–186.
Dowling, J. P., and Gea-Banacloche, J. (1995). *Phys. Rev. A* **52**, 3997–4003.
Dutta, G. S., and Agarwal, G. S. (1995). *Opt. Commun.* **115**, 597–605.
Einstein, A. (1917). *Phys. Z.* **18**, 121–128; reprinted in English *in* "Sources of Quantum Mechanics" (B. L. Van Der Waerden, ed.), 63–77. Dover, New York, 1968.
Ekert, A. (1995). *In* "Proceedings of the 14th International Conference on Atomic Physics" (D. J. Wineland, C. E. Wieman, and S. J. Smith, eds.), pp. 450–466. Amer. Inst. Phys. Press, Woodbury, NY.
Esslinger, T., Weidemuller, M., Hemmerich, A., and Hänsch, T. W. (1993). *Opt. Lett.* **18**, 450–452.
Feron, S., Reinhardt, J., Le Boiteux, S., Gorceix, O., Baudon, J., Ducloy, M., Robert, J., Miniatura, C., Nic-Chormaic, S., Haberland, H., and Lorent, V. (1993). *Opt. Commun.* **102**, 83–88.
Feron, S., Reinhardt, J., and Ducloy, M. (1994). *Phys. Rev. A* **49**, 4733–4741.
Feynman, R. P. (1982). *Int. J. Theor. Phys.* **21**, 467–488.
Feynman, R. P. (1985). *Opt. News* **11**, 11–46.
Feynman, R. P. (1986). *Found. Phys.* **16**, 507–531.
Glasgow, S., Meystre, P., Wilkins, M., and Wright, E. M. (1991). *Phys. Rev. A* **43**, 2455–2463.
Gordon, J. P., and Ashkin, A. (1980). *Phys. Rev. A* **21**, 1606–1617.
Gradshteyn, I. S., and Ryzhik, I. S. (1965). "Table of Integrals, Series, and Products." Academic Press, London.
Hajnal, J. V., and Opat, G. I. (1989). *Opt. Commun.* **71**, 119–124.
Hajnal, J. V., Baldwin, K. G. H., Fisk, P. T. H., Bachor, H.-A., and Opat, G. I. (1989). *Opt. Commun.* **73**, 331–336.
Hall, J. L., Zhu, M., and Buch, P. (1989). *J. Opt. Soc. Am. B* **6**, 2194–2205.
Harris, D. J., and Savage, C. M. (1995). *Phys. Rev. A* **51**, 3967–3971.

Helmerson, K., Rolston, S. L., Goldner, L., and Phillips, W. D. (1992). *In* "Optics and Interferometry with Atoms" (J. Mlynek and O. Carnal, eds.), Book Abstr., 97th WE-Heraeus Semin, Konstance University Press, Konstance, 1992.
Henkel, C., Kaiser, R., Westbroom, C., Courtois J.-Y., and Aspect, A. (1994). *Laser Phys.* **4**, 1042–1049.
Hori, H., Ohtsu, M., and Ohsawa, H. (1992). *Int. Quantum Electron. Laser Sci. Conf. Tech. Dig.* **9**, 48–49.
Iengo, R., and Lechner, K. (1992). *Phys. Rep.* **213**, 179–269.
Ito, H., Sakaki, K., Nakata, T., Jhe, W., and Ohtsu, M. (1995). *Opt. Commun.* **115**, 57–64.
Ito, H., Nakata, T., Sakaki, K., Ohtsu, M., Lee, K. I., and Jhe, W. (1996). *Phys. Rev. Lett.* **76**, 4500–4503.
Jackson, J. D. (1975). "Classical Electrodynamics," 2nd ed. Wiley, New York.
Jhe, W., Ohtsu, M., and Friberg, S. R. (1994). *Jpn. J. Appl. Phys., Part 2* **33**, L1680–L1682.
Kaiser, R., Levy, Y., Vansteenkiste, N., Aspect, A., Seifert, W., Leipold, D., and Mlynek, J. (1994). *Opt. Commun.* **104**, 234–240.
Kasevich, M. A., and Chu, S. (1991). *Phys. Rev. Lett.* **67**, 181–184.
Kasevich, M. A., Weiss, D. S., and Chu, S. (1990). *Opt. Lett.* **15**, 607–609.
Kawata, S., and Sugiura, T. (1992). *Opt. Lett.* **17**, 772–774.
Kazantsev, A. P., Ryabenko, G. A., Surdatowich, G. I., and Yakovlev, V. P. (1985). *Phys. Rep.* **129**, 75.
Kazantsev, A. P. Surdutovick, G. I., and Yakovlev, V. P. (1990). "Mechanical Action of Light on Atoms." World Scientific, Singapore and Teaneck, NJ.
Keith, D. W., Schattenburg, M. L., Smith, H. I., and Pritchard, D. E. (1988). *Phys. Rev. Lett.* **61**, 1580–1583.
Keith, D. W., Ekström, C. R., Turchette, Q. A., and Pritchard, D. E. (1991a). *Phys. Rev. Lett.* **66**, 2693–2696.
Keith, D. W., Soave, R. J., and Rooks, M. J. (1991b). *J. Vac. Sci. Technol., B* [2] **9**, 2846–2850.
Ketterle, W., and Pritchard, D. E. (1992). *Appl. Phys. [Part] B* **B54**, 403–406.
Korsch, H. J., and Lang, J. (1991). *J. Phys. A* **24**, 45–52.
Lee, K. I., Kim, J. A., Noh, H. R., and Jhe, W. (1996). *Opt. Lett.*, in press.
Lehtihet, H. E., and Miller, B. N. (1986). *Physica D (Amsterdam)* **21**, 93–104.
Lenz, G., Meystre, P., and Wright, E. M. (1994). *Phys. Rev. A* **50**, 1681–1691.
Liston, G. J., Tan, S. M., and Walls, D. F. (1995a). *Appl. Phys. [Part] B* **B60**, 211–227.
Liston, G. J., Tan, S. M., and Walls, D. F. (1995b). *Phys. Rev. A* **52**, 3057–3073.
Lofting, H. (1922). "The Voyages of Doctor Dolittle." Lippincott, Philadelphia.
Mabuchi, H., and Kimble, H. J. (1994). *Opt. Lett.* **19**, 749–751.
Marcatile, E. A. J., and Schmeltzer, R. A. (1964). *Bell Syst. Tech. J.* July, pp. 1783–1809.
Marksteiner, S., Savage, C. M., Zoller, P., and Rolston, S. L. (1994). *Phys. Rev. A* **50**, 2680–2690.
Marti, O., and Balykin, V. I. (1993). *In* "Near Field Optics" (D. W. Pohl and D. Corjon, eds.), pp. 121–130. Kluwer Academic Press, Amsterdam.
McCall, S. L., Levi, A. F. J., Slusher, R. E., Pearson, S. J., and Logan, R. A. (1992). *Appl. Phys. Lett.* **60**, 289–291.
Meystre, P., and Stenholm, S., eds. (1985). *J. Opt. Soc. Am. B* **2**, Spec. Issue.
Minnhagen, P. (1987). *Rev. Mod. Phys.* **59**, 1001–1066.
Minogin, V. G., and Letokhov, V. S. (1987). "Laser Light Pressure on Atoms." Gordon & Breach, New York.
Mlynek, J., Balykin, V., and Meystre, P., eds. (1992). *Appl. Phys. [Part] B* **B54**, Spec. Issue, 319–485.

Monroe, C., Meekhof, D. M., King, B. E., Itano, W. M., and Wineland, D. J. (1995). *Phys. Rev. Lett.* **75**, 4714-4717.
Murphy, J. E., Goodman, P., and Smith, A. E. (1993). *J. Phys: Condens. Matter* **5**, 4665-4676.
Murphy, J. E., Hollenberg, C. L., and Smith, A. E. (1994). *Phys. Rev. A* **49**, 3100-3103.
Murphy, J. E., Goodman, P., and Sidorov, A. (1995). *Opt. Commun.* **117**, 83-89.
Ol'Shanii, M. A., Ovchinnikov, Yu, B., and Letokhov, V. S. (1993). *Opt. Commun.* **98**, 77-79.
Ovchinnikov, Yu. B., Shul'ga, S. V., and Balykin, V. I. (1991). *J. Phys. B: At., Mol. Opt. Phys.* **24**, 3173-3178.
Ovchinnikov, Yu. B., Söding, J., and Grimm, R. (1995a). *JETP Lett. (Engl. Transl.)* **61**, 21-25; *Pis'ma Zh. Eksp. Teor. Fiz.* **61**, 23-27 (1995).
Ovchinnikov, Yu. B., Laryushin, D. V., Balykin, V. I., and Letokhov, V. S. (1995b). *JETP Lett. (Engl. Transl.)* **62**, 113-118; *Pis'ma Zh. Eksp. Teor. Fiz.* **62**, 102-107 (1995).
Pangaribuan, T., Yamada, K., and Jiang, S. (1992). *Jpn. J. Appl. Phys., Part 2* **31**, L1302-L1304.
Pauli, W. (1950). *Prog. Theor. Phys.* **5**, 526-543.
Pellizzari, T., Gardiner, S. A., Cirac, J. I., and Zoller, P. (1995). *Phys. Rev. Lett.* **75**, 3788-3791.
Pfau, T., Adams, C. S., and Mlynek, J. (1993a). *Europhys. Lett.* **21**, 439-442.
Pfau, T., Adams, C. S., Sigel, M., and Mlynek, J. (1993b). *Phys. Rev. Lett.* **71**, 3427-3430.
Pillet, P., ed. (1994). *J. Phys. II* **4**, Spec. Issue, 1877-2089.
Raether, H. (1988). "Surface Plasmons." Springer, Berlin.
Reichel, J., Bardou, F., Ben Dahan, M., Peik, E., Rand, S., Salomon, C., and Cohen-Tannoudji, C. (1995). *Phys. Rev. Lett.* **75**, 4575-4578.
Renn, M. J., Montgomery, D., Vdovin, O., Anderson, D. Z., Wieman, C. E., and Cornell, E. A. (1995). *Phys. Rev. Lett.* **75**, 3253-3256.
Renn, M. J., Donley, E. A., Cornell, E. A., Wieman, C. E., and Anderson, D. Z. (1996). *Phys. Rev. A* **53**, R648-R651.
Roach, T. M., Abele, H., Boshier, M. G., Grossman, H. L., Zetie, K. P., and Hinds, E. A. (1995). *Phys. Rev. Lett.* **75**, 629-632.
Sakurai, J. J. (1985). "Modern Quantum Mechanics." Benjamin/Cummings, Menlo Park, CA.
Savage, C. M., Marksteiner, S., and Zoller, P. (1993). *In* "Fundamentals of Quantum Optics III: Proceedings of the Fifth Meeting on Laser Phenomena, University of Innsbruck, 1993" (F. Ehlotzky, ed.), pp. 60-74. Springer-Verlag, Berlin.
Savage, C. M. Gordon, D., and Ralph, T. C. (1995). *Phys. Rev. A* **52**, 4741-4746.
Scully, M. O., and Dowling, J. P. (1993). *Phys. Rev. A* **48**, 3186-3190.
Seifert, W., Adams, C. S., Balykin, V. I., Heine, C., Ovchinnikov, Yu., and Mlynek, J. (1994a). *Phys. Rev. A* **49**, 3814-3823.
Seifert, W., Kaiser, R., Aspect, A., and Mlynek, J. (1994b). *Opt. Commun.* **111**, 566-576.
Shimoda, K. (1989). *IEEE Trans. Instrum. Meas.* **38**, 150-155.
Sigel, M., and Mlynek, J. (1993). *Phys. World* **6** (February), 36-42.
Snyder, A. W., and Love, J. D. (1983). "Optical Waveguide Theory." Chapman & Hall, London.
Söding, J., Grimm, R., and Ovchinnikov, Yu. B. (1995). *Opt. Commun.* **119**, 652-662.
Solimeno, S., Crosignani, B. B., and Diponto, P. (1986). "Guiding, Diffraction, and Confinement of Optical Radiation." Academic Press, New York.
Steane, A., Szriftgiser, P., Desbiolles, P., and Dalibard, J. (1995). *Phys. Rev. Lett.* **74**, 4972-4975.
Stenlake, B. W., Littler, I. C. M., Bachor, H.-A., Baldwin, K. G. H., and Fisk, P. J. H. (1994). *Phys. Rev. A* **49**, R16-R19.
Tan, S. M., and Walls, D. F. (1994). *Phys. Rev. A* **50**, 1561-1574.

Treussart, F., Hare, J., and Haroche, S. (1994). *Opt. Lett.* **19**, 1651–1653.
Van Der Hulst, H. C. (1981). "Light Scattering by Small Particles." Dover, New York.
Wallis, H. (1995). *Phys. Rep.* **255**, 203–287.
Wallis, H., Dalibard, J., and Cohen-Tannoudji, C. (1992). *Appl. Phys. [Part] B* **B54**, 407–419.
Wilkens, M., Goldstein, E., Taylor, B., and Meystre, P. (1993). *Phys. Rev. A* **47**, 2366–2369.
Wiseman, H., Collet, M., Martins, A., and Walls, D. F. (1995). *Coherence Quantum Opt.* **7**, in press.
Zhang, W., and Walls, D. F. (1992). *Phys. Rev. Lett.* **68**, 3287–3290.
Zhang, W., and Walls, D. F. (1993). *Phys. Rev. A* **47**, 626–633.

OPTICAL LATTICES

P. S. JESSEN
Optical Sciences Center,
University of Arizona
Tucson, Arizona

I. H. DEUTSCH
Center for Advanced Studies
University of New Mexico
Albuquerque, New Mexico

I. Introduction 95
II. The 1D Lin⊥Lin Model System 97
   A. Atomic Motion in a 1D Lin⊥Lin Optical Lattice 97
   B. Spectroscopy in 1D Lin⊥Lin Optical Lattices 100
III. Crystallography of Optical Lattices 104
   A. Two-Dimensional Lattices 105
   B. Three-Dimensional Lattices 107
IV. Laser Cooling in Optical Lattices: Theory 109
   A. General Formalism 109
   B. Semiclassical Methods 112
   C. The Monte Carlo Wavefunction Technique 113
   D. Band Structure Formalism 115
   E. The Secular Approximation 116
V. Spectroscopy 119
   A. Spectroscopic Techniques 120
   B. Experimental Results 123
VI. New Developments 128
   A. Cooling by Adiabatic Expansion in Optical Lattices 128
   B. Bragg Scattering in Optical Lattices 130
   C. Outlook 133
   References 136

I. Introduction

The interference pattern of a set of intersecting laser beams can create a stable periodic potential for neutral atoms through the ac Stark shift (light shift), which can trap and thereby organize atoms in an ordered, crystal-like structure. Historically, interest in these "optical lattices" grew out of the study of laser cooling (Hänsch and Schawlow, 1975; Wineland and Dehmelt,

1975). In 1988 the group at NIST Gaithersburg (Lett et al., 1988) discovered that atoms are readily cooled below the so called "Doppler limit." Research at the Ecole Normale Supérieure (ENS) (Dalibard and Cohen-Tannoudji, 1989) and Stanford (Ungar et al., 1989) soon thereafter identified new mechanisms for laser cooling that occur for atoms with degenerate ground states and light fields with spatially varying polarization. The ENS group in particular proposed an elegant one-dimensional model system in which cooling results from optical pumping between a set of spatially varying light shift potentials that depend on the atomic ground state. This model system cannot, however, be easily compared with the standard laser cooling configuration, known as optical *molasses*, which involves six laser beams and results in a complicated three-dimensional laser field of spatiotemporally varying intensity and polarization. Considerable effort was therefore directed toward realizing the one-dimensional model system in the laboratory. The first experiments of this type immediately led to spectacular new physical insight, prominent examples of which are atom trapping in microscopic optical potential wells, quantized atomic center-of-mass motion, and the generation of atomic samples with long range order imposed by light fields. At this point the concept of an optical *lattice* was coined, signifying a shift away from a physical image of laser-cooled atoms as a disordered vapor (optical molasses), toward a new physical image of an atomic sample whose properties are primarily determined by the periodic optical potential.

Aside from its role as a model system for laser cooling, the motion of atoms in an optical lattice is closely analogous to that of an electron in a solid state crystal. This situation offers unique opportunities to explore phenomena that were previously accessible only in condensed matter. In contrast to a solid, where the distance between atoms is a few angstroms, the interparticle spacing in an optical lattice can be many micrometers. Thus, the system can be essentially free from strong atom–atom interactions, which are difficult to model analytically and must be treated by approximate or numerical methods. Even at densities that are large by the standards of a typical laser cooling experiment, the interatomic interactions are sufficiently small that they can be studied using tractable, accurate theories. The optical lattice potential itself can be modeled exactly, and it can be produced in the laboratory without defects if appropriate attention is given to laser beam quality. Parameters characterizing the lattice, such as lattice constants/symmetry and potential well depth/shape, are easily adjusted through choices of the laser geometry, polarization intensity and frequency. Furthermore, the dissipative aspect of the dynamics, arising from spontaneous emission, is in principle controllable through adjustments of the laser frequency and intensity, and its description is within range of an abinitio theory. These considerations suggest that the atom/lattice system provides an attractive combination of tractable theory and well-controlled experiments, and so may prove a valuable testing ground for new developments in condensed matter physics. Examples of problems that may be addressed range from quantum transport phenomena to the physics of systems that exhibit long-range, but not short-range, order.

In this chapter we will review experimental and theoretical developments in the study of optical lattices. In Section II we introduce our subject by discussing the key one-dimensional model system for laser cooling and trapping in optical lattices. In Section III we discuss crystallography of various lattice geometries, and in Section IV we cover the basic laser cooling theory in optical lattices. Section V reviews experimental work that probes optical lattices through spectroscopy, and finally, in Section VI we discuss new developments in the field and possible future directions.

II. The 1D Lin⊥Lin Model System

In this section we discuss a one-dimensional (1D) model system identified by Dalibard and Cohen-Tannoudji (1989). This system consists of an atom with angular momenta $J_g = \frac{1}{2}$ in the ground state and $J_e = \frac{3}{2}$ in the excited state, moving in an optical lattice formed by a pair of linear and cross-polarized counterpropagating plane waves; in the following, we refer to this configuration as "1D lin⊥lin." Our use of the term 1D refers to the single dimension along which the field polarization and/or intensity is varying. In the remaining two dimensions, the atoms are neither trapped nor cooled; motion in this plane is, to a good approximation, not probed by experiments and therefore is ignored.

A. Atomic Motion in a 1D Lin⊥Lin Optical Lattice

Consider the 1D lin⊥lin optical lattice shown in Fig. 1(b), composed of two counter-propagating plane waves with orthogonal, linear polarization $\hat{\mathbf{x}}$ and $\hat{\mathbf{y}}$, traveling in the directions $\pm\hat{\mathbf{z}}$. The total electric field is $\mathbf{E}_L(z,t) = \frac{1}{2}\mathbf{E}_L(z)e^{-i\omega t} + c.c.$, where the spatial part, by convenient choice of phase and quantization axis along $\hat{z}$, can be written as

$$\mathbf{E}_L(z) = -E_0\left[\hat{\mathbf{x}}e^{ikz} + i\hat{\mathbf{y}}e^{-ikz}\right] = \sqrt{2}\,E_0\left[\hat{\mathbf{e}}_+\cos(kz) - i\hat{\mathbf{e}}_-\sin(kz)\right] \quad (1)$$

Equation (1) shows that the total field can be decomposed into two standing waves of σ_+ and σ_- polarization, offset by $\lambda/4$ so that the antinodes of one coincide with the nodes of the other. Note that the light intensity $I_L \propto |E_L(z)|^2$ is everywhere constant; only the field polarization changes from circular to linear and back to circular as one moves a distance $\lambda/4$ along $\hat{z}$.

Figure 1(a) shows the level structure of an atom with ground and excited state angular momenta $J_g = \frac{1}{2}$ and $J_e = \frac{3}{2}$. Since the electric field contains only $\sigma_\pm$ components, the atom–laser interaction consists of two independent "V-systems,"

$$\{|e, -\tfrac{1}{2}\rangle, |g, \tfrac{1}{2}\rangle, |e, \tfrac{3}{2}\rangle\} \quad \text{and} \quad \{|e, -\tfrac{3}{2}\rangle, |g, -\tfrac{1}{2}\rangle, |e, \tfrac{1}{2}\rangle\}$$

coupled only through $\Delta m = 0$ spontaneous decay of the $|e, \pm\tfrac{1}{2}\rangle$ states. In the limit of low saturation, the optical potentials for the $|g, \pm\tfrac{1}{2}\rangle$ states are

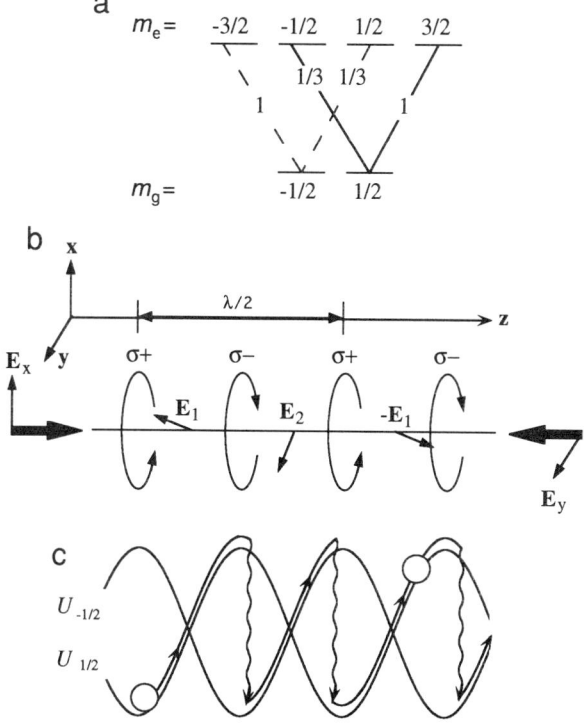

FIG. 1. Sisyphus cooling. (a) Level structure of a $J_g = \tfrac{1}{2} \to J_e = \tfrac{3}{2}$ atom, including the square of the Clebsch–Gordan coefficients. (b) 1D lin⊥lin lattice configuration. (c) Example of a trajectory followed by an atom in the bipotential associated with the two ground state sublevels. Optical pumping predominately transfers an atom from the hills of one optical potential to the valleys of the other, and the atom's kinetic energy is dissipated.

(Dalibard and Cohen-Tannoudji, 1989)

$$U_{1/2}(z) = \tfrac{2}{3}U_0 \cos^2(kz) + \tfrac{1}{3}U_0, \qquad U_{-1/2}(z) = \tfrac{2}{3}U_0 \sin^2(kz) + \tfrac{1}{3}U_0 \tag{2}$$

where $U_0 = \hbar \Delta s_0/2$ is the maximum value of the light shift (negative for $\Delta < 0$). In these expressions, $s_0 = 2\Omega^2/(4\Delta^2 + \Gamma^2)$ is the saturation parameter for the $|g, \tfrac{1}{2}\rangle \leftrightarrow |e, \tfrac{3}{2}\rangle$ transition at a point where the polarization is purely σ_+, with associated Rabi frequency Ω and detuning $\Delta = \omega_L - \omega_A$ of the lattice light from atomic resonance.

In a red-detuned lattice, $\Delta < 0$, laser cooling occurs because of optical pumping between the atomic ground state sublevels. In the following, we regard the center-of-mass position and momentum of the atom as classical variables in order to visualize how kinetic energy is dissipated. Consider an atom moving along $\hat{z}$ as illustrated in Fig. 1(c). At some time, the atom is in the state $|g, \tfrac{1}{2}\rangle$ at an antinode of the σ_+ standing wave. This corresponds to a location at the bottom of a well in the potential $U_{1/2}$. As the atom moves away from the σ_+ antinode, it expends kinetic energy by climbing out of the $U_{1/2}$ potential well. At first, it is unlikely that the atom will optically pump to the state $|g, -\tfrac{1}{2}\rangle$ because the local polarization is mostly σ_+, but as the atom enters a region where the local polarization is mostly σ_-, the probability of optical pumping increases. Optical pumping is most likely to occur at the antinode of the σ_- standing wave, transferring the atom from the top of a hill of the potential $U_{1/2}$ to the bottom of a well in the potential $U_{-1/2}$. The atom then expends more kinetic energy by climbing out of this potential well, only to be optically pumped to the bottom of another $U_{1/2}$ potential well, etc. On average the atom climbs more hills than it descends, and its kinetic energy decreases with time. This powerful physical image is known as *Sisyphus cooling* (Dalibard and Cohen-Tannoudji, 1989).

Based on a semiclassical model, Dalibard and Cohen-Tannoudji (1989) estimated spatially averaged friction and diffusion coefficients, and they predicted atomic temperatures corresponding to a mean kinetic energy of order the modulation depth of the lattice potential. Although this simple treatment explicitly ignores atomic localization on the scale of a wavelength, the result suggests that a substantial fraction of the atoms may be trapped in individual optical potential wells. In that case, one can approximate the atomic motion near the bottom of the wells by a simple harmonic oscillator that is thermally excited. Expansion of the optical potential around the minimum of a potential well yields the oscillation frequency,

$$\hbar \omega_{osc} = 2\sqrt{\tfrac{2}{3}U_0 E_R}, \qquad E_R = \frac{\hbar^2 K^2}{2M} \tag{3}$$

where E_R is the single photon recoil energy and M the atomic mass. Note that it is meaningful to talk about oscillatory atomic motion only if an atom resides in a given potential well for a time at least comparable with the oscillation frequency. This is precisely the regime in which one must describe the atomic center-of-mass motion quantum mechanically, as we will see in Section IV, and also the regime that has been found experimentally to lead to the lowest temperatures (Lett. *et al.*, 1988; Weiss *et al.*, 1989; Salomon *et al.*, 1990).

B. Spectroscopy in 1D Lin⊥Lin Optical Lattices

Early experimental work on six-beam optical molasses measured the distribution of atomic momenta (Lett *et al.*, 1988), which contains no direct evidence of atomic localization or oscillatory motion. In the period 1989–1992, several spectroscopic methods were pursued as alternative means of probing the laser cooling process. In the following, we briefly introduce the two most successful techniques, deferring a more detailed discussion to Section V.

Figure 2(a) shows schematically the technique of optical heterodyne spectroscopy of resonance fluorescence developed by the group of Phillips (Westbrook *et al.*, 1990). Fluorescence emitted by a sample of laser-cooled atoms is collected, mixed with a local oscillator beam, and detected by a photodiode. The resulting beat signal is analyzed with a radiofrequency spectrum analyzer, thereby producing a measurement of the power spectrum of the atomic fluorescence. An alternative approach was pioneered by the groups of Salomon, Grynberg, and Kimble (Grison *et al.*, 1991; Tabosa *et al.*, 1991). This probe spectroscopy technique, Fig. 3(a), measures the attenuation of a weak probe beam passing through the atomic sample as its frequency is scanned around the lattice light frequency. Figures 2(c) and 3(c) show fluorescence (Jessen *et al.*, 1992) and probe absorption spectra (Verkerk *et al.*, 1992), respectively, measured for samples of alkali atoms cooled and trapped in a 1D lin⊥lin optical lattice. In both experiments an effective 1D geometry was accomplished by keeping the measurement direction at a very small angle with respect to the lattice axis. Because of considerable laser heating in the plane transverse to the lattice axis, a 1D lattice can be achieved only transiently. Thus, the experiments of Verkerk *et al.* (1992) and Jessen *et al.* (1992) alternated between measurement periods of a few milliseconds duration with only the 1D lattice present, and periods where an additional superposed 3D molasses was used to cool the transverse atomic motion. As will be discussed, this transient measurement scheme restricts the information available in 1D spectra.

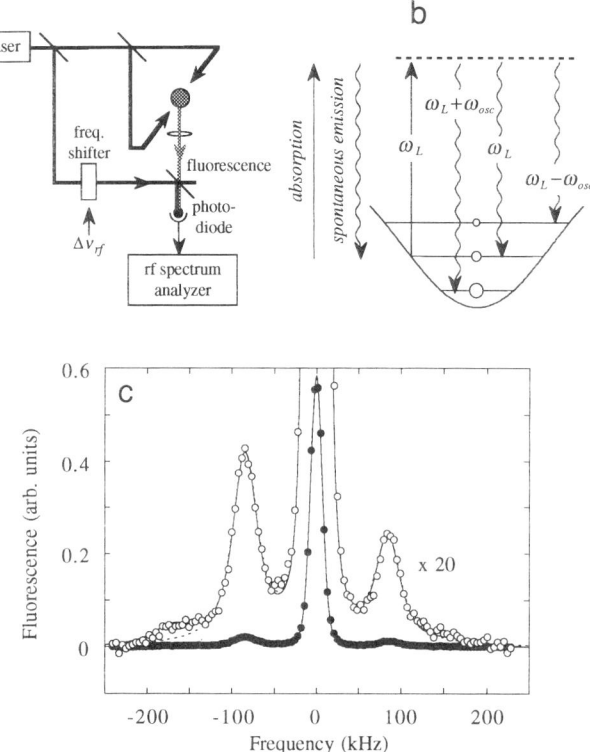

FIG. 2. Fluorescence spectroscopy. (a) Schematic experimental setup. (b) Spontaneous Raman transitions between vibrational eigenmodes of the atomic motion give rise to a Rayleigh line at the optical lattice frequency ω_L and Raman sidebands at $\omega_L \pm \omega_{osc}$. (c) Spectrum of fluorescence measured for ^{85}Rb atoms in a 1D lin⊥lin optical lattice (from Jessen et al., 1992).

Each set of features in both the fluorescence and probe absorption spectra can be interpreted in related physical terms. We first discuss the fluorescence spectrum shown in Fig. 2(c) under the assumption that the atomic center-of-mass motion can be described classically. For low saturation, light scattering is predominantly elastic with respect to the atomic internal degrees of freedom, so that the power spectrum of the scattered electric field becomes a delta function at the laser frequency ω_L, subject to broadening by the atomic motion. An observer seeing the light radiated by the atom would detect a scattered field of the form $\mathbf{E}_s(t) = \frac{1}{2}\mathbf{E}_s e^{-i\omega_L t} e^{-i\Delta k z(t)} + c.c.$, where $z(t)$ is the oscillating atomic position and Δk the change in photon wavevector upon scattering. If the atom is

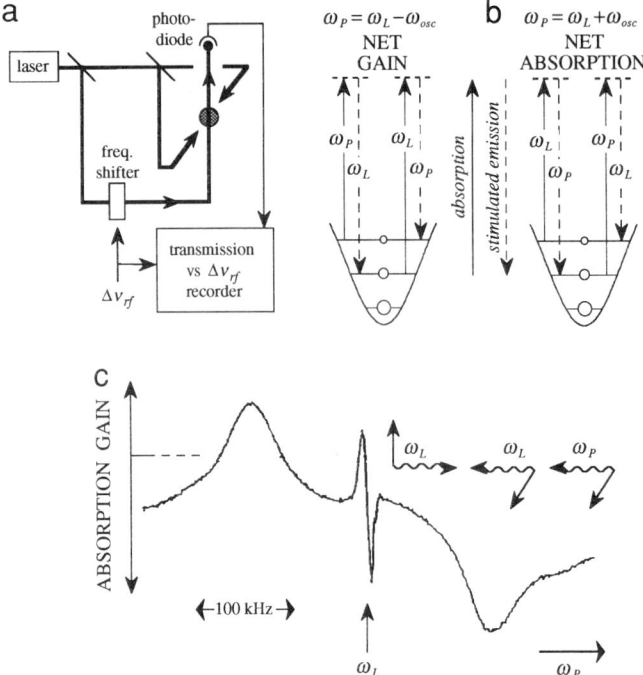

FIG. 3. Probe transmission spectroscopy. (a) Schematic experimental setup. (b) Stimulated Raman Transitions between vibrational eigenmodes of the atomic motion redistribute photons between the probe laser at frequency ω_p and the optical lattice at frequency ω_L. Competition between these processes results in net gain for $\omega_P < \omega_L$ and net absorption for $\omega_P > \omega_L$. (c) Probe transmission spectrum measured for ^{133}Cs atoms in a 1D lin⊥lin optical lattice (from Verkerk et al., 1992, reprinted with permission).

harmonically bound, the power spectrum of the radiated light is that of a phase-modulated carrier at ω_L with modulation index $\Delta k z_a$, where z_a is the classical oscillation amplitude. Neglecting damping of the atomic motion, the corresponding spectrum is

$$S(\omega) = \sum_{n=-\infty}^{\infty} J_n(\Delta k z_a)^2 \delta[\omega_L + n\omega_{osc}] \qquad (4)$$

where $J_n(x)$ is the Bessel function of order n. If the modulation index $\Delta k z_a \ll 1$, it follows from the properties of the Bessel functions that the amplitude in the nth sideband is of order $(\Delta k z_a)^n$. Therefore, the spectrum of Fig. 2(c), which exhibits only one pair of strongly suppressed sidebands, is strong evidence that the radiating atoms are well localized in optical potential wells of the 1D lin⊥lin optical lattice. Such tight localization is comparable with that achieved in ion trapping (Diedrich *et al.*, 1989) and is commonly known as the Lamb–Dicke limit. Because individual lines in the spectrum are well resolved, one can further conclude that the phase diffusion rate of the oscillatory atomic motion is comparable with or less than the oscillatory frequency. In other words, an atom undergoes, on the average, at least one full oscillation before the phase of oscillation is randomized by photon scattering or interrupted by optical pumping of the atom to a different internal state. The spectrum shown in Fig. 2(c) also shows an asymmetry in the power of the red and blue sidebands that cannot be accounted for in the purely classical model, Eq. (4).

In a quantum theory, the sidebands in the fluorescence and probe absorption spectra of Figs. 2(c), and 3(c) can be interpreted in terms of transitions between quantized vibrational states of the atom's center-of-mass motion in a harmonic potential. The fluorescence spectrum contains sidebands at frequencies $\omega_L + \omega_{osc}$ and $\omega_L - \omega_{osc}$ that arise from *spontaneous* Stokes and anti-Stokes Raman transitions of the type illustrated in Fig. 2(b), which change the atom vibrational quantum number by $\Delta n = \pm 1$. For this reason, we refer to these spectroscopic features as "Raman sidebands." In the probe absorption spectrum, sidebands arise due to *stimulated* Stokes and anti-Stokes Raman transitions of the type illustrated in Fig. 3(b). In such transitions a photon can be either absorbed from or scattered into the probe beam if the Raman resonance condition $\omega_p = \omega_L \pm \omega_{osc}$ is fulfilled. The competition between these two processes determines whether the probe beam experiences net gain or absorption.

Transition rates between vibrational states are readily calculated using Fermi's golden rule (Wineland and Itano, 1979), yielding $\gamma_{n \to n+\Delta n} \propto \gamma_s (k_L z_0)^{2|\Delta n|}$ for $(k_L z_0)^2 = E_R/\hbar \omega_{osc} \ll 1$, where $\gamma_s \approx \Gamma s_0/2$ is the total photon scattering rate and z_0 is the rms spread of the harmonic oscillator ground state. As in the classical picture, we conclude that emission into the sidebands is suppressed if the atoms are well localized on the scale of an optical wavelength (the Lamb–Dicke effect). The observation of well-resolved Raman sidebands is due largely to the strong localization, which suppresses the rate of decay of the vibrational populations and coherences.

The asymmetry between red and blue sidebands evident in the fluorescence spectrum, is readily accounted for in the quantum model in terms of population differences between vibrational levels. Consider a pair of vibrational states $|n\rangle$ and $|n+1\rangle$ in an optical potential well. In steady state the distribution of population over states is close to thermal, and for any n we have $\Pi_{n+1}/\Pi_n = f_B$, where Π_n is the population of state $|n\rangle$ and $f_B = \exp(-\hbar \omega_{osc}/k_B T)$ is the Boltzmann factor. Since the rates $\gamma_{n \to n+\Delta n}$ and $\gamma_{n+\Delta n \to n}$ are identical, it follows that Stokes Raman transitions are more frequent than anti-Stokes Raman transitions by a factor f_B. Therefore, the oscillator temperature T can be extracted from ω_{osc} and the measured sideband asymmetry. These parameters allow us to determine the mean vibrational excitation $\bar{n}$ and localization Δz of the oscillator. Typical numbers extracted from data such as that of Fig. 2(c) are $\bar{n} \approx 1$ and $\Delta z \approx \lambda/15$, consistent with atom trapping in the Lamb–Dicke regime.

The central feature at the laser frequency in the fluorescence spectrum arises from Rayleigh scattering, i.e., Raman transitions that begin and end in the same or a nearly degenerate quantum state. We will defer further discussion of this feature until Section V. Here we are simply pointing out that in a 1D lattice experiment the width of the Rayleigh line is dominated by Doppler broadening, which occurs because of a small component of the transverse atomic velocity along the measurement direction.

III. Crystallography of Optical Lattices

An optical lattice of any dimension can be characterized by its crystal structure in a manner analogous to a solid. An extensive discussion of the crystallography of optical lattices for various geometries is given in Petsas *et al.* (1994); here, we summarize some of the salient features of a few representative lattices. In general, a lattice is defined by a discrete set of vectors $\{\mathbf{R}_i\}$, such that the potential is invariant under translations, $U(\mathbf{x} + \mathbf{R}_i) = U(\mathbf{x})$. The potential can then be written as a Fourier sum

$$U(\mathbf{x}) = \sum_{\mathbf{K}_j} \tilde{U}_{\mathbf{K}_j} e^{i\mathbf{K}_j \cdot \mathbf{x}} \quad (5)$$

The set of vectors, $\{\mathbf{K}_j\}$, defines the reciprocal lattice according to the condition $\mathbf{R}_i \cdot \mathbf{K}_j = m\, 2\pi\, \delta_{ij}$. Although the optical lattice is normally used to trap a single atomic species, the resulting periodic potential is somewhat analogous to the NaCl crystal structure. This subtlety arises from the vector nature of the atom–photon interaction; different atomic "species" correspond to the different internal states of the atom when it is optically pumped by σ_+ or σ_- light. Thus, in general, an optical lattice must be described by a Bravais lattice with a basis (Ashcroft and Mermin, 1976). The reciprocal lattice determines the full periodicity of the light field (both intensity *and* polarization); the basis determines the sites of σ_+ and σ_- within the primitive cell.

In the low saturation limit, the light shift is equal to the potential energy stored in a linearly polarizable particle, $U(\mathbf{x}) = -\mathbf{E}_L^{(-)}(\mathbf{x}) \cdot \overleftrightarrow{\alpha} \cdot \mathbf{E}_L^{(+)}(\mathbf{x})/2$, where $\overleftrightarrow{\alpha}$ is the real part of the atomic polarizability tensor operator and $\mathbf{E}_L^{(+)}(\mathbf{x})$ is the spatial dependence of the laser field's positive frequency component. Given a lattice created by a set of plane waves with equal frequencies, but with arbitrary polarization, direction of propagation, amplitude, and phase, we have

$$\mathbf{E}_L(\mathbf{x}) = \frac{1}{2} \sum_j E_j \vec{\varepsilon}_j e^{i(\mathbf{k}_j \cdot \mathbf{x} - \omega_L t + \phi_j)} + c.c. \qquad (6)$$

The optical potential then takes the form

$$U(\mathbf{x}) = -\frac{1}{4} \sum_{j,j'} E_{j'}^* E_j \left(\vec{\varepsilon}_{j'}^* \cdot \vec{\alpha} \cdot \vec{\varepsilon}_j \right) e^{i(\phi_j - \phi_{j'})} e^{i(\mathbf{k}_j - \mathbf{k}_{j'}) \cdot \mathbf{x}} + c.c \qquad (7)$$

The potential in Eq. (2) is a special case of (7). In general, the potential is completely specified by the electric field and the atomic polarizability. It is evident by comparison of Eqs. (5) and (7) that the differences between the wavevectors of the laser beams are reciprocal lattice vectors. It also follows from Eq. (7) that a change in the relative phase between two of the beams, $\phi_j - \phi_{j'}$, is equivalent to an overall translation of the lattice, $\mathbf{x} \to \mathbf{x} + \mathbf{x}_0$. In general the relative phases of the laser beams will fluctuate with time, unless specific steps are taken to prevent this. For b laser beams, there are $b - 1$ independent relative phases. Thus, if one wants to create an optical lattice in n dimensions with b laser beams such that the topography of the lattice is constant (apart from n overall translations), one must choose $b = n + 1$. This is the approach originally proposed by Grynberg et al. (1993). Typically, the fluctuations in the relative phases are slow compared with the atomic response times, so that the trapped atoms adiabatically follow the lattice translations. One can also construct a lattice with time independent topography by using an arbitrary number of beams with stabilized relative phases; this is the approach used by Hemmerich and Hänsch (1993). In either case, a good optical lattice for trapping atoms will be one where the points of maximum light shift are points of pure σ polarization. In addition, if the lattice is to trap atoms and is tuned near resonance, the net radiation pressure from the lasers should be zero near the lattice potential minima.

A. Two-Dimensional Lattices

Two possible realizations of 2D Grynberg-style lattices are shown in Figs. 4(a) and (b). In both cases, the reciprocal lattice is the same, determined only by the differences in the laser wavevectors. The primitive basis for the reciprocal lattice is given by the wavevector differences of maximum magnitude. Assuming $\theta \leq \pi/3$ gives $\mathbf{b}_1 = \mathbf{k}_2 - \mathbf{k}_1 = K_x \hat{\mathbf{x}} + K_y \hat{\mathbf{y}}$ and

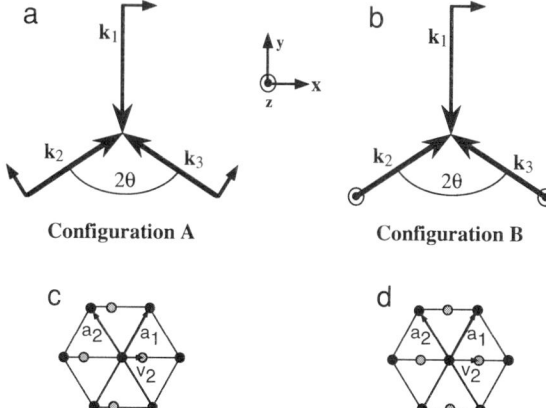

FIG. 4. two-dimensional optical lattice configurations of the Grynberg type, with $\theta = \pi/3$. (a) and (b) show the configurations of lattice wavevectors and polarization. (c) and (d) show the lattices defined by points of pure σ_+ polarization (black dots) and σ_- polarization (gray dots) for configurations (a) and (b), respectively. The lattices are characterized by a "Bravais lattice with a basis": a primitive cell spanned by $\mathbf{a}_1$ and $\mathbf{a}_2$, and a basis consisting of a σ_+ site at the origin and a σ_- site at $\mathbf{v}_2$.

$\mathbf{b}_2 = \mathbf{k}_3 - \mathbf{k}_1 = -K_x \hat{\mathbf{x}} + K_y \hat{\mathbf{y}}$, where $K_x \equiv k \sin\theta$ and $K_y \equiv k(\cos\theta + 1)$. In direct space, the primitive cell is spanned by

$$\mathbf{a}_1 = (\pi/K_x)\hat{\mathbf{x}} + (\pi/K_y)\hat{\mathbf{y}} \quad \text{and} \quad \mathbf{a}_2 = -(\pi/K_x)\hat{\mathbf{x}} + (\pi/K_y)\hat{\mathbf{y}}$$

The different polarization choices for these two geometries imply that the basis within a primitive cell, i.e., the relative positions of the σ_+ and σ_- sites, will be different. For configuration A, the positive frequency component of the electric field is given by

$$E_L(\mathbf{x}) = \frac{E_0 e^{-iky}}{\sqrt{2}} \left[-\mathbf{e}_+ \left\{ 1 + 2e^{iK_y y} \cos(K_x x) \right\} \right.$$
$$\left. + \mathbf{e}_- \left\{ 1 + 2e^{iK_y y} \cos(K_x x - 2\theta) \right\} \right] \quad (8)$$

where the quantization axis is chosen along $\hat{\mathbf{z}}$, perpendicular to the x–y plane, and the relative phases are chosen such that there is a local maximum of the σ_+ polarized field component at the origin. A basis associated with each primitive cell then consists of a σ_+ site at $\mathbf{v}_1 = 0$ and a σ_- site at $\mathbf{v}_2 = (2\theta/K_x)\hat{\mathbf{x}}$. This 2D configuration will only be a "good lattice," i.e., have maximum light shift at the points of pure σ polarization, for one choice of angle, $\theta = \pi/3$. Also, the radiation pressure near the potential minima will vanish for this symmetric geometry. Grynberg et al. (1993) used this lattice to trap Cs atoms in the Lamb–Dicke regime.

Consider now configuration B, shown in Fig. 4(b). We regard this lattice as a 2D generalization of the 1D lin⊥lin geometry, where one of the beams is split into two beams by an angle 2θ. Obtaining the maximum light shift at points of pure σ polarization requires unequal amplitudes of the three plane waves. Choosing $E_2 = E_3 = E_0/2$ and a convenient relative phase, the electric field for this geometry is

$$\mathbf{E}_L(\mathbf{x}) = \frac{E_0 e^{-iky}}{\sqrt{2}} \left[-\mathbf{e}_+ \left\{ 1 + e^{iK_y y} \cos(K_x x) \right\} + \mathbf{e}_- \left\{ 1 - e^{iK_y y} \cos(K_x x) \right\} \right] \tag{9}$$

Here the quantization axis is chosen in the x–y plane, along $\mathbf{k}_1$. Unlike configuration A, the radiation pressure at the potential minima is nonzero because of the imbalance in the intensities of the three beams, and so this geometry is not optimal for cooling and trapping in the typical near-resonance configuration. However, this geometry always has pure σ polarization at positions of maximal intensity for an arbitrary angle θ, and therefore it retains the possibility of varying the lattice constants through a change in θ while still maintaining a "good lattice." From Eq. (9) we see that the basis within a primitive cell has σ_+ at $\mathbf{v}_1 = 0$ and σ_- at $\mathbf{v}_2 = (\pi/K_x)\hat{\mathbf{x}}$. Figures 4(c) and (d) show the lattice structures for configurations A and B with $\theta = \pi/3$.

Hemmerich and Hänsch (1993) employed a four-beam geometry to create a 2D lattice from the interference pattern of an $\hat{\mathbf{x}}$ polarized standing wave along $\hat{\mathbf{y}}$ and a $\hat{\mathbf{y}}$ polarized standing wave along $\hat{\mathbf{x}}$. As we have discussed, this lattice requires phase stabilization to maintain a constant topography. If the relative phase of oscillation of the $\hat{\mathbf{x}}$ and $\hat{\mathbf{y}}$ polarized standing waves is $\pm\pi/2$ at the antinodes, this represents a "good" lattice with a square primitive cell, reciprocal lattice vectors $\mathbf{b}_1 = k(\hat{\mathbf{x}} + \hat{\mathbf{y}})$ and $\mathbf{b}_2 = k(\hat{\mathbf{x}} - \hat{\mathbf{y}})$, and electric field

$$\mathbf{E}_L(\mathbf{x}) = -\sqrt{2}\, E_0 [(\cos kx + \cos ky)\mathbf{e}_+ + (\cos kx - \cos ky)\mathbf{e}_-] \tag{10}$$

B. Three-Dimensional Lattices

As in the two-dimensional case, there are numerous geometries that one can employ to obtain a "good" three-dimensional lattice. Here we discuss a four-beam "Grynberg style" lattice that can be viewed as a three-dimensional generalization of the 1D lin⊥lin geometry (Verkerk et al., 1994). Other configurations were discussed by Petsas et al. (1994).

In a manner analogous to the 2D lattice shown in Fig. 4(b), we imagine starting with a 1D lin⊥lin lattice and "splitting" each of the beams into two, all with equal intensity and in such a way that all polarizations lie in the x–y plane, as shown in Fig. 5(a). Beams 1 and 2 create an $\hat{\mathbf{x}}$ polarized standing wave along $\hat{\mathbf{y}}$ with a traveling wave component in the $\hat{\mathbf{z}}$ direction, whereas beams 3 and 4 create a $\hat{\mathbf{y}}$ polarized standing wave along $\hat{\mathbf{x}}$ with a traveling wave component in the $-\hat{\mathbf{z}}$ direction. The overall field thus consists of a three-dimensional array of σ_+ and σ_- sites (where the quantization axis is chosen along $\hat{\mathbf{z}}$), at positions where the relative phase between the $\hat{\mathbf{x}}$ and $\hat{\mathbf{y}}$ polarized standing waves is $\pm\pi/2$ (Fig. 5b). Because all beams contribute in equal amplitude to the points of σ polarization, there will be no radiation pressure near the potential minima. There will, however, be an imbalance in the radiation pressure where the lines of nodes associated with one of the standing waves intersects the lines of antinodes of the other standing wave. This imbalance leads to "escape channels" for untrapped atoms, far from the potential minima.

For simplicity, let us consider $\theta_1 = \theta_2 \equiv \theta_L$. The primitive reciprocal lattice vectors are a linearly independent set of laser wavevector differences with the largest magnitude, e.g., $\mathbf{b}_1 = \mathbf{k}_1 - \mathbf{k}_3$, $\mathbf{b}_2 = \mathbf{k}_2 - \mathbf{k}_4$, and $\mathbf{b}_3 = \mathbf{k}_3 - \mathbf{k}_2$. In direct space, the spatial dependence of the electric

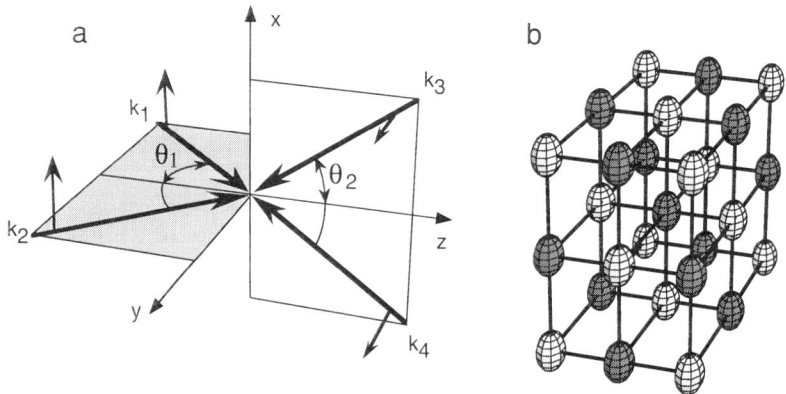

FIG. 5. (a) Grynberg-type 3D lin⊥lin lattice. One pair of laser beams travel in the $\hat{\mathbf{y}}$–$\hat{\mathbf{z}}$ plane ($\mathbf{k}_1$ and $\mathbf{k}_2$), separated by an angle $2\theta_1$, and polarized along $\hat{\mathbf{x}}$. They interfere with another pair that travel in the $\hat{\mathbf{x}}$–$\hat{\mathbf{z}}$ plane ($\mathbf{k}_3$ and $\mathbf{k}_4$), separated by an angle $2\theta_2$, and polarized along $\hat{\mathbf{y}}$. (b) The resulting centered tetragonal lattice of interleaved σ_+ and σ_- sites.

field is

$$\frac{\mathbf{E}_L(\mathbf{x})}{\sqrt{8}\,E_0} = \left(\hat{\mathbf{e}}_+ \left\{\cos(k_\parallel z)\left[\frac{\cos(k_\perp y) + \cos(k_\perp x)}{2}\right]\right.\right.$$
$$\left. + i\sin(k_\parallel z)\left[\frac{\cos(k_\perp y) - \cos(k_\perp x)}{2}\right]\right\}$$
$$+ \hat{\mathbf{e}}_- \left\{\cos(k_\parallel z)\left[\frac{\cos(k_\perp y) - \cos(k_\perp x)}{2}\right]\right.$$
$$\left.\left. + i\sin(k_\parallel z)\left[\frac{\cos(k_\perp y) + \cos(k_\perp x)}{2}\right]\right\}\right) \quad (11)$$

where $k_\parallel \equiv k_L \cos\theta_L$ and $k_\perp \equiv k_L \sin\theta_L$. Note that in the limit $\theta_L \to \pi/2$, this geometry reduces to the "Hänsch-style" 2D lattice, and Eq. (10) is recovered from Eq. (11); as $\theta_L \to 0$, we regain the 1D lin⊥lin geometry. Thus, we see that in the x–y plane the lattice is square with lattice constant $a_\perp = 2\pi/k_\perp$. Along the z axis, the lattice resembles the 1D lin⊥lin lattice, with a larger lattice constant $a_\parallel = \pi/k_\parallel$. In the nomenclature of crystallography, this lattice is centered tetragonal.

IV. Laser Cooling in Optical Lattices: Theory

The theory of laser cooling in optical lattices has seen enormous progress over the last few years, involving a synergism of new and old techniques. The close analogy between the periodic potential seen by an electron moving in a solid crystal and the optical lattice potential seen by a laser-cooled atom allows us to apply condensed matter formalism to the atomic physics problem. In attempting to solve these models, numerical techniques have been developed, which are not only powerful computational tools, but also give a new understanding of the quantum physics of dissipative systems. In this section, we review some of the theoretical models and methods that have been used to study laser cooling in optical lattices and to give interpretations to the experimental results.

A. GENERAL FORMALISM

The general theory of laser cooling of multilevel atoms was discussed in detail by Cohen-Tannoudji (1991) in the Les Houches Summer School. Here we highlight some of the central results. Consider a monochromatic laser field nearly resonant with an atomic transition with ground and excited state angular momenta J_g and J_e. The general description of the system involves the dynamics of $2(J_g + J_e + 1)$ states that evolve coherently through the coupling to the laser field, and dissipatively through the coupling to the vacuum. In the limit of low intensity or large detuning, one can reduce the problem significantly; fortunately, these are the parameters that lead to the lowest temperatures. In this regime, the atomic saturation is small and there is little excited state population. Mathematically, there will then be a separation of time scales—the spontaneous emission time and the optical pumping time—the latter being much longer than the former in the low saturation regime. The excited state populations and atomic coherences between excited and ground states then relax rapidly, and they adiabatically follow the evolution of the ground state manifold. It follows that the excited state variables can be "adiabatically eliminated," leading to a reduced Hamiltonian that acts only on the atoms's external coordinates and internal ground state manifold (Castin and Dalibard, 1991):

$$H = \frac{\mathbf{P}^2}{2M} + U_0\left(\vec{\varepsilon}_L(\mathbf{x}) \cdot \hat{\mathbf{d}}\right)^\dagger \left(\vec{\varepsilon}_L(\mathbf{x}) \cdot \hat{\mathbf{d}}\right) \tag{12a}$$

where

$$\hat{\mathbf{d}} \equiv \sum_{m_g, q} c_{m_g}^{m_g+q} |e; J_e, m_g + q\rangle\langle g; J_g, m_g| \mathbf{e}_q^* \tag{12b}$$

Here, $\mathbf{P}$ and $\mathbf{x}$ refer to the atom's center of mass, $\vec{\varepsilon}_L(\mathbf{x})$ is the laser polarization at $\mathbf{x}$, $c_{m_g}^{m_e}$ is shorthand for the Clebsch–Gordan coefficient coupling the states $|J_g, m_g\rangle$ and $|J_e, m_e\rangle$, and $\mathbf{e}_q$ are the spherical basis vectors ($q = 0, \pm 1$).

For geometries where all lattice beam polarization vectors lie in a plane, the field can be decomposed into solely $\sigma_\pm$ components (quantization axis perpendicular to the plane). Consider first an atom with a $J_g = \frac{1}{2} \to J_e = \frac{3}{2}$ transition as discussed in Section II. This system is particularly simple because the $m_g = \pm \frac{1}{2}$ ground states are not connected by the laser field; this allows us to consider the coherent atomic motion as taking place on one of two separate *scalar* potentials. This physical picture is too simple if $J_g \geq 1$. Consider, therefore, an atom with a $J_g = 2 \to J_e = 3$ transition as shown in Fig. 6(a). The light shift operator now contains both diagonal and off-diagonal terms in the basis $|g; J_g, m_g\rangle$; these correspond to absorption and stimulated emission processes with $\Delta m_g = 0$ and $\Delta m_g = \pm 2$ respectively. The $\Delta m_g = \pm 2$ stimulated Raman transitions couple states in families that have m_g either even or odd, as indicated as solid or dashed

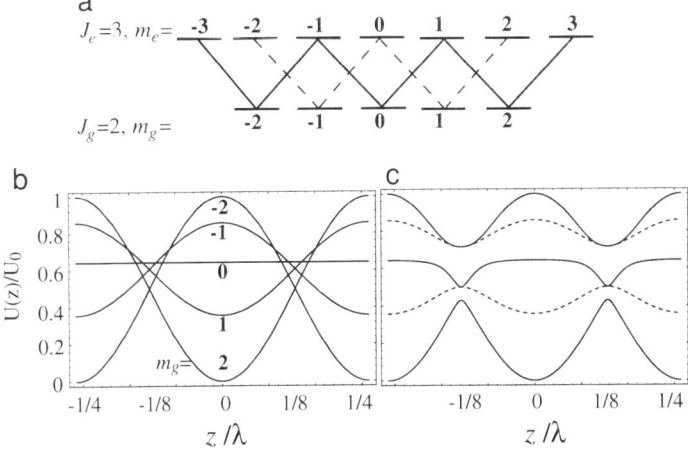

FIG. 6. (a) Level structure of a $J_g = 2 \to J_e = 3$ atom. The σ_+ and σ_- polarized components of the lattice light field couple the ground state sublevels in families with m_g even (solid lines) and m_g odd (dashed lines). (b) Light shift eigenvalues as a function of position in the 1D lin⊥lin field of Fig. 1(a), ignoring coupling between different m_g states. This yields the diabatic potentials. Numbers next to curves indicate the m_g value. (c) When $\Delta m_g = \pm 2$ coupling between m_g states is included one obtains the adiabatic potentials. The solid curves represent eigenvalues for the coupled "even family" of m_g-states ($m_g = 2, 0, -2$), and the dotted curves for the "odd family" ($m_g = 1, -1$).

lines in Fig. 6(a). If we neglect off-diagonal elements, we obtain "diabatic" potentials, whereas if we diagonalize the light shift operator, we obtain "adiabatic" potentials representing the light shift eigenvalues of a stationary atom localized at a particular position. Figures 6(b) and (c) show the diabatic and adiabatic potentials for a $J_g = 2 \to J_e = 3$ atom in a 1D lin⊥lin lattice. In general, one cannot view the atom as feeling the influence of any single adiabatic or diabatic potential; both are approximations. Slow atoms will follow a single adiabatic potential if the time spent in the crossing region of the diabatic potentials allows stimulated Raman transitions to occur; fast moving atoms will follow the diabatic potentials. Near the bottom of the optical potential, where the polarization is purely $\sigma_\pm$, there is little difference between the adiabatic and diabatic potentials.

The Hamiltonian in Eq. (12a) governs only the coherent evolution of the atom. To treat laser cooling, one must include dissipative processes, i.e., optical pumping between ground state sublevels. The starting point for most treatments of this problem is the master equation for the atomic density operator ρ following adiabatic elimination of the excited state

(Castin and Dalibard, 1991):

$$\frac{d\rho}{dt} = \frac{1}{i\hbar}[H,\rho] - \frac{1}{2}\gamma_s\{\Lambda,\rho\}$$
$$+ \gamma_s \sum_h \int d^2k_s N_h(\mathbf{k}_s)(W_h(\mathbf{k}_s)\rho W_h^\dagger(\mathbf{k}_s)) \quad (13)$$

The first term is the coherent Schrödinger evolution governed by the Hamiltonian, and the last two are the dissipative processes. The second term represents population decay, i.e., depletion of each ground state due to optical pumping into all others, with $\{,\}$ the anticommutator and the operator $\Lambda \equiv (\vec{\varepsilon}_L(\mathbf{x}) \cdot \hat{\mathbf{d}})^\dagger(\vec{\varepsilon}_L(\mathbf{x}) \cdot \hat{\mathbf{d}})$. The third term is a population feeding term arising from optical pumping replenishing the populations of each of the ground states. Here, the operator $W_h(\mathbf{k}_s) = (e^{i\mathbf{k}_s \cdot \mathbf{x}}\mathbf{e}_h \cdot \hat{\mathbf{d}})^\dagger(\vec{\varepsilon}_L(\mathbf{x}) \cdot \hat{\mathbf{d}})$ represents absorption of a lattice photon, followed by emission of a fluorescence photon with wavevector $\mathbf{k}_s$ and helicity h along the quantization axis. The function $N_h(\mathbf{k}_s)$ is the probability distribution for a fluorescence photon with helicity h to propagate along the direction $\mathbf{k}_s$. If one scales all energies by the recoil energy, we see that the master equation depends solely on two dimensionless parameters, $\overline{U}_0 \equiv U_0/E_R$, and $\overline{\gamma}_s \equiv \hbar\gamma_s/E_R$. The first parameter determines the time scale for the coherent processes such as the oscillation of atoms deeply bound at the bottom of the wells, and the second determines the time scale for the dissipative processes such as optical pumping. In principle, all information is contained within Eq. (13), which is an exact description of the atom–laser dynamics in the low saturation limit. Solutions to this equation are not, however, obtained trivially. In the remainder of this section, we outline some of the various approaches, their advantages and limitations, and the physical insight they give.

B. Semiclassical Methods

The master equation (13) is a complete quantum description of the atomic dynamics, including external degrees of freedom. In certain circumstances, one can simplify the problem by treating the center-of-mass coordinates as classical variables. This *semiclassical* approximation is valid when the spread in Doppler shifts due to the quantum uncertainty in momentum is small compared with the natural linewidth, and when the spatial coherence length of the atomic wavefunction is small compared with the optical wavelength. In addition, the internal degrees of freedom must relax much faster than the external degrees of freedom so that one can treat the atoms as a classical particle experiencing an instantaneous force. For multilevel atoms in optical lattices, the semiclassical approximation typically breaks down, not because the atomic wavefunction has spatial coherence over distances larger than the optical wavelength, but because the internal time scale τ_{int} becomes much longer than the external time scale τ_{ext}. Consider an atom tightly bound in a σ_+ potential well. The atomic localization reduces the overlap of the atomic center-of-mass distribution with the σ_- polarized standing wave by a factor of order the Lamb–Dicke parameter $E_R/\hbar \omega_{osc}$. This reduces the rate of optical pumping, so that $\tau_{int} \approx [(E_R/\hbar \omega_{osc})\gamma_s]^{-1} \gg \omega_{osc}^{-1} \approx \tau_{ext}$, where the inequality is fulfilled as soon as $\Delta/\Gamma \gg 1$. This parameter regime is known as the *oscillating regime*, whereas the reverse situation, $\tau_{int} \ll \tau_{ext}$, is known as the *jumping regime*. Various procedures have been employed to recover an approximate semiclassical equation of motion in the oscillating regime (Castin *et al.*, 1991; 1994). Some system properties can be deduced using such models, but they fail to account for purely quantum effects such as tunneling. For this reason a full quantum theory, including the external degrees of freedom, is necessary.

C. The Monte Carlo Wavefunction Technique

A brute force solution of the master equation would involve expanding the density operator in an appropriate basis, including $2J_g + 1$ internal states and a large number of external states. In principle, the external degrees of freedom have a continuous spectrum, but they can be restricted to a grid consistent with the desired resolution in momentum space. Even for a typical 1D laser cooling problem, the dimension of the Hilbert space, N, can be on the order of 10^3–10^4. The corresponding density matrix has $N^2 = 10^6$–10^8 elements, and numerical solution of Eq. (13) becomes intractable even for sparse matrices. Given these difficulties, the groups of Mølmer *et al.* (1993) and of Zoller (Marte *et al.*, 1993a) have applied a Monte Carlo wavefunction simulation technique to the problem. The essence of this method is to view the density operator as an ensemble average of "quantum trajectories." Because wavefunctions have only N elements, compared with the N^2 elements of the density matrix, the technique leads to considerable savings in computer memory. The trade-off is that extra computer time is needed to simulate a large number of trajectories. Each trajectory consists of the deterministic evolution of a wavefunction punctuated by quantum jumps at random times. The stochastic evolution represented by quantum jumps originates from the coupling of the small quantum system at hand to a large reservoir. For our case, the small system consists of an atom, and the role of the reservoir is played by vacuum fluctuations. Not only is the Monte Carlo method a useful computational tool, it gives physical insight into the quantum mechanics of dissipative systems. Many other problems in quantum optics have been studied by similar formulations in which the master equation is replaced by stochastic equations (Gardiner *et al.*, 1992; Gisin and Percival, 1992; Carmichael, 1993).

In a beautiful application of this technique, Marte *et al.* (1993b) calculated the fluorescence spectrum of Rb atoms laser-cooled in a 1D lin⊥lin lattice, and they found spectacularly close quantitative agreement with experiments. The calculated spectrum shows well-resolved Raman sidebands that are suppressed by the Lamb–Dicke effect, as discussed in Section II, and reproduces the sideband asymmetry that reflects the distribution of population over vibrational states. Castin *et al.* (1994) have used the Monte Carlo simulations to calculate the steady state properties of $J_g = \frac{1}{2} \rightarrow J_e = \frac{3}{2}$ atoms in a 2D lattice, in general agreement with other methods described in what follows. Castin and Mølmer (1995) have also applied this technique to a 3D optical molasses in the standard six-beam configuration. Their system was effectively an optical lattice, since a particular phase relation between the beams was chosen for the simulation.

In translating the master equation into a set of stochastic equations of motion, the basis of decomposition should be chosen appropriately for the application at hand. For studying velocity distributions, the natural basis elements are energy eigenstates, which are delocalized. On the other hand, if one studies spatial diffusion of atoms in a lattice, the natural basis would consist of localized states. Such an approach was initiated by Holland *et al.* (1996), using wavefunctions that represent the diffraction-limited images of an atom viewed through the photons it emits into a lens with a large aperture (the Heisenberg microscope). They applied this technique to calculate the diffusion of atoms in a one-dimensional optical lattice. The spatial trajectory of the atom shows long periods in which the atom is trapped near a given lattice site, punctuated by long "flights" over many wavelengths when energy fluctuations allow the atom to "boil" out of a well. The spatial diffusion coefficient was calculated for atoms driven on a $J_g = \frac{1}{2} \rightarrow J_e = \frac{3}{2}$ transition, and it was found to be an order of magnitude larger than for atoms driven on a $J_g = 3 \rightarrow J_e = 4$ transition. This indicates that atoms with large angular momenta are trapped for longer times in the wells; more research is, however, necessary to extend the standard model of Sisyphus cooling and to understand the mechanism by which this trapping occurs. The nature of the trajectories has interesting statistical features, which have been analyzed in terms of Lévy flights by Marksteiner *et al.* (1996).

D. Band Structure Formalism

A full quantum theory of laser cooling in optical lattices starts with the Hamiltonian in Eq. (12a), where now the external coordinates ($\mathbf{x}, \mathbf{P}$) are quantum operators. Because the Hamiltonian is invariant under translations by any lattice vector $\mathbf{R}$, as described in Section III, the energy eigenstates must satisfy Bloch's theorem. This is analogous to the description of an electron moving in the periodic potential formed by the ionic cores in a solid. The resulting energy spectrum consist of bands, with a large bandgap separating the tightly bound states and a quasicontinuous spectrum for free states. In the oscillating regime, laser cooling can be analyzed by first diagonalizing the laser–atom Hamiltonian and then treating the vacuum as a perturbation. This procedure is somewhat analogous to the dressed-state treatment of resonance fluorescence of atoms in intense laser fields (Cohen-Tannoudji *et al.*, 1992). The master equation (13) then governs the populations in Bloch states and coherences between them. In a near-resonant lattice, spontaneous emission will destroy the spatial coherence of the atomic wavefunction over distances large compared with the optical wavelength, and therefore the steady state density matrix will be a statistical mixture of Bloch states with a limited coherence length.

We begin with the diagonalization of the atom–laser Hamiltonian, Eq. (12a). The physical problem differs somewhat from the standard solid state problem, in part because the optical potential depends strongly on the internal state of the particle, and in part because the internal and external degrees of freedom are not generally separable. As a consequence the eigenstates are "entangled states" of these variables rather than product states. In general, one can write the state of the atom as a $(2J_g + 1)$-component spinor. Each wavefunction is then expressed in Bloch form as

$$|\Psi\rangle = \sum_{m_g} e^{i\mathbf{q}\cdot\mathbf{x}} |u_q^{m_g}\rangle \otimes |m_g\rangle \qquad (14)$$

where $|u_q^{m_g}\rangle$ has the periodicity of the lattice. A special, simple case is that of a 1D lin⊥lin lattice driving a $J_g = \frac{1}{2} \to J_e = \frac{3}{2}$ transition. The energy eigenstates are diagonal in the basis of m_g states as previously described, and the problem reduces to two independent scalar problems. The band structure for this system was calculated by Castin and Dalibard (1991).

Consider now a 1D lin⊥lin lattice driving a $J_g = 2 \to J_e = 3$ transition as shown in Fig. 6(a). A brute force approach would be to substitute the spinor representation, Eq. (14), into the Hamiltonian and diagonalize it to obtain eigenenergies and eigenstates. This task is simplified by careful consideration of the symmetries of the Hamiltonian (Castin, 1992; Courtois, 1993). The basic symmetry $T_{\lambda/2}$, consisting of a translation by $\lambda/2$, is built into the Bloch states. In addition, one immediately finds that the Hamiltonian is block diagonal, with even m_g in one block, and odd m_g in the other. This separation arises from a rotation symmetry of the system about the z axis by 180°. The unitary operator associated with this symmetry is $R = \exp\{-i\pi J_z\}$, with eigenvalues $\eta = (-1)^{m_g} = \pm 1$ corresponding to the two "families" of m_g states. We can now solve the problem for each family η separately. Diagonalizing the Hamiltonian for a given η, one finds that for each quasi-momentum q the eigenvalues come in pairs, which are nearly degenerate for tightly bound states and have increased energy separation for higher lying states. For integer angular momentum, these pairs are associated with a third symmetry of the problem, $\tau \equiv T_{\lambda/4} \otimes \Sigma_{m_g} |-m_g\rangle\langle m_g|$, corresponding to translation by $\lambda/4$ and a mapping $m_g \to -m_g$. Since $\tau^2 = T_{\lambda/2}$, one sees that τ has eigenvalues $\zeta e^{iq\lambda/4}$, where $\zeta = \pm 1$. Each pair of eigenfunctions represents even and odd parity solutions of the wavefunction between neighboring wells of opposite helicity. The energy eigenstates $|n, q, \eta, \zeta\rangle$ are thus simultaneous eigenfunctions of the four mutually commuting operators $\{H, T_{\lambda/2}, R, \tau\}$. There is yet a forth symmetry of the Hamiltonian: parity together with a rotation by 90° about the z axis, $I = P \otimes e^{-i\pi J_z/2}$. I does not commute with the translation operator; however, one can show $I|n, q, \eta, \zeta\rangle = \pm\sqrt{\eta}\,|n, -q, \eta, \eta\zeta\rangle$, from which it follows that the energy bands satisfy $E_{n,q,\eta,\zeta} = E_{n,-q,\eta,\eta\zeta}$.

The diagonalization procedure involves choosing a value for q, η, and ζ, and substituting the representation Eq. (14) into the Hamiltonian. The band structure for $U_0 = 200E_R$ is shown in Fig. 7. Compared with the $J_g = \frac{1}{2} \to J_e = \frac{3}{2}$ transition, the energy bands exhibit substantial curvature for states closer to the minima of the potential wells. This curvature arises because atoms with angular momentum $J_g \geq 1$ can tunnel between neighboring wells of opposite polarization. The band curvature becomes substantial as soon as the energy rises above the top of the lowest adiabatic potential well (Fig. 6c). Such bands do not, however, have completely free particle character, indicating the mixed adiabatic/diabatic nature of the potential.

E. THE SECULAR APPROXIMATION

Given the energy band structure just described, one can expand the master equation, Eq. (13), in the basis of the Bloch states. If we are interested only in the steady state solution, some simplifications can be made. We

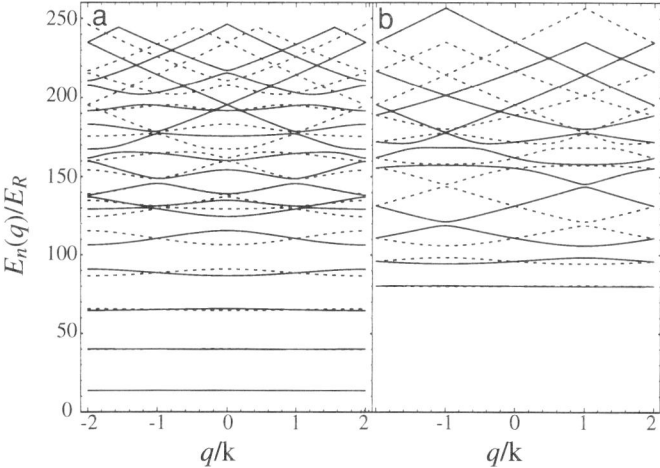

FIG. 7. Band structure in the first two Brillouin zones for a $J_g = 2 \to J_e = 3$ atom driven by a 1D lin⊥lin optical lattice, with a maximum light shift $U_0 = 200 E_R$. (a) The energy eigenvalues associated with the family of even m_g states ($\eta = 1$). (b) The energy eigenvalues associated with the family of odd m_g states ($\eta = -1$). Solid and dashed lines represent bands with parity eigenvalue $\zeta = +1$ or -1, respectively.

first note that all of the symmetries of the Hamiltonian, $T_{\lambda/2}$, R, τ, I, are also symmetries of the dissipative part of the master equation, implying that the steady state density operator must be invariant under these unitary operations. Therefore, no off-diagonal matrix elements between states of different q, η, or ζ survive. Consider the steady state master equation for a matrix element between two states of different energy bands, all other eigenvalues being the same:

$$\dot{\rho}_{nn'} = -(i/\hbar)(E_n - E_{n'})\langle n| \rho |n'\rangle + \gamma_s \langle n| \mathscr{L}_{relax}[\rho]|n'\rangle = 0 \quad (15)$$

where $\mathscr{L}_{relax}[\rho]$ is the dissipative part of the master equation. For the tightly bound states, the energy spectrum closely approximates a simple harmonic ladder with energy spacing $E_{n+1} - E_n = \hbar \omega_{osc}$. In the "oscillating regime," $\omega_{osc} \gg 1/\tau_{pump}$, the energy spacing is much larger than the incoherent linewidth. Then the first term in Eq. (15) will be much larger than the second, and therefore in steady state we must have $\langle n| \rho |n'\rangle \approx \rho_{nn'} \delta_{nn'}$ for tightly bound states. In this so-called *secular approximation*, the master equation reduces to a set of rate equations for population transfer between the Bloch states (Castin, 1992):

$$\dot{\Pi}_{nq\eta\zeta} = -\gamma_{nq\eta\zeta} \Pi_{nq\eta\zeta} + \sum_{n'q'\eta'\zeta'} \gamma_{n'q'\eta'\zeta' \to nq\eta\zeta} \Pi_{n'q'\eta'\zeta'} \quad (16)$$

where $\gamma_{nq\eta\zeta} = \gamma_s \langle nq\eta\zeta | \Lambda | nq\eta\zeta \rangle$ is the total departure rate from the state $|nq\eta\zeta\rangle$, and

$$\gamma_{n'q'\eta'\zeta' \to nq\eta\zeta} = \gamma_s \int_{-k}^{k} dk_s N(k_s) \sum_h |\langle nq\eta\zeta | W_h(k_s) | n'q'\eta'\zeta' \rangle|^2 \quad (17)$$

is the rate of optical pumping between states. Multiplying γ_s is the Franck–Condon factor, accounting for conservation of both linear and angular momentum in the transition. Conservation of the total population is ensured by the fact that $\gamma_{nq\eta\zeta} = \sum' \gamma_{n'q'\eta'\zeta' \to nq\eta\zeta}$.

The symmetries of the system determine certain selection rules for optical pumping, which in turn limit the number of nonvanishing transition matrix elements that must be calculated. For example, the transition matrix element between two Bloch states in a 1D lattice is nonvanishing only for $q - q' = \pm k_L + \mathbf{k}_s \cdot \hat{\mathbf{z}}$. This is a statement of momentum conservation. Since photons can be spontaneously emitted along any direction, the recoil along the z axis spans all values and therefore couples all quasi-momenta q within a band. This complete description makes the problem numerically intractable, and one typically simplifies the spontaneous emission pattern to match the symmetry of the lattice. Solving for the steady state populations in the various Bloch states is then straightforward. One chooses a finite but sufficient number of bands and calculates the transition rates between them according to Eq. (17). Given the conservation of the total population, one can then invert the rate equation matrix when the left hand side is set to zero.

Figure 8(a) shows the steady state population of the first five energy bands for a $J_g = \frac{1}{2} \to J_e = \frac{3}{2}$ transition as a function of total light shift, with a total of 80 bands included in the rate equations, as calculated by Castin and Dalibard (1991). The population of the lowest band reaches a maximum value of 34% of the total population, with only 20% of the population in the quasi-free bands. This suggests that atoms are indeed strongly localized in the optical potential wells. The rate equation approach has also been used to calculate properties of higher angular momentum atoms in a 1D lin⊥lin lattice, with the more complex band structure described previously. Figure 8(b) shows the steady state populations for cesium ($J_g = 4 \to J_e = 5$), as calculated by Courtois (1993). Sharp resonances appear in the populations of the lowest band at specific values of the light shift. These have been interpreted as arising from strong transfer of population to the ground state when energy bands associated with one of the high lying adiabatic potentials matches that of the deepest potential. Such resonances have also been predicted to occur in a 2D

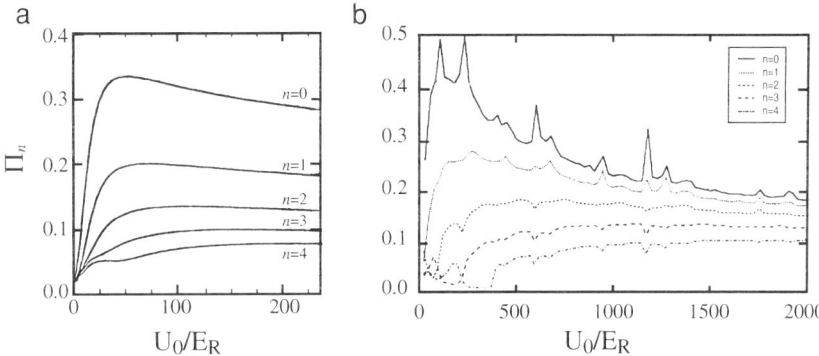

FIG. 8. (a) Steady state populations in the first five energy bands for a $J_g = \frac{1}{2} \to J_e = \frac{3}{2}$ transition (from Castin, 1992, reprinted with permission). (b) Steady state population in the first five bands for a $J_g = 4 \to J_e = 5$ transition (from Courtois, 1993, reprinted with permission). Both figures are for a 1D lin⊥lin optical lattice, and both show populations as a function of the maximum light shift U_0.

optical lattice, when bound states of neighboring wells are resonant (Berg-Sørensen et al., 1993).

The rate equation approach to laser cooling in optical lattices has been applied to numerous other systems. Doery et al. (1994) have used band theory to study the velocity distribution of $J_g = \frac{1}{2} \to J_e = \frac{3}{2}$ atoms in both a 1D lin⊥lin lattice and in a 1D σ_+ standing wave with a magnetic field. Castin et al. (1994) have calculated the steady properties of $J_g = \frac{1}{2} \to J_e = \frac{3}{2}$ atoms in a 2D "Hänsch-style lattice" described in Section III, and Berg-Sørensen (1994) has done a similar calculation for a "Grynberg-style" 2D lattice. The minimum temperatures in these lattices is on the same order as in the 1D lin⊥lin case, with the added feature that tunneling resonances may appear for certain parameters. To date, no calculations of this type have been performed on large angular momentum atoms in higher dimensional lattices.

V. Spectroscopy

Spectroscopy has been the method of choice for probing the quantized atomic motion and cooling dynamics in optical lattices. Section II briefly introduced resonance fluorescence and probe transmission spectroscopy; in addition, phase conjugation spectroscopy (Lounis et al., 1993; Hemmerich et al., 1994a), which measures the frequency dependent phase conjugate reflection of a probe beam, has been employed in the study of optical lattices. These techniques probe atoms in an optical lattice through the interaction between the atomic dipole moment **d** and the electromagnetic field. In fluorescence spectroscopy, one measures the spectrum of fluctuations of **d**, driven by the coupling of the atomic dipole to the lattice field and vacuum modes. In probe absorption or phase conjugation spectroscopy, an external probe field drives the system, resulting in energy exchange between the probe and lattice fields.

In Section IV, we showed that the evolution of the atomic system is described by the master equation (13). We can write Eq. (13) in matrix form as $\dot{\mathbf{S}} = \mathscr{L}\mathbf{S}$, where **S** is the Bloch vector containing the elements of the density matrix, and $\mathscr{L}$ is the super-operator governing the evolution of **S**. A representation can be chosen in which $\mathscr{L}$ is diagonal, in which case the evolution of **S** is trivially determined. It follows that the normal modes associated with the dynamical evolution of the system are eigenvectors of $\mathscr{L}$, and that the kth normal mode evolves as $e^{\lambda_k \tau}$, where λ_k is the corresponding eigenvalue. Optical lattice parameters are often close to the secular regime, so that the normal modes of the system remain close to the basis states of Eq. (14) and the system dynamics can be discussed in terms of transitions between those states. As outlined in Section II, this often allows a simple physical interpretation of the spectrum in terms of Raman and Rayleigh processes.

A. Spectroscopic Techniques

1. *Fluorescence Spectroscopy*

The normalized power spectrum of the fluorescence electric field is, by definition, the Fourier transform of the normalized first order correlation function $g_s^{(1)}(\tau) = \langle \hat{\varepsilon}_D^* \cdot \mathbf{E}_s^{(-)}(t+\tau)\, \hat{\varepsilon}_D \cdot \mathbf{E}_s^{(+)}(t)\rangle / I_s$, where $\hat{\varepsilon}_D$ is the detection polarization and I_s is the scattered intensity. The scattered field, $\mathbf{E}_s^{(\pm)}(t)$, is proportional to $\mathbf{d}^{(\pm)}(t - r/c)$, where r is the atom–detector distance. It follows that $g_s^{(1)}(\tau)$ is also the normalized correlation function of the atomic dipole. The quantum regression theorem states that the evolution of averages of the type $g_s^{(1)}(\tau)$ is governed by the same equation of motion as that of the Bloch vector **S** (Cohen-Tannoudji et al., 1992), so that

$$g_s^{(1)}(\tau) = e^{i\omega_L \tau} \sum_k c_k e^{\lambda_k \tau}; \quad \tau \geq 0 \qquad (18)$$

In this expression, $\{\lambda_k\}$ are the eigenvalues of the super-operator $\mathscr{L}$, and $\{c_k\}$ are complex amplitudes that are in principle calculable. Equation (18) shows that in the secular regime the spectrum is the sum of Lorentzian lines with amplitudes $\text{Re}[c_k]$, center frequencies $\omega_L + \text{Im}[\lambda_k]$, and halfwidths $\text{Re}[\lambda_k]$ that are small compared with the line separations. In this limit, one can associate each line with transitions between a particular set of states, and determine the amplitudes $\text{Re}[c_k]$ from rate equations. Some general information about the nature of the spectrum is available from the properties of $\mathscr{L}$. Because a representation can be chosen in which $\mathscr{L}$ is real, the eigenvalues are either real or come in pairs $\lambda_k = \lambda_{k'}^*$. Consequently, the spectrum contains components at $\omega_L \pm \text{Im}[\lambda_k]$ with identical widths, but since possibly $c_k \neq c_{k'}^*$, the spectrum need not be symmetric around ω_L. If the system is closed, $\text{Tr}(\rho) = 1$, a representation may be chosen in which one of the elements of **S** is constant. The corresponding eigenvalue is zero, and the corresponding eigenvector is the normal mode associated with the steady state; this is the source of the coherent component in the spectrum.

The group of Phillips have demonstrated an optical heterodyne technique for spectrum analysis of resonance fluorescence (Westbrook *et al.*, 1990; Jessen *et al.*, 1992). The technique is illustrated in Fig. 2(a) and was discussed briefly in Section II. Light radiated by atoms in the optical lattice is combined with a local oscillator beam, and it interferes to produce a total electric field $\mathbf{E}_{tot}(t) = \hat{\varepsilon}_D E_{LO} \cos(\omega_{LO} t) + \mathbf{E}_s(t)$. For the moment, we consider these fields to be classical. The total intensity of the interfering fields is measured with a photodiode, resulting in a current

$$i(t) \propto |\mathbf{E}_{tot}(t)|^2 \propto |E_{LO}|^2 + \text{Re}\left[\mathbf{E}_s(t)^* \cdot \hat{\varepsilon}_D E_{LO} e^{-i\omega_{LO} t}\right] \quad (19)$$

where terms that are second order in $|\mathbf{E}_s(t)|$ and terms that oscillate at optical frequencies have been omitted. The physical significance of the remaining terms is as follows: The term $|E_{LO}|^2$ produces a dc component in the spectrum, as well as shot noise due to the quantization of the photocurrent; the term $\mathbf{E}_s(t)^* \cdot \hat{\varepsilon}_D E_{LO} e^{-i\omega_{LO} t}$ is the signal of interest. The power spectrum of the current for $\omega \geq 0$ is

$$P_i^{meas}(\omega) = \frac{ei_{LO}}{\pi} + 2i_{LO}^2 \delta(\omega) + \frac{i_{LO}\langle i_s(t)\rangle}{\pi}$$
$$\times \int_{-\infty}^{\infty} \exp(i\omega_{LO}\tau) g_s^{(1)}(\tau) \exp(-i\omega\tau) d\tau \quad (20)$$

where i_{LO} and i_s are the separate currents generated by the local oscillator and signal light. Equation (20) shows that the measured power spectrum of the photo-current contains three contributions: a frequency independent shot-noise background, a dc current generated by the local oscillator, and a signal term that is an exact replica of the power spectrum of $\hat{\varepsilon}_D \cdot \mathbf{E}_s(t)$, translated to an rf intermediate frequency $|\omega_{LO} - \omega_L|$.

The primary technical concerns in optical heterodyne spectroscopy are frequency resolution and signal-to-noise ratio (S/N). Westbrook *et al.* derived both the local oscillator and optical lattice beams from the same laser, in which case the measurement setup of Fig. 2(a) is equivalent to a Mach–Zender interferometer, with atoms in the lattice playing the role of a mirror. It is well known that such an interferometer has good fringe visibility as long as the optical path difference in the interferometer is much smaller than the laser coherence length. Typical laser sources employed in laser cooling therefore contribute negligibly to the frequency resolution, which in practice is limited by mechanical mirror vibration. Obtaining a useful S/N in a measurement of the spectrum is usually a much more serious concern. Jessen (1993) has performed a detailed analysis of S/N issues in optical heterodyne fluorescence spectroscopy; here, we note only that since the rate of photon scattering $\gamma_s \propto 1/\Delta^2$ a useful S/N has been achieved only for moderate detunings $|\Delta| \leq 10\Gamma$.

Photon correlation or self-heterodyne spectroscopy measures the power spectrum of the photocurrent generated by the fluorescence without a local oscillator beam. In that case, the photocurrent correlation function is a replica of the second order correlation function $g_s^{(2)}(\tau)$ of the scattered field. If the photon statistics of the source is Gaussian, one can use the relation $g_s^{(2)}(\tau) = 1 + |g_s^{(1)}(\tau)|^2$ to partly recover $g_s^{(1)}(\tau)$. Because the phase of $g_s^{(1)}(\tau)$ is not available, the power spectrum of the photocurrent is centered on dc, and information about the asymmetry of the spectrum is lost. Considerations of resolution and S/N are essentially the same as for heterodyne detection. Photon correlation spectroscopy has been applied recently to an atomic sample in a magneto-optic trap (Jurczak *et al.*, 1995).

2. *Probe Spectroscopy*

By applying a probe to the system at frequency ω_p, one can measure either the transmitted signal or the phase-matched phase conjugate reflection (via four-wave mixing) as a function of $\delta_p = \omega_p - \omega_L$. Typically, the probe is weak, and its only effect is to excite particular normal modes of the unperturbed system (Courtois and Grynberg, 1992). As discussed, the normal modes are the eigenvectors of the super-operator $\mathscr{L}$ that governs the evolution of the atom–lattice system in the absence of the probe. The transfer of energy between the probe field and the atom–lattice system is most efficient when δ_p corresponds to a particular dynamical evolution frequency Im[λ_k]. Thus, the information contained in the probe transmission and phase conjugation spectra is largely equivalent to the information in the fluorescence spectrum. In the secular regime, one can associate features in the spectra with stimulated Raman transitions between system eigenstates that lead to redistribution of photons among lattice and probe beams. As discussed in Section II, the net amount of gain and absorption at a given frequency is determined by the competition between stimulated Raman processes that involve lattice and probe photons in opposite order, as illustrated in Fig. 3(b). Thus, the gain/absorption of a probe beam passing through the lattice is proportional to the total population difference between the coupled states. It follows that the pair of Raman lines at frequencies $\omega_L \pm \text{Im}[\lambda_k]$ is antisymmetric; a gain of α at frequency $\omega_L - \text{Im}[\lambda_k]$ is accompanied by an absorption of $1/\alpha$ at frequency $\omega_L + \text{Im}[\lambda_k]$. In general, the probe transmission through a medium without population inversion will exhibit gain for $\omega_p < \omega_L$ and absorption for $\omega_p > \omega_L$, as is the case in Fig. 3(c).

The frequency resolution that can be achieved in probe spectroscopy is subject to the same limitations as fluorescence spectroscopy, whereas the problem of achieving useful S/N is less severe. The difference in the S/N ratio of these two techniques can be explained as follows. In probe spectroscopy, emission of a probe photon is *stimulated* by the lattice/probe beams, whereas the fluorescence is *spontaneous* emission. Viewed classically, in probe spectroscopy the atomic motion is coherently driven by the interference of probe and lattice fields, and therefore the light scattered from each atom constructively interferes. In fluorescence spectroscopy, the atomic motion is incoherently driven by the lattice and vacuum fields, and light emitted from different atoms does not interfere constructively. This results in a much weaker scattered field.

B. EXPERIMENTAL RESULTS

The first spectroscopic measurements on laser-cooled atoms were performed by Westbrook *et al.* (1990), who found evidence of localization of Na atoms in the optical potential wells of a 3D molasses. This work was followed by that of Verkerk *et al.* (1992) and Jessen *et al.* (1992), who studied 1D lin⊥lin optical lattices as discussed in Section II. At this time, Gupta *et al.* (1992) were independently studying the 1D lin⊥lin lattice configuration by rf spectroscopy; physical understanding of this type of measurement was, however, complicated by the high rf fields necessary to drive the electric-dipole forbidden one-photon transitions between bound states of the lattice. Insight gained from these experiments also permitted the extension to 2D and 3D lattice configurations as reported by Grynberg *et al.* (1993) and Hemmerich and Hänsch (1993). Since then, spectroscopy has been performed on a wide range of lattice configurations. Generally, one finds closely analogous features in fluorescence and probe spectroscopy, i.e., Raman sidebands and a Rayleigh line. In the following, we organize our discussion so as to address the experimental findings for each of these features in turn.

1. *Raman Sidebands*

As discussed in Section II, resolved Raman sidebands in the fluorescence and probe spectra are characteristic of a well-designed optical lattice. These spectroscopic features may be interpreted in terms of spontaneous or stimulated Raman transitions between eigenstates of the atomic center-of-mass motion. In the following, we restrict the discussion to the secular regime, where the positions of Raman lines in the spectrum correspond to transition frequencies of the system. Strictly speaking, the eigenstates of the system are the Bloch states, Eq. (14). However, because long range coherence is destroyed by spontaneous emission, localized states form a more appropriate basis in which to discuss the spectra. Consider, therefore, a highly localized state in a σ_+ potential well of the lowest adiabatic potential of Fig. 6(c), belonging to a 1D lin⊥lin lattice. Tightly bound atomic states approximately factorize as $|J_g, m_g\rangle \otimes |n\rangle \otimes |\mathbf{p}_\perp\rangle$, where m_g is the maximum magnetic quantum number, $|n\rangle$ is a harmonic oscillator state, and we characterize the free atomic motion transverse to the lattice by the state $|\mathbf{p}_\perp\rangle$. Optical pumping transfers most of the population to these states, and light scattering then occurs mainly on the closed $|J_g, m_g\rangle \leftrightarrow |J_e, m_g + 1\rangle$ transition. This suggests that the spectra may reasonably be interpreted in terms of a harmonically bound two-level atom. The transition rate, Eq. (17), due to scattering of lattice light into the direction $\mathbf{k}_s$ is approximately

$$\gamma_{n, \mathbf{p}_\perp \to n', p'_\perp} \approx \gamma_s |\langle n', \mathbf{p}'_\perp | e^{i\mathbf{k}_s \cdot \mathbf{r}} \cos(\mathbf{k}_L \cdot \mathbf{r}) | n, \mathbf{p}_\perp \rangle|^2$$

$$\approx \gamma_s \big(\delta_{n,n'} \delta(\mathbf{p}'_\perp - \mathbf{p} - \hbar \Delta \mathbf{k}) + \langle n', \mathbf{p}'_\perp | \mathbf{k}_s \cdot \mathbf{r} | n, \mathbf{p}_\perp \rangle^2$$

$$\times \delta_{n, n' \pm 1} \delta(\mathbf{p}'_\perp - \mathbf{p} - \hbar \Delta \mathbf{k}) + \cdots \big) \quad (21)$$

where $\hbar \Delta \mathbf{k}_\perp$ is the momentum transfer in the transverse plane. The first term in Eq. (21) corresponds to Doppler-broadened Rayleigh scattering, whereas the second represents an inelastic process changing the vibrational quantum number by one. The Taylor expansion of the photon modes is valid when the spatial extent of the atom is small compared with the optical wavelength, as is true for tightly bound atoms. Note that to lowest order in the Lamb–Dicke parameter, $E_R/\hbar \omega_{osc}$, inelastic scattering arises from the photon recoil associated only with the *spontaneously* emitted photon. This is because the absorbed lattice photon is a σ polarized standing wave, which gives no recoil (to lowest order) to the atom. Therefore, for highly localized atoms, fluorescence spectroscopy will probe the atoms' vibrational motion only in the direction that the fluorescence photons are detected. This result generalizes to probe spectroscopy and to lattices of higher dimensionality.

Raman sidebands occur in fluorescence and probe spectra from lattices of any dimension, as illustrated in Figs. 2(c), 3(c), and 9(a)–(d). The atomic motion near the bottom of the potential wells is approximately separable, and the states are products of two or three 1D oscillator states. For the four-beam 3D lin⊥lin lattice described in Section III.B, the optical potential for tightly bound atoms can be approximated by an anisotropic harmonic oscillator with oscillation frequencies $\hbar \omega_{osc,z} = 2 \cos \theta_L \sqrt{U_0 E_R}$ and $\hbar \omega_{osc,x} = \hbar \omega_{osc,y} = \sin \theta_L \sqrt{U_0 E_R}$. Only one set of Raman sidebands will be observed, centered at the corresponding oscillation frequencies, when photons are detected along these Cartesian axes. Verkerk et al. (1994) report that sidebands corresponding to the oscillation along directions other than the observation axis are in fact seen in probe spectra; this has been interpreted as a breakdown of the harmonic oscillator model (G. Grynberg, private communication, 1995). In principle, one can use the amplitude asymmetry of a pair of Raman sidebands in the fluorescence spectrum of higher dimensional lattices to extract the "temperature" of that degree of freedom, as discussed in Section II for the 1D case. However, in 3D lattices the atom density may be sufficiently large that Raman gain/absorption of the spontaneously emitted photons can distort the spectrum, thus preventing a measurement of the temperature (S. L. Rolston, private communication, 1995).

Let us finally address the width of the observed Raman sidebands. It can be shown that a harmonic oscillator coupled to the electromagnetic field is characterized solely by the frequency of oscillation and the energy damping rate (Cohen-Tannoudji et al., 1992). For the conditions leading to the spectrum of Fig. 2(c) with the appropriate Lamb–Dicke suppression factor, this rate is approximately 2.7×10^4/sec, which should lead to a sideband width of ≈ 4 kHz. The apparent width of the sidebands is nearly 10 times larger and indicates that a simple harmonic oscillator model is not sufficient. The Raman sideband width seen in the spectra is in fact almost completely determined by anharmonicity of the optical potential. A perturbation calculation shows that neighboring Raman lines, corresponding to pairs of levels $|n\rangle$, $|n+1\rangle$, are separated by a frequency $E_R/\hbar$. A full band structure calculation confirms that this anharmonic shift indeed accounts for the major frequency deviations from the simple harmonic oscillator (SHO) model, and both models agree well with the experimentally measured sideband frequency (Jessen et al., 1992).

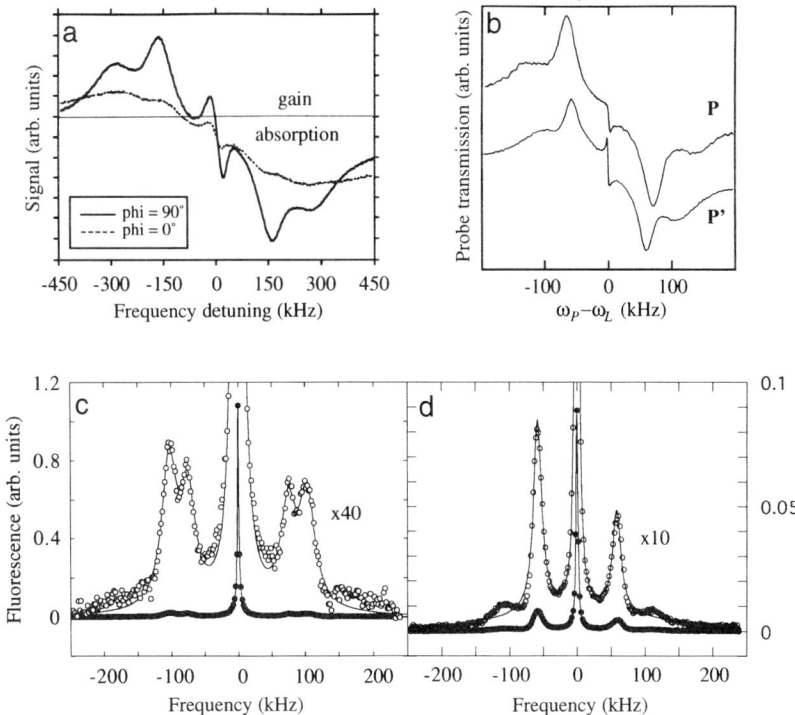

FIG. 9. (a) Probe transmission spectrum from a two-dimensional lattice of the Hänsch type. Solid and dotted lines are data measured for different relative phases of the two standing waves forming the lattice (from Hemmerich and Hänsch, 1993, reprinted with permission). (b) Probe transmission spectrum from a three-dimensional lattice of the Grynberg type. Data is shown for different probe propagation directions P and P' (from Grynberg et al., 1993, reprinted with permission). (c) Fluorescence spectrum measured with a lattice of the type shown in Fig. 5(a). The spectrum is measured along $\hat{x} + \hat{z}$, so that two separate oscillation frequencies are observed. (d) Same as (c), but measured along $\hat{x}$, so that only one oscillation frequency is observed.

Hemmerich et al. have measured probe transmission spectra using an intense probe beam (Hemmerich et al., 1994b). In that case, the atom–lattice–probe interaction cannot be treated by lowest order perturbation theory, and multiphoton transitions between vibrational states become important. The result is the appearance of subharmonics in the spectrum at frequencies ω_{osc}/n, for processes involving n probe photons and n lattice photons. These subharmonics have been seen in probe transmission spectra for n up to 5.

2. The Rayleigh Component

Fluorescence spectra typically contain a coherent component associated with the steady state of the system under study. In 1D and 2D lattices, the atomic density in the detection volume decays rapidly due to transverse heating, and it is clear that a steady state does not exist in the strict sense; as a result, the coherent component vanishes (Cooper and Ballagh, 1978). Nevertheless, normal modes other than the steady state also contribute to the Rayleigh feature centered at the lattice frequency, but such contributions are broadened according to their finite evolution rates. In the secular regime, the Rayleigh line is associated with scattering events that begin and end in the same, or a nearly degenerate, quantum state. In spectra from 1D (2D) lattices, these contributions may be Doppler-broadened due to motion in the unbound dimensions if the direction of measurement is not parallel (coplanar) with the lattice beams. In a deep 3D lattice, the atomic density decays much more slowly because there is cooling in all dimensions, and an approximate steady state exists on time scales up to a few hundred milliseconds. Furthermore, most atoms occupy states that are tightly bound in all directions, and Doppler broadening is negligible. The experimental spectrum of Figs. 9(c) and (d) show a correspondingly narrow Rayleigh line. Indeed, the width of the Rayleigh features in fluorescence spectra from 3D lattices has generally been close to the instrumental resolution, in which case it is doubtful that physical information about the lattice dynamics can be extracted.

The Rayleigh feature observed in probe transmission spectra is associated with a similar class of normal modes as the Rayleigh feature in the fluorescence spectra, except that probe transmission spectra do not contain a contribution from the steady state of the system (Courtois and Grynberg, 1992). For 1D (2D) lattice spectra probed in nonparallel (nonplanar) configurations, the dominant contribution to the Rayleigh feature in probe spectroscopy is a so-called 'recoil-induced resonance" (Courtois et al., 1994; Guo, 1994), which is closely analogous to Doppler-broadened Rayleigh scattering in the fluorescence spectrum. Consider the case of a 1D optical lattice, where atoms move freely in the plane perpendicular to the 1D axis. Stimulated Raman transitions then couple pairs of eigenstates that differ only with respect to the transverse atomic momentum. Because the transverse momentum distribution is Maxwell–Boltzmann, there will be a population difference between such states. Raman transitions of the type analogous to those of Fig. 3(b) occur when $\delta_p = \Delta \mathbf{k} \cdot \mathbf{v}_\perp$, where $\Delta \mathbf{k} = \mathbf{k}_p \pm \mathbf{k}_L$ is the change in photon wavevector due to the scattering event, and $\mathbf{v}_\perp$ is the atom velocity in the transverse plane. The magnitude of the probe gain or absorption is determined by the population difference between the initial and final states, and the recoil-induced resonance line is therefore the derivative of the Doppler lineshape in the fluorescence spectrum. This interpretation is confirmed by a close examination of the central feature of Fig. 3(c) (Courtois *et al.*, 1994).

In deep 3D lattices, the majority of atoms occupy tightly bound states, and a recoil-induced resonance is unlikely to occur. In this situation, it is nontrivial to identify the system modes associated with the Rayleigh feature. Experimental 3D probe transmission spectra contain Rayleigh features having widths of a few hundred hertz (Hemmerich *et al.*, 1993), which do not scale with the photon scattering rate as expected based on the master equation, Eq. (13) (S. L. Rolston, private communication, 1995). It is presently unresolved what dynamical evolution modes of the atom–lattice system might be responsible for such long time scales and unexpected scaling laws. Courtois and Grynberg (1992) have proposed a mechanism that involves a wave mixing process, in which either the probe beam or one of the lattice beams is scattered off a grating of atomic density or magnetization. Guo (1995a, b) has shown that the observed feature cannot be fully explained in this fashion and has proposed a mechanism involving a recoiled-induced resonance between quasi-continuum states. Further study is necessary, however, to establish how the small population in such states might account for the large observed signal.

IV. New Developments

A. Cooling by Adiabatic Expansion in Optical Lattices

A near-resonant optical lattice will localize atoms with a center-of-mass distribution that is much narrower than the separation between potential wells. If the optical lattice is used primarily as a source of cold atoms, this microscopic atomic localization is of no direct advantage, and it is lost when the lattice beams are extinguished and the atoms released. It is therefore advantageous to trade the atomic localization for a decrease in temperature, e.g., by adiabatic expansion in the optical potential wells. Adiabatic cooling has been demonstrated in an intense, blue-detuned 1D lattice by Chen *et al.* (1992), and in a 3D lattice of the type shown in Fig. 5 by Kastberg *et al.* (1995). To illustrate the basic principle, we discuss adiabatic expansion for a harmonically bound atom, with a thermal distribution of population characterized by a Boltzmann factor f_B. If the oscillator frequency is decreased, e.g., by lowering the lattice light intensity, the energy separation between vibrational states will decrease. We assume that the change in the oscillator frequency is slow enough to be adiabatic, the condition being that $(d\omega_{osc}/dt) \times (1/\omega_{osc}) \ll \omega_{osc}$, yet fast enough that dissipative processes can be ignored. In that case, the vibrational populations remain unchanged, and the system temperature has obviously been reduced. Because the Boltzmann factor remains unchanged during the expansion, we have $f_B = e^{-\hbar \omega_i/k_B T_i} = e^{-\hbar \omega_f/k_B T_f}$, and therefore $T_f = T_i \omega_f/\omega_i$, where ω_i, ω_f and T_i, T_f are the initial/final oscillation frequencies and temperatures. For a true harmonic oscillator, there is no lower bound on the oscillator frequency and, therefore, no lower limit for adiabatic cooling.

In an optical lattice, the microscopic potential wells have finite depth, and adiabatic cooling therefore stops when the lattice becomes shallow enough to allow atoms to escape from their original potential wells either by tunneling or above-barrier motion. Kastberg *et al.* (1995) proposed a simple band theory for adiabatic expansion in a one-dimensional scalar potential, from which a cooling limit can be derived. In this model, atoms are initially assumed to occupy tightly bound states, and the band structure, Fig. 10(a), is well known from the tight binding approximation of solid state physics (Ashcroft and Mermin, 1976). During the adiabatic expansion process, this band structure adiabatically evolves into that of free momentum states, Fig. 10(b). Thus, the final momentum distribution relates in a particularly simple fashion to the initial distribution of population over Bloch states, as illustrated in Fig. 10(c). One can then show that the final momentum distribution and temperature depends only on the lattice constant R and the Boltzmann factor f_B that initially characterized the population distribution over bound states. In a 1D lattice, one finds

$$\frac{T}{T_R} = \left(\frac{2\pi/R}{k}\right)^2 \frac{1 + 4f_B + f_B^2}{12(1-f_B)^2} \qquad (22)$$

where $k_B T_R/2 = E_R$. This result can be generalized to a cubic 3D lattice, where we again find $T/T_R \propto (2\pi/Rk)^2$. This simple theory was found to be in reasonable quantitative agreement with the experiment of Kastberg *et al.* (1995).

Equation (22) suggests that the adiabatic cooling limit might be improved if atoms are trapped in an optical lattice with a very large lattice constant. Adiabatic cooling in a far-off-resonant lattice configuration with arbitrarily large R in all directions has been demonstrated by Anderson *et al.* (1994). However, since a large optical potential well has closely spaced bound states and therefore $f_B \approx 1$, even for atoms that are initially very cold, this approach has so far not been particularly successful. It is possible that Raman sideband cooling (Taïeb *et al.*, 1994) may be used to cool

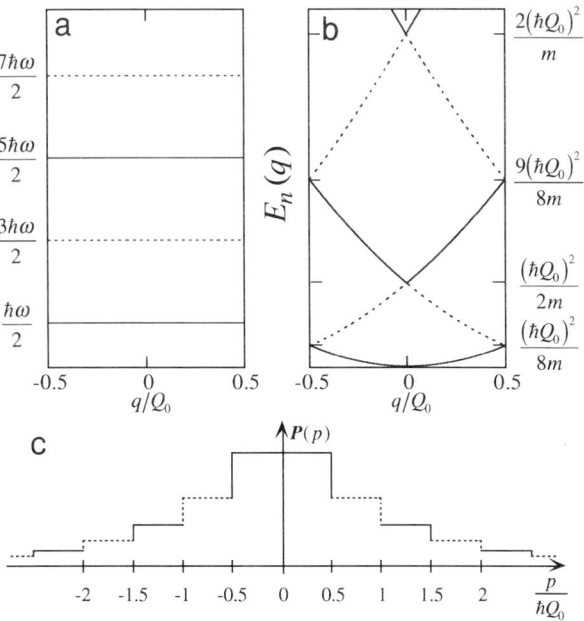

FIG. 10. (a) band structure in a scalar one-dimensional lattice potential in the tight binding regime. The lattice constant is $2\pi/Q_0$. (b) Band structure in the same potential, but in the free particle regime. During adiabatic cooling, the tight binding Bloch states evolve adiabatically into free particle Bloch states. One can show that, in the tight binding regime, all Bloch states within a band are equally populated; since the adiabatic evolution conserves all populations, this is true also for the free particle Bloch states. (c) The corresponding momentum distribution. Solid and dashed lines indicate odd and even numbered bands and their contributions to the final momentum distribution. From Kastberg et al. (1995).

atoms to the vibrational ground state of the lattice. In that case, adiabatic release from a lattice of the type shown in Fig. 5 would lead to a temperature as low as $T_R/4$, corresponding to 50 nK in the case of Cs atoms.

B. Bragg Scattering in Optical Lattices

The experiments discussed in Sections II and V have investigated the localization and quantized motion of atoms in optical lattices. As the lattice light field is a topologically stable configuration, the atoms are trapped at strictly periodic positions, and therefore the density distribution possesses long range correlations. Optical lattices studied so far have been only sparsely populated; for a 3D lattice of the type shown in Fig. 5, an atom density of 10^{11} cm^{-3} corresponds to a filling fraction of only $f \sim 0.01$. Even so, long range order has important consequences for the propagation of light through the atomic sample. In particular, waves scattered into well-defined directions other than forward will be phase-matched; i.e., the atomic order will give rise to Bragg reflection. Two factors make this Bragg scattering process markedly different from Bragg scattering in solid crystals. First, the optical lattice constant is of order 1 μm, compared with a few angstroms in solids, and Bragg scattering therefore occurs at optical rather than X-ray wavelengths. Second, we consider scattering of a weak probe beam tuned near an atomic resonance, where the absorption cross section is enhanced and the atomic response is highly dispersive.

Atoms arranged in a lattice described by reciprocal lattice vectors $\{\mathbf{K}\}$ will Bragg reflect a probe with wavevector $\mathbf{k}_p$ into the direction $\mathbf{k}_p + \mathbf{K}$ when one satisfies the Bragg condition, $2\mathbf{k}_p \cdot \mathbf{K} = -K^2$, written here in the von Laue form. As shown in Section III, the set of vectors $\{\mathbf{K}_j\}$ are given by differences between the lattice beam wavevectors $\{\mathbf{k}_i\}$. Bragg reflection therefore occurs for a probe beam propagating parallel to a lattice beam, $\{\pm\mathbf{k}_i\}$, into directions parallel to the remaining beams $\{\pm\mathbf{k}_{j \neq i}\}$.

A Bragg reflection experiment was performed by Birkl *et al.* (1995) on Cs atoms in a three-dimensional optical lattice (Fig. 5). In that work, atoms were first allowed to equilibrate and become spatially ordered. The lattice light field was then quickly extinguished, and a weak probe beam, tuned close to the atomic transition used to form the lattice, was applied along one of the directions for which Bragg reflection occurs. Performing the experiment in the absence of the lattice laser beams avoids a contribution from nearly degenerate four-wave mixing; however, once the lattice potential is removed, the center-of-mass distribution of the initially localized atoms starts expanding ballistically. By measuring the temporal dependence of the Bragg reflected signal, one gains information about the atomic localization and temperature. Weidemuller *et al.* (1995) have used an alternative approach, in which a probe resonant with a separate transition of much shorter wavelength was Bragg reflected in the presence of the lattice light field. In this situation the large frequency difference between lattice and probe beams prevents any contribution from four-wave mixing.

Physical information about the trapped atoms is available in the reflected intensity. In the following, we ignore diffraction effects and assume that the atom density is so low that multiple reflections can be ignored. Consider, then, an atomic sample distributed over n infinite planes, illuminated by a plane probe wave. The Bragg reflection coefficient for intensity is

$$R_n = n^2 R f_n(\Delta_P) \qquad (23)$$

where R is the single plane reflectivity and n^2R is the familiar Bragg result. One can show that $n^2R \sim \beta[N\sigma(\Delta_p)]^2$, where N is the number density or atoms in the lattice, $\sigma(\Delta_p)$ is the absorption cross section and Δ_P the detuning of the probe from atomic resonance. The parameter $\beta = \exp[-(2\pi\Delta x/d)^2]$ is known as the Debye–Waller factor. In this expression, d is the separation between lattice planes and Δx is the rms width of the atomic center-of-mass distribution perpendicular to the lattice planes. The factor $f_n(\Delta_P)$ in Eq. (23) accounts for loss and dispersion.

Figure 11 shows the measured Bragg reflection intensity as function of probe detuning, for various atomic densities. At low density, the reflection lineshape is a Lorentzian with natural linewidth Γ. As atomic density is increased, the lineshape is broadened, and it eventually develops a dip near resonance. This behavior is easily interpreted in terms of optical density of the atomic sample. At low density, the sample is optically thin, and the reflection spectrum is similar to a disordered vapor, but enhanced by a factor $\sim 10^5$. As the atom density is increased, both probe and reflected waves are attenuated, and the effective number of planes contributing to the Bragg scattering is reduced. The effect is most pronounced

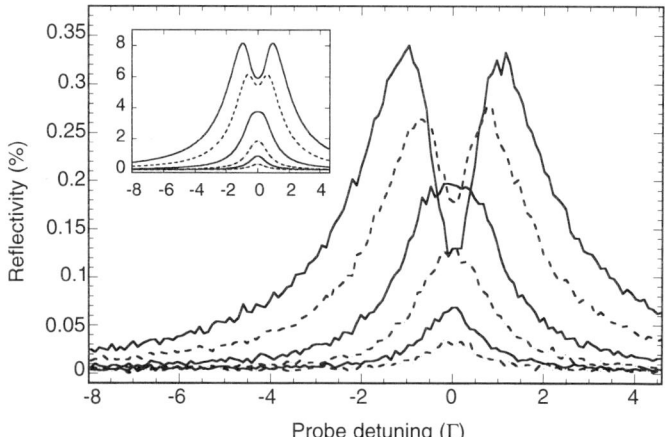

FIG. 11. Bragg reflectivity as a function of probe laser detuning. Atomic densities in the lattice range from 3×10^9 to 8×10^{10} cm^{-3}, increasing by roughly a factor of two between traces. The inset shows calculated spectra for the same parameters (from Birkl et al., 1995).

near resonance, where the optical density is maximum, and eventually a dip will develop. A calculation of the lineshape using the foregoing model of Bragg reflection (inset of Fig. 11) is in qualitative agreement with the experimental spectra. When the lattice light is extinguished, the atomic center-of-mass distribution expands ballistically and the Bragg reflectivity decays. The atom temperature can therefore be extracted from the time dependent Debye–Waller factor. These measurements of atom temperature and localization are in reasonable agreement with measurements based on spectroscopy or time of flight measurements of the atomic momentum distribution.

Bragg reflection is a new tool, which complements the spectroscopic techniques discussed in Section V. The strong dependence of Bragg reflectivity on atomic localization provides a unique probe of both equilibrium dynamics and driven atomic motion. One can also measure the change in the lattice constant with lattice detuning and atomic density (Birkl *et al.*, 1995) due to the back-action of the atomic polarizability on standing wave light fields (Deutsch *et al.*, 1995).

C. OUTLOOK

Most studies of optical lattices have so far been aimed at understanding the basic process of laser cooling and trapping, as well as the physics involved in fluorescence and probe spectroscopy. These efforts have been largely successful, and it appears likely that the main focus of both theory and experiment will shift in the near future. One area of considerable practical importance is the application of optical lattices to the control of atomic motion. Examples range from the use of optical lattices in preparing cold atomic samples for atomic fountain clocks, to the development of new lithographic techniques based on light-controlled deposition of atoms onto substrates.

To illustrate the potential of light-controlled atom deposition, consider that in the 1D optical lattice experiments discussed in Section II, Rb atoms were found to be localized in optical potential wells with an rms center-of-mass spread of $\lambda/15$, corresponding to a FWHM of ≈ 100 nm (Jessen *et al.*, 1992). If such atoms could be deposited on a surface, they would form lines with a resolution well below the current state of the art of optical lithography. In a practical version of this experiment, Timp *et al.* (1992) used a 1D optical lattice to control the deposition of an Na atomic beam on a surface. Atoms were first laser-cooled in the direction transverse to the atomic beam, and then channeled along planes defined by an optical lattice oriented parallel to the atomic beam axis. Immediately downstream from the lattice, the atoms struck a substrate, where they formed a regular array of lines a few tens of nanometers wide. McClelland et al. (1993) have demonstrated similar deposition schemes using 1D and 2D lattices and chromium atoms. The combination of nanometer resolution and the ability to deposit, in parallel, features over a very large area, makes atom lithography appear an attractive technology; much development still remains, however, before it can be demonstrated that light-controlled deposition of arbitrary nanoscale patterns is feasible.

One of the most interesting aspects of optical lattices is the close analogy between the motion of cold atoms in the periodic optical potential and systems in condensed matter physics. The flexibility with which we can modify the parameters of an optical lattice makes this an attractive model system for studying fundamental problems, e.g., transport in a periodic system. Such studies are crucial for the microscopic understanding of electrical and thermal conductivity in solids, and increasingly they are relevant to studies of biophysical processes such as energy transport in photosynthesis. Though the optical lattice is far removed from the highly complex systems of condensed matter, and thus no direct information can be inferred, its relative simplicity opens the door to well-controlled experiments that can be precisely modeled. In addition, cooperative phenomena may be studied in a realizable system that has long been an idealized "toy" model for condensed matter physicists—the dilute, weakly interacting gas. Preliminary studies of the effects of dipole–dipole interaction between atoms in a 3D lattice have been carried out by Goldstein et al. (1996). Practical realization of these model systems requires two important experimental advances: trapping atoms in the lattice in an essentially dissipation-free environment, and increased atomic densities.

The typical near-resonant optical lattice is not particularly well suited for such work, both because it is highly dissipative and because inelastic long range dipole–dipole interactions between atoms in the lattice, along with other types of light-assisted collisions, tend to limit the atomic density. The rate of dissipative processes, such as spontaneous emission and inelastic collisions, are proportional to the excited state population, and it is therefore apparent that the design of new types of lattices that minimize atomic excitation is of considerable interest. To this end, Grynberg et al. have proposed a "gray" optical lattice in which atoms are trapped near points where they are in a "dark state" (Grynberg and Courtois, 1994), and such that they only couple weakly to the field away from these equilibrium points. A similar, closely related lattice has been demonstrated experimentally by Hemmerich et al. (1995).

Another approach toward reducing the excited state population is to continue to trap atoms at points where they strongly couple to the field, but to increase both the lattice light detuning Δ and intensity I_L. In that regime, the photon scattering rate scales as $\gamma_s \propto I_L/\Delta^2$ whereas the potential well depth scales as $U_0 \propto I_L/\Delta$, so that dissipation can be made arbitrarily small while still maintaining deep wells in the optical lattice. One difficulty associated with these lattices is that the very absence of dissipation makes it difficult to load atoms into the optical potential wells. This problem is much more severe than for the "gray" lattices just discussed, in which bound states are efficiently populated by a combination of Sisyphus cooling and velocity-selective coherent population trapping (Aspect *et al.*, 1989). In a far-off-resonance lattice, loading may be achieved through a separate laser-cooling field, or perhaps through the adiabatic transfer of population from an initial near-resonance lattice.

Once atom trapping in a dissipation-free lattice is accomplished, it may be possible to implement new cooling mechanisms to improve the atomic localization, or to generate specific quantum states of the atomic center-of-mass motion. Raman sideband cooling might, for example, be implemented in a far-off-resonance lattice, in a manner analogous to that of trapped ions (Heinzen and Wineland, 1990). Theoretical studies by Taïeb *et al.* (1994) indicate that this method might allow one to cool atoms to the vibrational ground state of their individual potential wells. If this can indeed be accomplished, then such atoms might be used as a starting point for adiabatic cooling to sub-recoil temperatures, or for experiments with driven atomic motion. As an example, parametric driving of an atom initially in the vibrational ground state will produce an oscillating atomic wavepacket that is squeezed alternately in position and momentum space (Marksteiner *et al.*, 1995).

One of the most basic physical phenomena characterizing quantum motion is tunneling. Tunneling dynamics are expected to become important if the rate of dissipation in an optical lattice can be reduced sufficiently. In that case, an optical lattice may constitute an attractive and flexible model system in which to study the propagation of matter waves in periodic potentials. One of the most striking manifestations of tunneling in a periodic lattice is the existence of "Bloch oscillations" (Wannier, 1962) of the atomic momentum under the influence of a constant force. The observation of such oscillations in a crystal lattice is greatly complicated by strong interactions and extremely rapid decoherence of the electronic motion, and thus it has been accomplished only in semiconductor superlattice structures (Mendez and Bastard, 1993). An optical lattice constitutes a new system in which a conceptually simpler version of this physical phenomenon may be studied, i.e., noninteracting particles in a simple lattice free from dissipation, and with an external force applied through a constant acceleration of the lattice in the laboratory frame. Very recently, experiments have seen Bloch oscillations directly in the time domain

(Dahan et al., 1996), and performed spectroscopy of the "Wannier–Stark ladder" (Wilkinson et al., 1996). These developments illustrate how optical lattices offer a new "laboratory" in which one can study a wide variety of phenomena that combine concepts from atomic, optical and condensed matter physics.

Acknowledgments

The authors thank S. L. Rolston for helpful discussions. This work was supported in part by the National Science Foundation Contract No. PHY-9503259.

References

Anderson, B. P., Gustavson, T. L., and Kasevich, M. (1994). *Proc. Int. Quantum Electron. Conf.*, QPD2-1/3, 8–13 May, Anaheim, CA.
Ashcroft, N. W., and Mermin, N. D. (1976). "Solid State Physics." Holt-Saunders, Philadelphia.
Aspect, A., Arimondo, E., Kaiser, R., Vansteenkiste, N., and Cohen-Tannoudji, C. (1989). *J. Opt. Soc. Am. B* **6**, 2112.
Berg-Sørensen, K. (1994). *Phys. Rev. A* **49**, R4297.
Berg-Sørensen, K., Castin, Y., Mølmer, K., and Dalibard, J. (1993). *Europhys. Lett.* **22**, 663.
Birkl, G., Gatzke, M., Deutsch, I. H., Rolston, S. L., and Phillips, W. D. (1995). *Phys. Rev. Lett.* **75**, 2823.
Carmichael, H. J. (1993). "An Open Systems Approach to Quantum Optics." Springer, Berlin.
Castin, Y. (1992). Ph.D. Dissertation, Ecole Normale Supérieure, Université Paris VI.
Castin, Y., and Dalibard, J. (1991). *Europhys. Lett.* **14**, 761.
Castin, Y., and Mølmer, K. (1995). *Phys. Rev. Lett.* **74**, 3772.
Castin, Y., Dalibard, J., and Cohen-Tannoudji, C. (1991). *In* "Light Induced Kinetic Effects on Atoms, Ions, and Molecules" p. 5, (L. Moi, S. Gozzi, C. Gabbanini, E. Arimondo, and F. Strumia, eds.). ETS Editrice, Pisa.
Castin, Y., Berg-Sørensen, K., Dalibard, J., and Mølmer, K. (1994). *Phys. Rev. A* **50**, 5092.
Chen, J., Story, J. G., Tollet, J. J., and Hulet, R. G. (1992). *Phys. Rev. Lett.* **69**, 1344.
Cohen-Tannoudji, C. (1991). *In* "Fundamental Systems in Quantum Optics, Les Houches, Session LIII" (J. Dalibard, J. M. Raimond, and J. Zinn-Justin, eds.) Elsevier, Amsterdam.
Cohen-Tannoudji, C., Dupont-Roc, J., and Grynberg, G. (1992). "Atom-Photon Interaction," pp. 326–329, 384–386, 427–436. Wiley (Interscience), New York.
Cooper, J., and Ballagh, R. J. (1978). *Phys. Rev. A* **18**, 1302.
Courtois, J.-Y. (1993). Ph.D. Dissertion, Ecole Polytechique, Paris.
Courtois, J.-Y., and Grynberg, G. (1992). *Phys. Rev. A.* **46**, 7060.
Courtois, J.-Y., Grynberg., G., Lounis, B., and Verkerk, P. (1994). *Phys. Rev. Lett.* **72**, 3017.
Dahan, M. B., Peik, E., Reichel, J., Castin, Y., and Salomon, C. (1996). *Phys. Rev. Lett.* **76**, 4508.
Dalibard, J., and Cohen-Tannoudji, C. (1989). *J. Opt. Soc. Am. B* **6**, 2023.
Deutsch, I. H., Spreeuw, R. J. C., Rolston, S. L., and Phillips, W. D. (1995). *Phys. Rev. A* **52**, 1394.

Diedrich, F., Bergquist, J. C., Itano, W. M., and Wineland, D. J. (1989). *Phys. Rev. Lett.* **62**, 403.
Doery, M., Widmer, M., Bellanca, J., Vredenbregt, E., Bergeman, T. H., and Metcalf, H. (1994). *Phys. Rev. Lett.* **72**, 2546.
Gardiner, C. W., Parkins, A. S., and Zoller, P. (1992). *Phys. Rev. A* **46**, 4363.
Gisin, N., and Percival, I. C. (1992). *J. Phys. A* **25**, 5677.
Goldstein, E. V., Pax, P., and Meystre, P. (1996). *Phys. Rev. A* **53**, 2604.
Grison, D., Lounis, B., Salomon, C., Courtois, J.-Y., and Grynberg, G. (1991). *Europhys. Lett.* **15**, 149.
Grynberg, G., and Courtois, J.-Y. (1994). *Europhys. Lett.* **27**, 41.
Grynberg, G., Lounis, B., Verkerk, P., Courtois, J.-Y., and Salomon, C. (1993). *Phys. Rev. Lett.* **70**, 2249.
Guo, J. (1994). *Phys. Rev. A* **49**, 3934.
Guo, J. (1995a). *Phys. Rev. A* **51**, 2338.
Guo, J. (1995b). *Phys. Rev. A* **52**, 1458.
Gupta, R., Padua, S., Bergeman, T., and Metcalf, H. (1992). *In* "Laser Manipulation of Atoms and Ions" p. 345. (W. D. Phillips, E. Arimondo, and F. Strumia, eds.). North-Holland Publ., Amsterdam.
Hänsch, T., and Schawlow, A. (1975). *Opt. Commun.* **13**, 68.
Heinzen, D. J., and Wineland, D. J. (1990). *Phys. Rev. A* **42**, 2977.
Hemmerich, A., and Hänsch, T. W. (1993). *Phys. Rev. Lett.* **70**, 410.
Hemmerich, A., Zimmermann, C., and Hänsch, T. (1993). *Europhys. Lett.* **22**, 89.
Hemmerich, A., Weidemüller, M., and Hänsch, T. (1994a). *Europhys. Lett.* **27**, 427.
Hemmerich, A., Zimmermann, C., and Hänsch, T. W. (1994b). *Phys. Rev. Lett.* **72**, 625.
Hemmerich, A., Weidemüller, M., Esslinger, T., Zimmermann, C., and Hänsch, T. W. (1995). *Phys. Rev. Lett.* **75**, 37.
Holland, M., Marksteiner, S., Marte, P., and Zoller, P. (1996). *Phys. Rev. Lett.* **76**, 3683.
Jessen, P. S. (1993). Ph.D. Dissertion, University of Aarhus, Århus, Denmark.
Jessen, P. S., Gerz, C., Lett., P. D., Phillips, W. D., Rolston, S. L., Spreeuw, R. J. C., and Westbrook, C. I. (1992). *Phys. Rev. Lett.* **69**, 49.
Jurczak, C., Sengstock, K., Kaiser, R., Vansteenkiste, N., Westbrook, C. I., and Aspect, A. (1995). *Opt. Commun.* **115**, 480.
Kastberg, A., Phillips, W. D., Rolston, S. L., Spreeuw, R. J. C., and Jessen, P. S. (1995). *Phys. Rev. Lett.* **74**, 1542.
Lett. P. D., Watts, R. N., Westbrook, C. I., Phillips, W. D., Gould, P. L., and Metcalf, H. J. (1988). *Phys. Rev. Lett.* **61**, 169.
Lounis, B., Verkerk, P., Courtois, J.-Y., Salomon, C., and Grynberg, G. (1993). *Europhys. Lett.* **21**, 13.
Marksteiner, S., Ellinger, K., and Zoller, P. (1996). *Phys. Rev. A* **53**, 3409.
Marksteiner, S., Walser, R., Marte, P., and Zoller, P. (1995). *Appl. Phys. [Part] B* **B60**, 145.
Marte, P., Dum. R., Taïeb, R., and Zoller, P. (1993a). *Phys. Rev. A* **47**, 1378.
Marte, P., Dum, R., Taïeb, R., Lett, P. D., and Zoller, P. (1993b). *Phys. Rev. Lett.* **71**, 1335.
McClelland, J. J., Scholten, R. E., Palm, E. C., and Celotta, R. J. (1993). *Science* **262**, 887.
Mendez, E. E., and Bastard, G. (1993). *Phys. Today* **46**, 34.
Mølmer, K., Castin, Y., and Dalibard, J. (1993). *J. Opt. Soc. Am. B* **10**, 524.
Petsas, K. I., Coates, A. B., and Grynberg, G. (1994). *Phys. Rev. A* **50**, 5173.
Rolston, S. L. (1995). Private Communication.
Salomon, C., Dalibard, J., Phillips, W. D., Clairon, A., and Guellati, S. (1990). *Europhys. Lett.* **12**, 683.

Tabosa, J. W. R., Chen, G., Hu, Z., Lee, R. B., and Kimble, H. J. (1991). *Phys. Rev. Lett.* **66**, 3245.
Taïeb, R., Dum, R., Cirac, J. I., Marte, P., and Zoller, P. (1994). *Phys. Rev. A* **49**, 4876.
Timp, G., Behringer, R. E., Tennant, D. M., Cunningham, J. E., Prentiss, M., and Berggren, K. K. (1992). *Phys. Rev. Lett.* **69**, 1636.
Ungar, P. J., Weiss, D. S., Riis, E., and Chu, S. (1989). *J. Opt. Soc. Am. B* **6**, 2058.
Verkerk, P., Lounis, B., Salomon, C., Cohen-Tannoudji, C., Courtois, J.-Y., and Grynberg, G. (1992). *Phys. Rev. Lett.* **68**, 3861.
Verkerk, P., Meacher, D. R., Coates, C. B., Courtois, J.-Y., Guibal, S., Lounis, B., Salomon, C., and Grynberg, G. (1994). *Europhys. Lett.* **26**, 171.
Wannier, G. H. (1962). *Rev. Mod. Phys.* **34**, 645.
Weidemüller, M., Hemmerich, A., Gorlitz, A., Esslinger, T., and Hänsch, T. W. (1995). *Phys. Rev. Lett.* **75**, 4583.
Weiss, D. S., Riis, E., Shevy, Y., Ungar, P. J., and Chu, S. (1989). *J. Opt. Soc. Am. B* **6**, 2072.
Westbrook, C. I., Watts, R. N., Tanner, C. E., Rolston, S. L., Phillips, W. D., and Lett, P. D. (1990). *Phys. Rev. Lett.* **65**, 33.
Wilkinson, S. R., Bharucha, C. F., Madison, K. W., Niu, Q., and Raizen, M. G. (1996). *Phys. Rev. Lett.* **76**, 4512.
Wineland, D. J., and Dehmelt, H. (1975). *Bull. Am. Phys. Soc.* [2] **20**, 637.
Wineland, D. J., and Itano, W. M. (1979). *Phys. Rev. A* **20**, 1521.

CHANNELING HEAVY IONS THROUGH CRYSTALLINE LATTICES

HERBERT F. KRAUSE AND SHELDON DATZ

Physics Division,
Oak Ridge National Laboratory,
Oak Ridge, Tennessee

I. Introduction	139
II. Channeling Trajectories and Interaction Potentials	144
III. Planar Channeling	145
IV. Axial Channeling	146
V. Experimental Methods	150
VI. Charge Changing Collisions	152
VII. Radiative Electron Capture	153
VIII. Electron Impact Ionization	158
IX. Dielectronic Excitation and Recombination and Resonant Transfer with Excitation	161
X. Resonant Coherent Excitation	166
References	176

I. Introduction

In recent years, considerable progress has been made in the study of collisions involving multiply charged ions and electrons. Areas that have been opened to investigation include excitation, ionization, radiative recombination, and dielectronic recombination. All of these processes occur when swift ions impinge on gases and solids. They are contributors to the dynamics of highly charged ion sources, storage rings, and hot plasmas. In these environments, the collisions of short-lived excited ionic states with electrons also play a significant role and, until now, have not been amenable to study. In the past 15 years, many groups have used the channeling effect that occurs in crystalline solids to create conditions under which such processes may be studied relatively easily at electron densities in excess of $10^{23}/cm^3$.

Before discussing the recent advances in this field, we will summarize the essential points that are necessary to define the "channeling effect" and the unique collision environment for channeled particles. This summary will be done in a rather intuitive way because a more rigorous treatment is contained in a review up to 1973 (Gemmell, 1974). That review contains more than 700 references; however, most of the early work employed proton and helium ion beams at low energy ($E < 1$ MeV) to study properties of the crystal being penetrated. In fact, results of those channeling studies have been applied in the study of lattice disorder, surfaces and epitaxial growth, location of dopant and impurity atoms, and sputtering and in the measurement of nuclear lifetimes.

Heavy ion channeling work done between 1965 and 1982, where the focus was the projectile, was reviewed by Datz and Moak (1983). That review addressed questions of projectile energy loss in axial and planar channels that allow one to (i) test scattering potentials for the particle–solid interaction and (ii) study screening effects and higher order corrections for electronic stopping power. Charge changing and excitation collisions involving the projectile were also discussed.

Most recently, channeling techniques have been used in the study of atomic collision phenomena involving very energetic and heavier multiply charged projectile ions, even uranium projectiles moving at relativistic energy. For heavier ions moving at much higher speed, cross sections for electron impact phenomena are greatly reduced. In this case, X-ray techniques are also more easily applied to probe specific excited states of heavier projectiles. Here, we focus on studies that have occurred within the past 15 years, and we emphasize the study of motion, atomic collisions, and ionic electronic states as ions move through crystal channels.

Swift heavy ions penetrating solids undergo a rapid series of collisions that degrade their energy and cause excitation, ionization, and capture involving both the projectile ion and the atoms in the solid. It is usually difficult to study isolated atomic collision processes in dense media and follow the evolution of the charge states or electronic states of ions as they penetrate amorphous solids, because of the chaotic nature and the short time between relatively violent collision events. Although target thickness changes can be made in particle–amorphous solid studies, the basic nature of the collision environment is independent of the beam–target orientation.

The situation can be quite different when the solid is monocrystalline. When an energetic ion enters the lattice close to a major axial direction or in a planar direction, it undergoes a set of correlated small-angle collisions that tend to "channel" its motion so as to avoid close collisions with lattice atoms (i.e., relatively large impact parameter collisions). The interaction can be visualized using the face-centered cubic (fcc) lattice model shown in Fig. 1, which represents the atomic arrangement in Ag or Au crystals. Atomic rows parallel to the [001] direction are shown in Fig. 1(a). Any four neighboring rows shown are the boundaries of a $\langle 100 \rangle$ axial channel. The motion becomes channeled because the projectile will be slightly deflected by the repulsive potential of the first atom, again by the second, etc., until the ion is moving more nearly parallel to the atomic string. A one-dimensional illustration of the concept is shown in Fig. 2. The crystal appears to be highly transparent when it is viewed along a direction of low Miller index, such as a $\langle 100 \rangle$ direction shown in Fig. 1(a). In fact, for the fcc lattice shown in Fig. 1(a), a $\langle 100 \rangle$ direction is the second most transparent direction after a $\langle 110 \rangle$ direction. Rotating the crystal of Fig. 1(a) about the [100] direction leads to a view of a {100} planar channel shown in Fig. 1(b). For planar channeling, it is the repulsive sheet potential that deviates the incident projectile's direction. Once the particle becomes channeled, it tends to remain channeled. (Throughout this chapter, [uvw] and (hkl) represent a particular axis and plane, while $\langle uvw \rangle$ and $\{hkl\}$ represent the sets of all crystallographically equivalent axes and planes.)

In an axial direction, entering ions are guided two-dimensionally through the crystal by repulsive interactions with the closest parallel atomic strings. For planar channeling (Fig. 1b), the guided motion is predominately one-dimensional in a direction perpendicular to the densely packed atomic sheets. Ions entering a single crystal in a random direction (the majority of possible orientations) cannot be guided by atomic rows and planes, and thus, in a random orientation, the crystal acts like an amorphous solid.

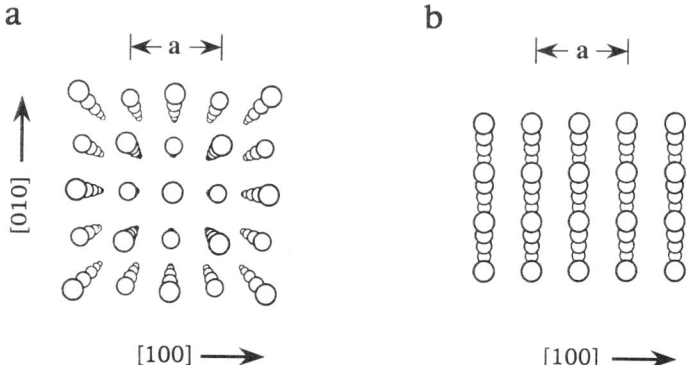

FIG. 1. (a) Perspective view of an fcc lattice, such as Ag or Au, that shows $\langle 100 \rangle$ axial channels. (b) View of {100} planar channels produced by rotating the crystal about the [100] direction. The unit cell dimension, indicated by **a**, is 4.078 Å for Au.

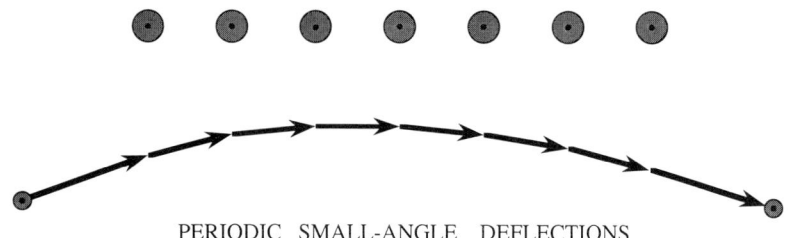

FIG. 2. Effect of correlated collisions when a penetrating ion interacts with a row of atoms in a crystal and is channeled.

Because the transparency of the crystal is orientation dependent, the yield in physical collision processes involving channeled particles must be strongly orientation dependent. These directional effects were actually anticipated by Stark (1912) and Stark and Wendt (1912) and rediscovered in computer calculations of Robinson and Oen (1963a, b) about 50 years later.

Lindhard et al. (1963) demonstrated that the correlated collisional sequences for positive ion channeling can be accurately described by classical mechanics. Assuming a screened Coulomb potential and using a continuum string approximation for the potential of a row of atoms, they showed that, for high velocity particles entering close to an axial direction (e.g., Fig. 1a), hard collisions with lattice atoms are avoided if the incidence angle with an atomic row is less than

$$\psi_1 = \left(\frac{2Z_1 Z_2 e^2}{Ed} \right)^{1/2} \quad (1)$$

where Z_1 and Z_2 are the atomic numbers of the projectile and target atoms, E is the ion energy, and d is the distance between atoms in the row. For incidence at the angle ψ_1, the distance of closest approach to an atomic string is approximately equal to the Thomas–Fermi screening length

$$a = 0.8853 a_0 Z_2^{-1/3} \quad (2)$$

where a_0 is the Bohr radius (0.52917×10^{-8} cm). Strictly speaking, Eq. (2) gives the screening length appropriate to the potential distribution in an isolated Thomas–Fermi atom. However, one would expect approximately the same distribution if $Z_1 = 1$ and $Z_2 \gg 1$. In fact, this length has been experimentally shown also to apply even when multiply charged projectile ions are fully stripped or nearly stripped (Robinson, 1969; Krause et al., 1994).

For incidence very close to the atomic string direction, where the beam's transverse energy can be made very small by a high degree of beam collimation, the closest approach to a string for ions entering near the center of the channel is usually much larger than a.

The maximum channeling impact parameter with respect to any atomic string or plane depends on the axial channel. Figure 3 illustrates the geometric arrangement of atoms for $\langle 100 \rangle$, $\langle 110 \rangle$, and $\langle 111 \rangle$ axial channels of an fcc lattice, such as Ag and Au. The area lying within the shaded region at the center of each axis shown indicates the full range of two-dimensional impact parameters for the channel. Figure 3 shows that, in a more closed $\langle 111 \rangle$ axis (shown at the right), the range of impact parameters is smaller than for a $\langle 100 \rangle$ axis (shown at the left). These impact parameter ranges can be compared with that for the most open $\langle 110 \rangle$ axis (shown in the center). In planar channels, the most distant impact parameter is half the spacing between adjacent atomic planes. Ions that undergo the "softest" collisions, corresponding to most distance impact parameters, are sometimes called the *best channeled ions*. In atomic collision experiments, one chooses the axial or planar direction that is most appropriate for the physics being studied. Electron density and electric field strengths inside the channel depend strongly on the distance from any atomic string or plane. Therefore, the range of impact parameters, the distance between atoms, and in some cases, the specific arrange-

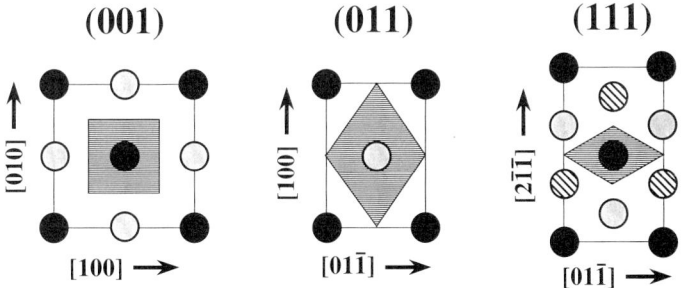

FIG. 3. Projections of an fcc lattice in several crystallographic directions that illustrate, from left to right, $\langle 100 \rangle$, $\langle 110 \rangle$, and $\langle 111 \rangle$ axial channels. The effective impact parameters for each channel are contained within the shaded boundary around each central atomic string. The range of channeling impact parameters depends on the channeling direction. The $\langle 110 \rangle$, $\langle 100 \rangle$, and $\langle 111 \rangle$ axial channels are the first, second, and fifth most open axes in an fcc lattice.

ment of atoms along a row are sometimes important variables in atomic collision studies.

The primary advantage of using channeled ions in atomic collision studies is to prevent the small impact parameter and more violent collisions from occurring. These ions also experience the least energy loss and suffer the least deflection. When the channeling condition is met, nuclear reactions, Rutherford scattering, and nuclear stopping are eliminated and inner shell excitation and ionization are greatly reduced. A channeled ion samples only slow valence and conduction electrons in its path. Even relatively low-Z multiply charged ions in the ground state, where the cross sections for excitation and ionization are the largest, can pass through the crystal with a reasonable probability of not being ionized (i.e., the incident charge state remains "frozen"). In addition, the channel milieu offers some unique possibilities. Because the electron density is so high, short-lived excited states that would normally decay before undergoing collisions become amenable to study. The periodicity of the lattice atoms coupled to the ion velocity can create perturbing frequencies that can cause coherent excitation of swift channeled ions. Channeled ions experience relatively strong electric fields, partly due to the potential gradient between planes or rows, and partly due to a dynamic wake field set up by an enhanced electron density trailing behind the swift ion. These fields can Stark mix otherwise degenerate states and can lead to the production of selected substates. Since 1962, studies of these directional effects (e.g., ion trajectories, ranges, and energy losses) have done much to elucidate the detailed nature of atomic collision processes experienced by heavy ions.

II. Channeling Trajectories and Interaction Potentials

Particle trajectories are related to the transverse electrostatic potentials through which the particle travels. The potential is determined by the spatial distribution of charges in the crystal. The energy loss of the particle is also related to this distribution of charges inside the crystal in the vicinity of the particle trajectory. The interrelation of trajectories, potential, and particle stopping power allow one to test various models for the potentials in the crystal.

Particle trajectories can be accurately computed in planar or axial channeling, using a general Monte Carlo trajectory computer code such as LAROSE (Barrett, 1971, 1990). The calculation begins by randomly selecting the starting point of each trajectory and an entrance ray lying within the beam divergence having a Gaussian distribution. Deflection of the ion is computed classically in the impulse approximation, including the most important atomic interactions as the ion moves through the lattice at each atomic layer. The effects of thermal vibration and mosaic spread in the crystal and of projectile energy loss and electron multiple scattering are also usually simulated.

The interaction energy between the projectile ion and a lattice atom has often been assumed to be Molière's approximation to the Thomas–Fermi function. With this general screening function, the interaction potential is

$$U(r) = \frac{Z_1 Z_2 e^2}{r} [0.35 \exp(-br) + 0.55 \exp(-4br) + 0.10 \exp(-20br)] \quad (3)$$

where $b = 0.3/a$, $Z_1 e$ and $Z_2 e$ are the nuclear charges of the projectile and target species, respectively, and a is a screening length given in Eq. (2). Potentials, such a Hartree–Fock or the Ziegler–Biersack–Littmark (ZBL) potential (Ziegler et al., 1985), can be used in trajectory calculations to more accurately model important atomic physics parameters such as electron density or include shell effects (Dygo and Turos, 1987; Smulders et al., 1992). Experience has shown, though, that the simple Molière potential is a good starting point because it yields accurate trajectories for the majority of channeling impact parameters in low order channels.

III. Planar Channeling

In planar channels, the projectile executes predominantly one-dimensional oscillatory motion in a direction perpendicular to the bounding atomic planes. Three one-dimensional trajectories for planar channeled positive ions initially moving parallel to closely packed lattice planes are illustrated in Fig. 4. At sufficiently high projectile energy, where the stopping power is very small, the transverse motion is that of an undamped anharmonic oscillator. Ions entering close to the channel midplane (path a) have comparatively low sinusoidal oscillation amplitude, and the wavelength is independent of amplitude. Those entering closer to an atomic plane (paths b and c) are repelled more strongly and have paths of greater amplitude and shorter wavelength. For any realistic potential describing the atomic plane the anharmonicity increases with increasing amplitude. Those entering closer to the atomic plane have a wavelength that depends on their entrance point or impact parameter (Robinson, 1969). Particles having the least amplitude also interact least with the atomic plane and, thus, experi-

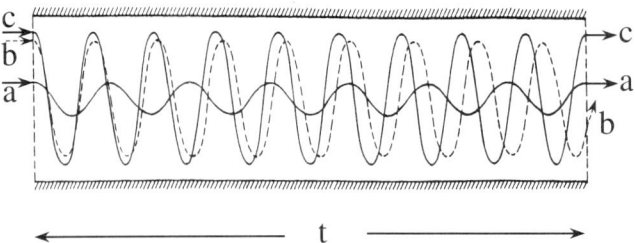

FIG. 4. Planar channeling trajectories for three entrance points. The ion executes one-dimensional simple harmonic motion. Trajectories b and c, which start farthest from the center of the channel, are the most anharmonic.

ence the least electron density and energy loss. The sinusoidal oscillation frequency for a given lattice and planar channel is proportional to $(q/m_p)^{1/2}$, where q and m_p are the projectile charge and mass, respectively. In fact, energy loss measurements performed on planar channeled particles have clearly resolved ions that have undergone a different number of integral oscillations in passing through the crystal (Lutz *et al.*, 1966; Gibson *et al.*, 1968; Datz *et al.*, 1969; Appleton *et al.*, 1971), and extensive data of this type has been used to test ion–atom potentials (Robinson, 1971). In atomic physics measurements, energy loss is often measured in coincidence with other parameters so that only trajectories having the minimum energy loss are included in the measurements, often simplifying interpretation of experimental results.

IV. Axial Channeling

Axial channeling trajectories are much more complex than planar trajectories. Three calculated trajectories for ions channeled in the [001] direction of Au are shown in Fig. 5. The projectile executes transverse oscillations in two Cartesian directions perpendicular to the direction of motion; in an axial direction, different atomic planes always cross. If the projectile has insufficient transverse energy to cross the electrostatic saddle point between strings, it will be confined within the boundary formed by a single set of atomic strings, and this special case of axial channeling is called *hyperchanneling* (Appleton *et al.*, 1972). The acceptance angle for hyperchanneled ions is smaller than the angle in ordinary channeling. Because these ions sample larger average impact parameters, their energy loss is also the smallest. This special case is illustrated in the upper left corner of

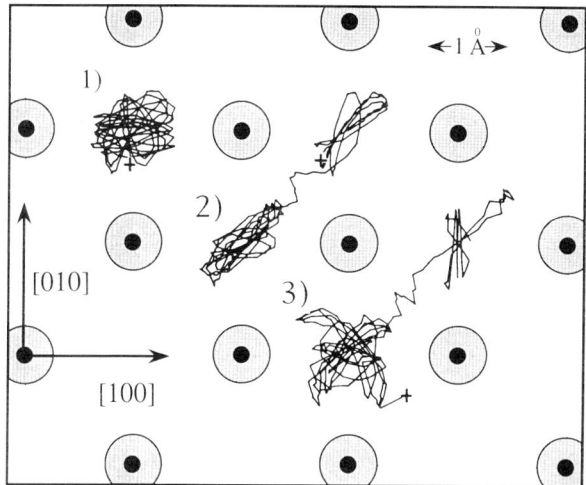

FIG. 5. Computer simulation showing the transverse motion of particles moving in axial channels of Au. The particle motion indicated by (1), confined to a single set of four atomic strings, is called *hyperchanneling*. Particles having sufficient transverse energy to surmount the saddle-point potential energy barrier between adjacent axial channels are indicated by (2) and (3).

Fig. 5. The two-dimensional oscillations can lead to a focusing of the flux inside each axial channel in the incident beam direction.

Despite the complex motion that occurs in axial channels, there is still a remarkable underlying trajectory simplicity for the majority of incident impact parameters, especially if the crystal is very thin (i.e., a thickness of several hundred atomic layers). Transmitted ions have highly structured two-dimensional angular distributions that depend sensitively on the projectile's velocity and incident charge state and the target material, its thickness, and azimuthal orientation (Krause *et al.*, 1986a, b; Nešković, 1986a). For a given lattice and axial channel, the trajectories and hence the transmitted angular distribution have been shown to scale simply (Krause *et al.*, 1994). In contrast to planar channeling, the atomic strings surrounding each channel focus the flux two-dimensionally at periodic depths in the channeling direction, a phenomenon called *flux peaking*. In fact, realistic trajectory calculations indicate that the flux can be enhanced about a factor of five or more in the vicinity of a flux peak. Interesting, axial channeling angular distributions emerging from thin crystals have also been discussed in terms of catastrophe theory (Nešković and Perović, 1987; Nešković *et al.*, 1993, 1994).

A basic description of trajectories and the emergent angular distribution for axially channeled ions is relevant to dealing with the practical matters of actually aligning a crystal's axis to the incident beam direction and the effective size of the detector located after the crystal, which will analyze the emergent ions. Knowledge of the universal scaling of axial angular distributions can be used to predict the depth at which flux peaking occurs, provide a clue about the family of trajectories relevant to the chosen experimental conditions, and help interpret results in atomic physics experiments. A first order description of motion in a very thin crystal, which will be given here, has been derived from experiments involving Monte Carlo simulations of highly charged carbon ions axially channeled in a $\langle 100 \rangle$ direction of silicon (Krause et al., 1986a, b; 1994).

Projections of ion trajectories that lead to the most focused angular distribution are shown schematically in Fig. 6(a). Here, the length of the crystal is equal to one-half the effective wavelength of the oscillatory motion (i.e., where the normalized parameter $\Lambda = \frac{1}{2}$; see Eq. (4) following). Particles exiting the crystal with this trajectory have experienced the minimum emittance growth. The flux peak occurs halfway through the crystal ($\Lambda = \frac{1}{4}$), where the two trajectories cross (and, in general, at other thicknesses in the sequence $\frac{1}{4}, \frac{3}{4}, \frac{5}{4}, \ldots$). Based on experimental data and Monte Carlo trajectory calculations, it is possible to predict the general shape of axial channeling angular distributions in the first oscillation cycle for arbitrary crystal thickness and projectile conditions using the reduced crystal thickness Λ (Fig. 7), defined by

$$\Lambda = \frac{f(q, m_p)t}{v} \quad (4)$$

In Eq. (4), v is the average projectile speed in cm/sec, t is the [001] crystal thickness (cm), and f is the transverse oscillation frequency (Hz). The frequency for multiply charged ions in a $\langle 100 \rangle$ axial direction in silicon, for example, is closely approximated by $f(q, m_p) = 6.226 \times 10^{13}(q/m_p)^{1/2}$ Hz, where q is the projectile charge state (unit charge) and m_p is the projectile mass (amu). In the case where exit charge states are not dispersed, q can be taken to be the average exit charge, to obtain the exit charge-averaged Λ.

The two-dimensional angular distributions shown in Fig. 7, which have been shown to agree well with experimental measurements, depend sensitively on the normalized thickness Λ. The atomic strings bounding an axial channel can act as a strong atomic lens, which sometimes leads to a focusing in the direction of incident ion flux (e.g., $\Lambda = 0.50$). For other normalized thicknesses, the maximum emergent flux does not occur at zero

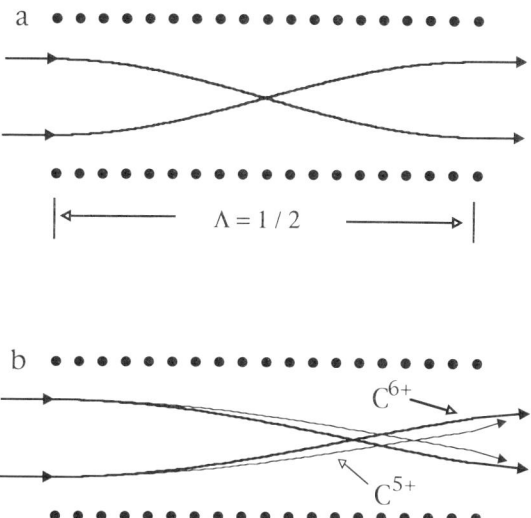

FIG. 6. Normalized trajectories that illustrate axial channel oscillations inside a very thin crystal. The two-dimensional trajectories have been projected in one dimension. Illustration (a) shows the condition when the beam emittance growth is at a minimum, that is, where the crystal thickness corresponds to $\Lambda = \frac{1}{2}$ defined in Eq. (4). Illustration (b), 30-MeV C^{q+} for $q = 5$ and 6, shows that the oscillation frequency depends on the projectile's charge state. The C^{5+} ions exiting the crystal are slightly more deflected than the C^{6+} ions because their wavelength is 10% longer.

scattering angle (e.g., $\Lambda = 0.20$ and 0.30). The trajectories shown in Fig. 6(b) illustrate the fact that, because the axial oscillation frequency depends weakly on the projectile's q/m_p, trajectories for different entrance charge states are slightly different. The normalized angular scale used in Fig. 7, $\alpha = 0.32\psi_1$, is proportional to the characteristic channeling angle. Therefore, virtually all the emergent flux can be detected when the detector of channeled ions has a radial angular acceptance of about α. Given Eq. (1), this angle can become very small for projectiles moving at very high energy (experimental situations discussed later in this chapter). In general, the best channeled ions, corresponding to trajectories that start close to the center of the axial channel, are deflected the least. Therefore, it is possible to limit the range of effective channeling impact parameters by limiting the angular acceptance in actual experiments; this selectivity can be achieved easily where the channeling angle ψ_1 is relatively large (i.e., at the lowest projectile velocity). Although the results just discussed are immediately

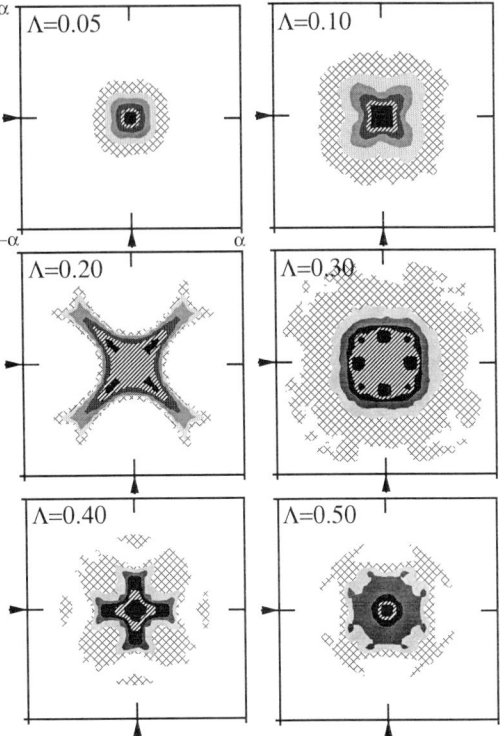

FIG. 7. Calculated two-dimensional angular distributions versus crystal thickness, applicable to any ion axially channeled along an Si $\langle 100 \rangle$ axis. The atoms bounding the channel are congruent with the corners of each angular distribution (i.e., the horizontal and vertical axes are in the [110] and [1$\bar{1}$0] directions, respectively). The two-dimensional contours, represented in increasing gray-scale density, are cuts through the distribution at 2%, 5%, 10% 20%, 50%, and 75% of the peak intensity. The reduced angular scale α corresponds to $0.32\psi_1$; the reduced crystal thickness, Λ, is defined in Eq. (4).

applicable only to a $\langle 100 \rangle$ axis of silicon (diamond lattice), trajectory codes can be used to obtain detailed results for other lattices and axial directions should details of the motion be needed in other investigations.

V. Experimental Methods

A typical channeling arrangement is shown schematically in Fig. 8. The beam incident on the setup has usually been charge-state and momentum analyzed. Typically, after this analysis, the beam is collimated so that its divergence is much less than the channeling angle ψ_1. At the highest energies that have been employed in recent experiments, where ψ_1 can be extremely small ($< 10^{-4}$ rad), achieving the necessary beam collimation can be challenging.

Most atomic physics experiments involving low-Z_1 projectiles traveling at nonrelativistic speeds require a very thin single crystalline target, where the thickness lies in the range 0.1–3 μm. Relevant cross sections for the physics under study dictate the thickness. A thickness of 1000 Å (0.1 μm), corresponding to about 200 atomic layers for most lattices, is approximately the minimum practical thickness for a self-supporting nearly "perfect" crystal (i.e., one having few dislocations and low mosaic spread). Single crystals are typically about 0.5–1.0 cm in diameter. The crystal is usually mounted in a three-axis goniometer so that initially the crystal axis normal to the surface can be precisely aligned to the incident beam direction. Two motions (ϑ and Φ) are used to choose the crystallographic direction. The third motion, δ, is perpendicular to ϑ. The ϑ and δ motions are adjusted together to align the normal axial direction of the crystal to the beam axis. The goniometer positioning mechanism must be precise compared with the required channeling angle. Whenever the beam–target relative collision energy exceeds the nuclear Coulomb barrier, the goniometer must be driven under remote control using stepping motors.

The channeled ions transmitted through the crystal (having a divergence half-angle $\leq \psi_1/2$) are sometimes postcollimated before they enter

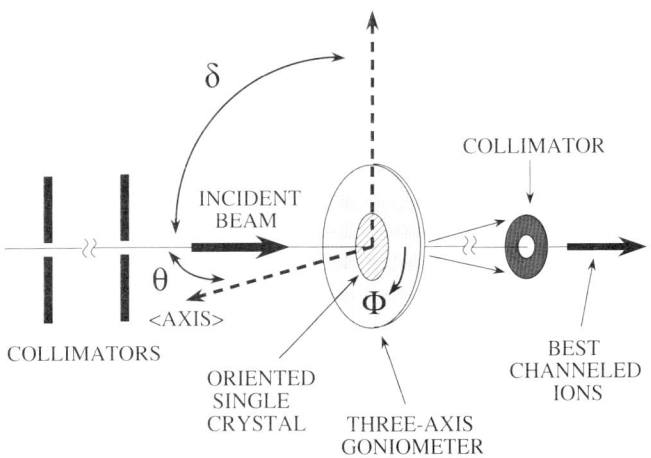

FIG. 8. Illustration of a typical channeling experiment.

an analyzing device or detector to select individual ionic charge states. Electrostatic or magnetic deflection can be used to analyze exit charge-state fractions. High resolution magnetic spectrometers of the Elbek or Enge design are usually used to measure projectile energy loss. One- and two-dimensional position-sensitive solid state detectors, gas-filled detectors, and channel plates are used to detect the energetic ions. X-ray radiation produced by the projectile target interaction can be detected by use of one or more Si(Li) detectors.

A ZnS phosphor located after the crystal that detects characteristic "star patterns" can be used to visually align the crystal to the incident beam axis when the projectile speed is well below the Coulomb barrier for the projectile–target system. The highly variable count rate from a detector (channeled vs unchanneled) can be used with a remotely controlled goniometer to align the crystal at the highest projectile speeds.

VI. Charge Changing Collisions

Energetic ions traveling through crystals at small angles to low index directions may be steered to avoid hard collisions and interact in close collisions only with loosely bound electrons. For ion velocities $v \gg v_f$, where v_f is a target Fermi electron velocity representative of the sampled electrons, the penetrating ion may be viewed as being bombarded by a flux of electrons moving at velocity v. The qualitative verity of this approximation was first demonstrated in 1972 (Datz *et al.*, 1972) when oxygen ions (2 MeV/u) channeled through thin crystals were shown to undergo electron capture and loss that could be characterized by interactions with only valence electrons. Since that time, the effect has been used in numerous and varied investigations involving heavier ions traveling at still higher velocities. It is the basis for all atomic physics investigations discussed in succeeding sections.

In an amorphous solid, both capture and loss processes are so rapid that the charge state distribution of the emergent ions is independent of the input charge state. This phenomenon is illustrated in Fig. 9 for 315-MeV Ti ions emerging from a 2.6-μm Si crystal oriented in a random direction. In the concave downward charge-state distribution shown, charge state equilibrium has been achieved. When the crystal is oriented to a $\langle 100 \rangle$ axial direction, the situation is quite different. Here, the exit charge-state distribution depends strongly on the input charge. Charge capture is limited to $\approx 10\%$ and charge loss is too small to indicate on the ordinate scale. This comes about because of the constriction that allows only interaction with valence electrons. The ion velocity, approximately

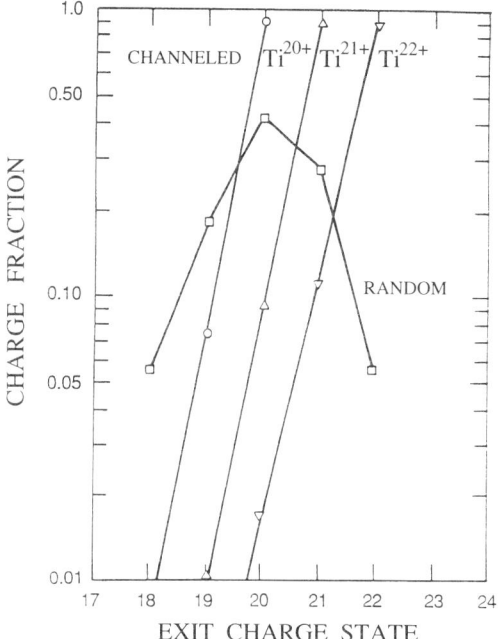

FIG. 9. Typical exit charge-state distributions for channeled and unchanneled ions. The distributions for channeled ions depend on the incoming charge state for 315-MeV Ti^{20+}, Ti^{21+}, and Ti^{22+} in a $\langle 100 \rangle$ channel of a 2.6-μm-thick Si crystal. The concave downward distribution shows that charge-state equilibrium occurred when the ions were traveling in a random crystal orientation.

7 MeV/u, corresponds to an electron velocity of about 3.5 keV, and the probability of capture of such semifree electrons is small. On the other hand, the ionization potential of Ti^{21+}, for example, is about 6.6 keV, so that the electron bombardment energy is much below threshold. Thus, by the use of channeling, a "thick" target for charge transfer has been converted into a "thin" target from which one can derive single-collision cross sections. In the channeling studies discussed in later sections, projectile ions were prepared under these "thin" target conditions.

VII. Radiative Electron Capture

Channeling techniques have been used to measure the radiative recombination processes in solids. Radiative electron capture (REC) is a recombination process whereby an ion captures a target electron into one of its shells and emits a photon. REC is closely related to the process called radiative recombination (RR), in which a free electron is transferred into a bound state of a positive ion, with simultaneous emission of a momentum and energy conserving photon acting as the necessary third body. The RR process is the exact inverse of the photoelectric effect. REC is distinguished from RR in that the active electron in REC initially occupies a bound state in the target. Therefore, REC occurs in ion–gas and ion–solid collisions, and RR occurs in ion–electron collisions. Knowledge about REC involving highly charged ions can be applied in numerous arenas such as materials research, storage rings, laboratory plasmas, and some astrophysical environments.

For fully stripped or hydrogen-like projectiles, the dominant contribution to REC involves capture into the projectile's K shell. For a projectile moving at the nonrelativistic energy E, the REC photon energy (centroid) $E_{h\nu}$ is given by

$$E_{h\nu} = E_p - E_t + \frac{m}{M_1} E \tag{5}$$

where E_p and E_t are the electron binding energies in the projectile and target, and the electron and projectile masses are m and M_1, respectively.

For a bare ion that captures an electron to produce a hydrogen-like ion in the K shell, the term $E_p - E_t$ becomes $(Z_1)^2$ Ry, where Ry is the Rydberg energy. The last term in Eq. (5) represents the kinetic energy of the electron in the rest frame of the ion (e.g., ~500 eV for an ion moving at 1 MeV/u). Thus, for swift, highly charged ions, the REC photon lies in the X-ray region. Because the shape of this X-ray peak reflects the velocity distribution of the electron captured, REC measurements performed in solids or gases probe the target electron's density and Compton profile. In channeling, the ions whose trajectory passes through the channel center sample only conduction or valence electrons in the target (Fermi electron distribution) with width of order 10 eV.

The process has been described theoretically by Bethe and Salpeter (1957). In the impulse approximation, the REC cross section per free electron captured into the K shell of a bare ion is (Stobbe, 1930; Bethe and Salpeter, 1957)

$$\sigma_{\text{REC}} = 9.1 \left(\frac{\kappa^3}{1 + \kappa^2} \right)^2 \frac{\exp(-4\kappa \operatorname{arc cot} \kappa)}{1 - \exp(-2\pi\kappa)} \times 10^{-21} \text{ cm}^2 \tag{6}$$

Here, $\kappa = Z_1 \alpha / \beta$ is the Sommerfield parameter, α is the fine structure constant, $\beta = v/c$, where v is the incident projectile speed and c is the speed of light. The parameter κ is simply related to the *adiabaticity* parameter η, with respect to the projectile K shell electron (Stöhlker et al., 1992):

$$\eta = \frac{E_{kin}}{E_B} = \kappa^{-2} \qquad (7)$$

Here, E_{kin} is the initial kinetic energy of the quasi-free electron in the projectile frame and E_B is the K binding energy of the captured electron.

Stöhlker et al. have demonstrated that measured K-REC cross sections obtained in ion–gas studies involving numerous projectile species are easily compared with theory (Eq. 6) using a plot of $\sigma(\eta)$. The adiabaticity parameter is also useful in the comparison of the ion–solid results, discussed in the following.

The first observation of REC, involving energetic O^{q+} ions in solids, was reported by Schnopper et al. (1972). Their study showed that the width of the REC peak is affected by a number of things. When heavy ions enter an amorphous solid, rapid ionization and capture lead to a distribution of charge states and, as shown in Eq. (5), each charge state can result in an REC photon of a different energy. Second, the intrinsic width of the REC line arising from a single charge state will reflect the velocity distribution of the target electrons captured by the moving ion. Even a narrow electron energy distribution for electrons in the target contribute significantly to the REC linewidth when it is transformed into the frame of the moving projectile. These effects are noticeable even when the REC photons emitted from the crystal are detected using a low resolution solid state detector. The REC process was also studied using atmospheric gas targets (Schnopper et al., 1974).

The use of channeling greatly simplified the study of REC in solids (Appleton et al., 1979). In the first study, 17- to 40-MeV oxygen ions were channeled in either Ag or Si, demonstrating several advantages. Channeling allows the projectile ions to remain predominately in a fixed charge state during their passage through the crystal. Channeling also greatly diminishes the prominent target X-ray background that can interfere with the emission in the REC process. Essentially, channeled ions interact only with the free or conduction electrons in the target because close encounter collisions involving bound target electrons are eliminated. Study of the single cross section and lineshape for K-REC in a solid became feasible.

Several investigations of the REC process, using solid and gaseous targets, have been reported since the last review. Advances in ion source technology and the availability of more energetic accelerators have allowed REC processes for higher-Z_1 projectiles to be studied. For these species at higher collision energy, the X-ray energy is substantially increased, so that the X rays emitted lie in a region where the detected signals are large, and the efficiency of windowed Si(Li) or Ge(Li) detectors is not changing rapidly in the vicinity of the REC peak. For high-Z_1 projectiles, it is easier not only to study K-REC but also to study the less probable end products where the electron is captured into the L and M shells. The quality of REC data has also improved by studying the REC emission in coincidence with the $(q - 1)+$ resultant charge state. Some X-ray signals that interfere with the low intensity REC peaks are eliminated using the coincidence technique. The use of a universal scaling law to represent total cross sections in both ion channeling and ion–gas investigations has simplified comparisons with theory and provided evidence for enhanced REC cross sections in ion–solid collisions because of a solid state effect. Detailed studies of the REC photon distribution and the absolute energy for REC photons have also been studied in ion–solid and ion–gas collisions.

Andriamonje et al. (1987) studied the effects of 25-MeV/u hydrogen-like Xe^{53+} ions channeled along a $\langle 110 \rangle$ axis in a 21-μm Si crystal, the most open axis in a diamond lattice. Their high energy channeled and random X-ray spectra were dominated by projectile X rays (e.g., centroid of K = REC = 55 keV) so that coincidence measurements were not necessary to distinguish REC emission. Using charge-state measurements and the X-ray spectra, they were able to determine cross sections corresponding to electron capture into the K, L, and M shells of the projectile. The cross sections are based on a self-consistent electron density calculation for a $\langle 110 \rangle$ channel (L'Hoir et al., 1990). Their REC cross sections per target valence electron for well-channeled Xe^{53+} were in agreement with the cross section given in Eq. (6). In fact, the results show that REC accounts for nearly all the electron capture by 25-MeV/u Xe ions channeled along a $\langle 110 \rangle$ axis. By adjusting the crystal's alignment with the beam axis, they showed that the K- and L-REC intensity and centroid depends on the alignment. These investigators provided qualitative arguments for shift of the REC centroid in terms of changes in the density and velocity of the electrons sampled by the projectile. However, no theoretical calculations have been reported that duplicate the measured shifts. Recently, K-, L-, and M-REC have also been studied for a number of high-Z_1 ions in targets of H_2 or He (Stöhlker et al., 1992).

Several ion channeling studies of REC have been performed by the Oak Ridge group. These studies involved energetic sulfur and titanium ions in silicon and gold (Datz et al., 1992). Spectra that illustrate the observation of K-REC when 315 MeV Ti^{q+} ions are channeled in a $\langle 100 \rangle$ channel of Si (2.6 μm) are shown in Fig. 10. The REC peaks are broader than other spectral features shown because of the Compton profile of the target electrons captured. Each spectrum was taken in coincidence with emer-

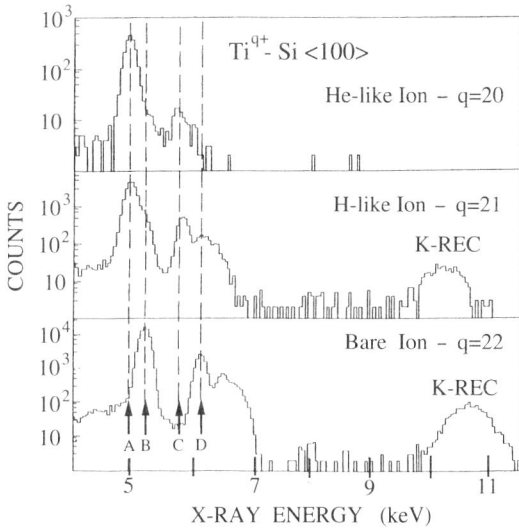

FIG. 10. X-ray spectra for 315-MeV Ti^{q+} incident ions axially channeled in a $\langle 100 \rangle$ direction of a 2.60μm-thick Si crystal. The spectrum in each panel was taken in coincidence with the emergent charge state ($q - 1$). The centroid of the K-REC X rays appear at 10.3 and 10.7 keV for the $q = 21$ and 22 incident charge states, respectively. Other radiative transitions identified in the spectra are: (A) $1s2l \rightarrow 1s^2$; (B) $2p \rightarrow 1s$; (C) $2s3l \rightarrow 1s^2$, and (D) $2p \rightarrow 1s$.

gent ions in the charge state $q - 1$; therefore, target X rays and Bremsstrahlung are nearly eliminated. A goal in these studies was to determine the absolute energy of REC photons and the shape of the photon distribution.

Another study involved 5.83-MeV/u fully stripped titanium ions (Ti^{22+}) channeled along a $\langle 100 \rangle$ axis in a very thin (0.263-μm) Au crystal (Vane et al., 1992). Here, Ti^{21+} ions were momentum analyzed using a high resolution magnetic spectrometer (Enge design). Using X-ray and energy loss data collected in the "list mode," the X rays could be rehistogrammed later by gating on the measured Ti^{21+} energy loss. An interesting finding is that ions having the minimum energy loss yielded REC photons having two components. The profile of the most narrow X-ray peak is consistent with that predicted by the Fermi free electron gas model (i.e., delocalized electrons in the solid). The profile of the second component, which is about twice as broad as the first component, is consistent with capture involving the more tightly bound $5d$ and $5p$ target electrons, the wavefunctions of which extend into the center of the channel. An X-ray spectrum generated by including only ions that have the most channeled energy loss even shows a small X-ray peak corresponding to Au $L\alpha$. These results are understandable by recalling that channeled ions that undergo the minimum energy loss have trajectories which remain close to the center of channel throughout their passage in the crystal. Also, ions having the maximum energy loss have trajectories that pass through the inner atomic shells of an atomic string.

The titanium channeling data also showed the interesting finding that the K-REC peak was slightly shifted to lower energy than expected from Eq. (5) (e.g., ≈ 80 eV at about 5.9 keV). Within experimental error, this shift can be approximately explained in terms of the dynamical screening and depolarizing effects for ions traveling in a solid medium (Pitarke et al., 1991). The dynamic screening is due to the electron density enhancement that occurs in the vicinity of the ions when highly charged ions move through the sea of conduction electrons in the solid (i.e., the wake). Additional data obtained for other projectiles in Si channels indicate similar shifts. We emphasize that these shifts in the REC centroid energy observed in the ion–solid experiments have not been observed in ion–gas studies of REC, at least for swift fluorine or oxygen ions in H_2 or He (Tawara et al., 1982; Vane et al., 1994).

Most recently, Tribedi et al. (1994) have studied K-REC using bare and hydrogen-like ions of C, O, F, and S channeled in a $\langle 100 \rangle$ direction of a very thin (0.17-μm) Si crystal. The ion energies in this study were in the range 3–7.5 MeV/u. Their measurements of the K-REC peak shift and the cross section also provide evidence of a solid state effect. In the case of S, for example, the peak was shifted slightly toward lower photon energy and the projectile velocity dependence of the shift scaled approximately like Z_1/v, where v is the projectile speed in units of the Bohr velocity. Also, the cross section for K-REC, universally scaled (σ vs η) was found to lie on the universal curve for channeled S ions (i.e., Eq. (7)), whereas most ion–gas REC cross sections lie about 50% below the theoretical curve (Stöhlker et al., 1992). The increased electron density in solids due to the wake, which depends on Z_1/v (Pitarke et al., 1991), may be the cause of both the increased REC cross section and the peak shift for some highly charged projectiles.

VIII. Electron Impact Ionization

Channeling techniques have been used to measure the electron impact ionization (EII) cross sections for few-electron uranium ions moving at a relativistic speed (Claytor et al., 1988). Specifically, they channeled 405-

MeV/u beryllium-like through hydrogen-like uranium ions along an open ⟨110⟩ axis of thick silicon crystals. As in the case of other atomic physics studies involving heavy ions, channeling allows the highly charged channeled ions to avoid close encounters with Si nuclei and, therefore, interact primarily with a dense electron gas. In these experiments, the equivalent energy is 222 keV in the projectile frame. The crystals were relatively thick (0.11 and 0.37 mm) because charge changing cross sections are very small at their relativistic energy. Because of the small channeling angle at their high energy ($\psi_1 = 0.011°$), excellent beam collimation and a precision goniometer were required in the experiments. Electron impact ionization studies of this type would not be possible using an amorphous target or in a random direction of a single crystal, because the cross section for ionization of uranium ions in collision with Si nuclei is much larger than for ionization in collisions with electrons. Measurements by other methods are only now becoming possible by the use of heavy-ion storage rings that have electron beam coolers or a high energy ion source such as the electron beam ion trap (EBIT) (Marrs et al., 1994).

Claytor et al. (1988) measured the electron density integrated along the paths of the channeled ions by comparing a cross section for electron capture by channeled ions with a previously measured capture cross section for ions when the crystal was oriented in a random direction (Anholt et al., 1987). This comparison is possible because the most significant capture process involved in both cases is radiative electron capture (REC). For relativistic ions in low-Z targets such as Si, REC has been shown to be the dominant electron capture mechanism, both for channeled ions (Andriamonje et al., 1987) and for ions in amorphous materials, where cross sections have been measured and agree with theory (Anholt et al., 1984, 1987; Gould et al., 1984; Meyerhof et al., 1985).

The cross sections for ionization were obtained using the charge state yield versus target thickness data and known capture cross sections. The U^{90+} and U^{91+} (K shell) ionization cross sections of 11.0 and 3.9×10^{-24} cm^2 were found to be about two to four times larger than K shell ionization cross sections extrapolated from Scofield (1978), or with the calculations of Younger (1980, 1981) or with the Lotz formula (Lotz, 1967). The cross sections obtained for L shell ionization were not sufficiently accurate to distinguish between different calculations. Claytor et al. concluded that the ionization of excited states, populated by electron excitation, did not make a significant contribution to the measured U^{90+} and U^{91+} cross sections because the radiative lifetime is too short to lead to a significant contribution from subsequent collisional ionization. Therefore, excitation-ionization cannot account for the experimental-theoretical discrepancy.

Although one might argue about whether the adduced ionization cross sections are high in Claytor's channeling work because of systematic uncertainty in the average electron density for a $\langle 110 \rangle$ channel that was obtained in their measurements, more recent work also indicates that there are considerable differences between experimental and theoretical values for the $1s$ ionization cross sections for high-Z hydrogen-like ions (Marrs et al., 1994; Moores and Reed, 1995). Moores found that by including the Møller interaction (i.e., the effects of magnetic interactions and retardation) and electron exchange into relativistic distorted-wave calculations for EII, the cross section for hydrogen-like uranium increased significantly in the near-threshold region. In fact, inclusion of the Møller interaction with exchange increased the EII cross section obtained by both the methods of Zhang and Sampson (1990) and Pindzola et al. (1989), so that theory and the EBIT measurements on hydrogen- and helium-like uranium (Marrs et al., 1994) now agree much more closely. Using generally similar physics but a different formulation, Fontes et al. (1995) also found improved agreement between theory and EBIT results for uranium.

Electron impact ionization cross sections have also been obtained in a study of the impact parameter dependence of the stopping power for 27-MeV/u Xe^{35+} ions channeled along a $\langle 110 \rangle$ axis of Si (Andriamonje et al., 1989; L'Hoir et al., 1990). Under these conditions, the incident charge state is far below the narrow equilibrium charge state in a random direction ($\langle q' \rangle = 49.3$) for the 20.7-$\mu$m-thick crystal used. The projectile moving at 7.2×10^9 cm/sec impinges on the electrons in the channel with an equivalent energy of 14.7 keV. The resultant very broad and nonequilibrium emerging charge-state distribution ($35 \leq q' \leq 53$) arises nearly only from electron impact ionization ($q' = 53$ corresponds to hydrogen-like ions), because the best channeled projectiles only interact with the low electron density located near the center of the channel. This expectation was verified by showing, in angular scans across a $\langle 110 \rangle$ channel, that the emerging Xe^{37+} ions are hyperchanneled. For example, the width of the emergent Xe^{37+} ions was about 7.5% of the channeling half-angle as measured by observing the Xe Lyman α photon yield. This was not the case for most of the channeled particles emerging in the $q' = 44+$ or $45+$ and higher charge states. By the use of Monte Carlo simulations and an iterative procedure, they derived the electron impact ionization cross sections $\sigma(q)$ for $35 \leq q \leq 45$.

The EII cross sections determined by Andriamonje et al. (1989) involve both direct and indirect ionizations (i.e., starting by excitation). The predictions of the Lotz and Thompson formulas (Barnett, 1989) gave results that are two to four times smaller. The cross sections measured by Donets (1983) are also up to three times smaller than those obtained in the channeling study. The higher cross sections obtained in the channeling experiment may be due to the occurrence of multistep ionization that cannot happen in the low electron flux experiment of Donets; the mean time interval between two collisions in the solid leading to excitation or ionization is about 3×10^{-14} sec, and this time may become comparable with the radiative lifetime for highly excited states. Cross sections for $q > 45$ could not be determined with high precision because nucleus-induced effects, which may dominate the ratio between EII and nucleus-induced charge-exchange cross sections, are only known approximately. In their analysis, a nucleus-induced ionization cross section that is 200 times greater than the corresponding EII cross section ($\sim Z_2^2$ times larger) was considered (Andriamonje et al., 1989). The use of REC measurements to determine the effective electron density (as in the case of Claytor et al.) is impractical in the case of Xe ionization because the measured REC cross sections (Andriamonje et al., 1987) are vanishingly small compared with EII.

IX. Dielectronic Excitation and Recombination and Resonant Transfer with Excitation

Dielectronic excitation is a resonant electron–ion collision process wherein the free electron excites a bound electron on the ion and, in so doing, the incident electron loses energy and is captured to a discrete state. The following equation illustrates the production of the dielectronically excited $2l2l'$ states in the case of a hydrogenic ion:

$$A^{(z-1)+}(1s) + e \rightarrow \{A^{(z-2)+}(2l2l')\}^{**} \tag{8}$$

In a dilute medium, this doubly excited intermediate state can relax either by an Auger transition, the reverse of Eq. (8), or by radiative stabilization leading to recombination:

$$\{A^{(z-2)+}(2l2l')\}^{**} \rightarrow \{A^{(z-2)+}(1s2l')\}^* + h\nu_1 \rightarrow A^{(z-2)+}(1s^2) + h\nu_2 \tag{9}$$

In electron spectroscopy, the reverse of Eq. (8) is called a *KLL* Auger event. The end result of collision sequence (9), dielectronic recombination (DR), is the production of an ion of reduced charge and two photons ($h\nu_1$ and $h\nu_2$, respectively).

The resonant excitation in Eq. (8) occurs below the energy required for direct excitation. For resonance in Eq. (8), the energy of the electron to be captured must match that of a *KLL* Auger transition in the rest frame of the moving ion. Thus, the resonant energy E_r is

$$E_r = E_K - 2E_L = \frac{(\mathbf{p} - m_e\mathbf{v})^2}{2m_e} = \frac{m_e v^2}{2} + \frac{\mathbf{p}^2}{2m_e} - \mathbf{p} \cdot \mathbf{v} \quad (10)$$

where E_K and E_L are the K and L shell binding energies of the highly charged ion, m_e is the electron mass, $\mathbf{v}$ is the ion velocity, and $\mathbf{p}$ is the electron momentum in the laboratory frame. In virtually all experimental situations, the last term in Eq. (10) determines the resonance width. The sharp resonances associated with DR involving highly charged ions have been studied at high resolution in very dilute configurations such as in merged electron-ion beam experiments and storage-ring electron beam coolers (Müller et al., 1991). The DR process is also known to occur in nuclear fusion plasmas. In the ion channeling manifestations of DR to be discussed, a small correction in Eq. (10) has been ignored; there is an additional small shift in energy of order 1%, because of the polarization potential induced by the projectile charge.

An ion-gas collision analogous to DR also occurs when the target electron in Eq. (8) is bound to a heavy particle (Tanis et al., 1986; Feagin et al., 1984). Here, the process is usually referred to as resonant transfer and excitation (RTE). In the field of ion channeling, both the DR and RTE terminology are used to describe the corresponding phenomena. The resonances observed in ion-gas or ion channeling are much broader than those observed in free electron collisions, due to the momentum distribution of the target's bound electron (i.e., the Compton profile). In the case of channeling, ions experience "almost free" electrons that have a momentum distribution characteristic of conduction or valence electrons. The ion channeling environment is unique because it allows dielectronic phenomena to be studied in a very dense electron gas ($\approx 10^{23}/cm^3$), the properties of which are calculable. The environment in laboratory fusion plasmas is less controlled.

Considering the high electron density in channels, simple DR phenomena can be studied only when the crystalline target is very thin (e.g., < 0.05 μm). For significantly thicker crystals, however, second and even third electronic collisions can occur before the dielectronically excited ion either radiates or escapes from the crystal.

Dielectronic excitation and recombination processes have been studied by the Oak Ridge group (Datz et al., 1989, 1992). These experiments involved the study of hydrogen-like S^{15+} (80–210 MeV), hydrogen-like Ca^{19+} (150–330 MeV), hydrogen-like Ti^{21+} (250–370 MeV), and helium-like Ti^{20+} (237–375 MeV). The S^{15+}, Ca^{10+}, and Ti^{21+} ions beams were channeled along a $\langle 110 \rangle$ axis of a 1.2-μm-thick silicon crystal. X-ray spectra were measured using a Si(Li) detector. The emerging charge-state distributions were measured using electrostatic deflection and a solid state position-sensitive detector. Coincidence spectra for X-rays and charge-exchanged ions were also measured. These studies are based on the principle that the penetrating ion may be viewed as being bombarded by a flux of electrons moving at velocity v when the speed is much greater than the target Fermi electron velocity, v_f. The fundamental fact that these channeled heavy ions also have a high probability for maintaining their incident charge state in moving through the crystal was shown at the outset (Fig. 9).

For the thicker crystal used, only those recombined excited ions formed near the end of the path will survive, whereas X rays formed all along the path will be recorded. We take as an example the *KLL* resonance in the formation of Li-like Ti:

$$\text{Ti}^{20+}(1s^2) + e \leftrightarrow [\text{Ti}^{19+}(1s2l2l')]^{**} \tag{11}$$

$$[\text{Ti}^{19+}(1s2l2l')]^{**} \rightarrow \text{Ti}^{19+}(1s^22l) + h\nu_1 \tag{12}$$

$$[\text{Ti}^{19+}(1s2l2l')]^{**} + e \rightarrow [\text{Ti}^{20+}(1s2l)]^* + 2e \tag{13}$$

$$[\text{Ti}^{20+}(1s2l)]^* \rightarrow \text{Ti}^{20+}(1s^2) + h\nu_2 \tag{14}$$

$$[\text{Ti}^{20+}(1s2l')]^* + e \rightarrow \text{Ti}^{21+}(1s) + 2e \tag{15}$$

In this case, the radiative rate for Eq. (12) exceeds the collisional rate for Eq. (13) by a factor of about eight. However, the stabilized recombined ion (Eq. 12) is susceptible to ionization via $1s^22l + e \rightarrow 1s^2 + 3l \rightarrow 1s^2 + 2e$ with a mean free path of about 1.5 μm. This effect was demonstrated by passing a $\text{Ti}^{20+}(1s^2)$ beam through a $\langle 100 \rangle$ channel in a Si crystal having a thickness of 2.6 μm. Because of the relatively low yield of surviving Ti^{19+} ions, coincidence measurements are necessary. Data for the coincidence of Ti^{19+} ions with Ti $K\alpha$ and $K\beta$ X rays are shown in Fig. 11. Direct excitation, in contrast to the dielectronic excitation process, rises sharply at its threshold energy. The solid curves are a computer simulation involving the steps indicated in Eqs. (11)–(15). Here, electron binding and transition energies have been reduced to approximate the Li-like doubly excited states, small backgrounds have been added, and the calculations for $K\alpha$ and $K\beta$ have been independently normalized to the data at 300 MeV. Clearly, these effects should be studied as a function of path length, the upper limits being determined by complications arising from energy loss of ions in the crystal.

Excitation of *KLL* and *KLM* was also observed in the case of injected hydrogen-like Ti^{21+} ions. Here, the absolute cross section for DR was

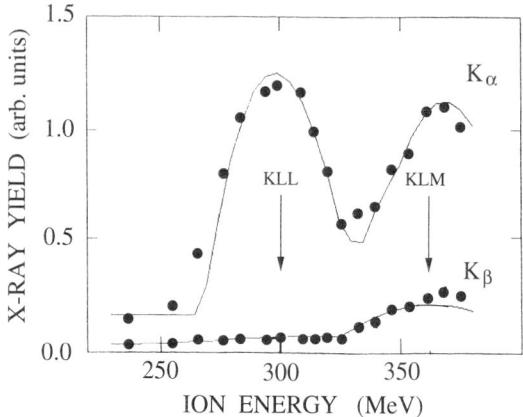

FIG. 11. Titanium $K\alpha$ and $K\beta$ X-ray yields for He-like (Ti^{20+}) incident ions axially channeled along a $\langle 100 \rangle$ direction of a 2.6-μm-thick Si crystal. The X rays were measured in coincidence with Ti^{19+} ions. Solid curves show normalized results of a computer simulation.

found to be 0.70×10^{-21} cm^2 per electron. Although no direct RTE measurement exists for this system, cross comparisons with ion–gas experiments were made. The RTE results for lithium-like Ti^{19+} (Reusch et al., 1987) and the RTE results for hydrogen-like, helium-like, and lithium-like Ca (Tanis et al., 1986) were used. The channeling method yielded cross sections that agree well with the ion–gas experiments where comparisons are possible (Datz et al., 1992). These and other experiments have demonstrated that a crystal channel provides an environment for electron–ion collision studies with $v_i \gg v_f$ that is qualitatively analogous to a dense electron gas with a near Fermi distribution of energies.

A high resolution study of dielectronic capture involving channeled Ti ions was reported (Belkacem et al., 1990). In this study, 267- to 320-MeV helium-like Ti^{20+} and lithium-like Ti^{19+} ions were channeled along a $\langle 110 \rangle$ axis of a very thin Au crystal (0.12 μm thickness). They reported only the 19+ exit charge-state fraction but isolated the best channeled ions via high resolution measurements of their energy losses. The high resolution energy measurements were accomplished by detecting the channeled ions with a position-sensitive parallel-plate avalanche detector positioned in the focal plane of an Enge split-pole magnetic spectrograph. Like Datz et al., they observed enhanced electron capture when the projectile speed corresponds to ejected *KLL* Auger electron energies. The analysis of their energy loss data also yielded a narrow resonance at a projectile energy of 293 MeV. When the analysis focused on the particles having the minimum energy loss, the observed peak width was at least five times narrower than any previously observed for RTE, a surprising result.

Band structure calculations for a $\langle 110 \rangle$ axis in Au predict an average electron density of about three electrons per Au atom, or 0.18 Å^{-3} (Anderson et al., 1993). A uniform gas of this density implies that a Compton profile of about 12 eV would be anticipated instead of the much narrower line observed. In a theoretical investigation, Feagin and Wanser (1991) claimed that the Compton profile of conduction electrons in Au is significantly narrower than that predicted by a simple free electron gas model (Wanser and Feagin, 1991). They suggested that the width of the DR resonance in a gold crystal aligned in a $\langle 110 \rangle$ direction could be as low as the value obtained by Belkacem et al. The surprising possibility stimulated several experimental studies in very thin Au, Ni, and Si crystals by the Oak Ridge and Chalk River laboratories (Dittner et al., 1992; Anderson et al., 1993; Forster et al., 1993, 1994). Another RTE study involving Xe^{52+} ions channeled in a much thicker Si crystal was also performed (Andriamonje et al., 1992).

The Oak Ridge group (Dittner et al., 1992) essentially duplicated the high resolution energy loss experiment of Belkacem et al. using helium-like Ti^{20+} ions. An Enge split-pole spectrograph was also employed to distinguish the best channeled ions. The only differences in the two experiments were that in the Oak Ridge experiment, ions were channeled along a $\langle 100 \rangle$ axis of a Au crystal and the crystal was about two times thicker than that used by Belkacem. The Ti^{19+} charge fraction exiting the crystal was measured as a function of ion energy in the range 280–310 MeV, where the *KLL* RTE resonance occurs. The width of the resonance peak including only the best channeled ions was found to be comparable with that obtained with an H$_2$ gas target (i.e., a Compton spread of order 10–12 eV). A narrow peak in the energy loss spectrum coincident with the particles having the least energy loss which had been reported by Belkacem et al., was not observed. In addition, the number of Ti $K\alpha$ X rays emitted by the Ti ions due to RTE (i.e., X rays in coincidence with Ti^{19+}), were measured at two energies. No difference in the X-ray yield was observed on and off the previously reported narrow resonance, as would have been expected.

The GANIL collaboration also investigated *KLL* resonant transfer and excitation using 33–43 MeV/u helium-like Xe^{52+} ions channeled along a $\langle 110 \rangle$ axis of a 21-μm-thick Si crystal (Andriamonje et al., 1992). Two-electron projectiles (52+) were used because they could be produced with a large intensity around the resonance energy, and because RTE rates involving helium-like ions should be larger than those involving hydrogen-like ions. Like the measurements of the Oak Ridge group, charge-state and energy analysis of the transmitted ions and projectile X-ray spectroscopy were performed. Their X-ray emission and charge-state fraction data collected for the *KLL* resonance yielded nearly the same results. Their theoretical estimates reproduced the experimental RTE capture probability with an accuracy better than 50%. Moreover, their analysis indicated that resonant *KLL* RTE and nonresonant *L*-REC captures are rather independent of the local electron density near the center of a $\langle 110 \rangle$ channel. These experiments involving a Si crystal also did not reveal an anomalous energy and width for the *KLL* resonance as reported by Belkacem *et al.* for a $\langle 110 \rangle$ axis in gold. If these anomalous effects in Si were as large as in Au, their experiment should have shown them. However, the width of the fine structure of doubly occupied *L* states in Xe^{51+}, together with the energy loss of the projectile ions, did not yield a precise Compton profile for RTE.

Several additional channeling studies of the *KLL* RTE resonance were performed at Chalk River in an attempt to observe the narrow RTE peak. Their studies involved helium-like Br^{33+} ions channeled along a $\langle 110 \rangle$ axis in Si (Forster *et al.*, 1993), Ni (Forster *et al.*, 1994), and Au (Anderson *et al.*, 1993). The ions had an energy range of roughly 12–18 MeV/u, and the energy loss was measured using their high resolution Q3D spectrograph. The crystal thicknesses were 0.99, 0.55, and 0.90 μm, respectively. In each case, they observed a broad *KLL* resonance including only the ions that underwent the minimum energy loss. They showed that both the width and strength of the resonance are consistent with theoretical predictions for a uniform and isotropic electron gas having an average electron density in a $\langle 110 \rangle$ channel of 2.6 electrons per atom. Coincidence X-ray data were also measured. All Chalk River results and the Oak Ridge results for Au $\langle 100 \rangle$ (Dittner *et al.*, 1992) strongly suggest that the anomalous results for the DR resonance involving Ti (Belkacem *et al.*, 1990) were incorrect. Results of the French group also indicate that the effect does not occur in Si.

X. Resonant Coherent Excitation

When an ion moves parallel to the atomic strings or an ordered atomic plane inside a crystalline solid, the ion experiences a variation of the static electric field as it passes each lattice atom. The spatial variation appears as a periodic time varying electric field in the projectile frame. The fundamental perturbation frequency ν experienced by the ion is v/d, where v is the ion speed and d is the distance between the perturbing atoms. The spatial electric field is not a pure sinusoid; therefore, in general, the ion is also perturbed at integral multiples of the fundamental frequency, $\nu = K(v/d)$, where $K = 1, 2, \ldots, N$. When one of these frequencies coincides with $\nu_r = \Delta E_{ij}/h$, where ΔE_{ij} is the energy difference between electronic states i and j of the ion, a fraction of the incident ground state ions can become excited provided that electric dipole selection rules are satisfied. This phenomenon is called resonant coherent excitation (RCE) (Datz et al., 1978; Moak et al., 1979).

The ion energy required for a resonance E_r is given by

$$E_r \text{ (MeV/u)} = 3.03 \, K^{-2} \, d_{\text{Å}}^2 \left[\Delta E_{ij} \text{ (keV)} \right]^2 \frac{2}{1 + \gamma} \tag{16}$$

where $\gamma = (1 - \beta^2)^{-1/2}$. For example, a second harmonic $(K = 2)$ resonance for the $n = 1 \to 2$ transition in N^{6+} in $\langle 100 \rangle$ Au requires 44.1 MeV.

Once the ion enters an excited state, two paths are possible. First, because the electron bombardment ionization cross section is much larger than for the ground state [e.g., for H-like ions, $\sigma_i(n = 2) \approx 10\sigma_i(n = 1)$], it may be ionized, and the signal for RCE is a loss of the "frozen" charge component. Second, if the ion exits the crystal in its excited state, it can radiate, and the signal is then expressed in photon emission.

The Oak Ridge group carried out the earliest successful RCE experiments by channeling low-Z ($Z = 5$–9) ions through thin Au crystals. An example is shown in Fig. 12 for the second harmonic of $n = 2$ excitation in $\langle 100 \rangle$ Au. The vertical axis is the ratio of N^{6+} to the sum of $N^{6+} + N^{7+}$, i.e., the surviving fraction of H-like ions. The horizontal axis, v/v_r, is in dimensionless velocity units, where v_r corresponds to the calculated $(1s$–$2p)$ resonant velocity in vacuum. As can be seen, there is a doublet split around the resonant frequency, which has been attributed to the static crystal field that removes the degeneracy between $2p_x$ and the $2p_z$ states (Datz et al., 1978; Moak et al., 1979, 1982; Crawford and Ritchie, 1979; Crawford, 1980; Nešković, 1986b). Experiments using the Holifield Heavy Ion Research Facility (HHIRF) at Oak Ridge (Miller et al., 1986) enabled the extension of work with low-Z ions to lower harmonics ($K = 2$ for F^{8+} and $K = 1$ for C^{5+}).

During this series of experiments at Oak Ridge, no higher order, i.e., $\Delta n > 1$, transitions were observed. This result was attributed to the ionization cross sections for states with $n > 2$ for these low-Z ions being so high that the lifetime of the upper state was too short to allow a measurement of a narrow resonance. Forster et al. (1996) at Chalk River have observed $\Delta n = 1, 2,$ and 3 transitions with Si^{13+} ions channeled in Si $\langle 111 \rangle$ and Si $\langle 112 \rangle$; an example is shown in Fig 13.

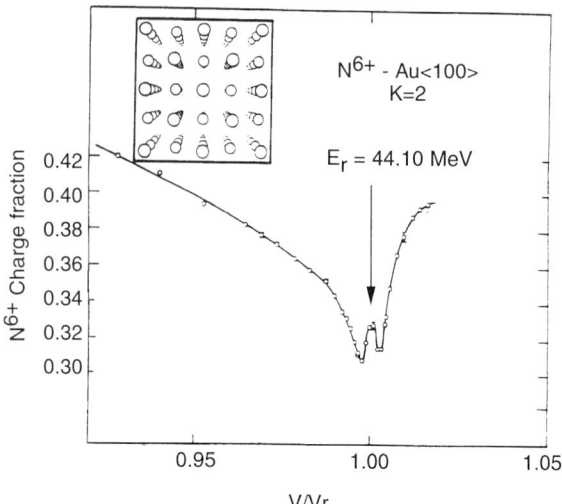

FIG. 12. Surviving N^{6+} fraction channeled through $\langle 100 \rangle$ Au as a function of the ion velocity divided by the velocity at which an RCE resonance should appear if the ion were in vacuum (indicated by an arrow).

Krause *et al.* (1993) demonstrated that under certain symmetry conditions in a $\langle 111 \rangle$ axis in diamond-like lattices, interferences could occur. As shown in Fig. 14, a $\langle 111 \rangle$ string in the fcc structure has atoms indicated by "1" (solid dots). The diamond structure has additional atoms indicated by "2" (open circles) that are displaced from atoms "1" by the distance $d/4$.

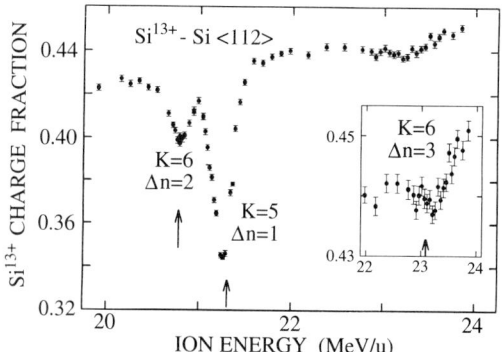

FIG. 13. Surviving Si^{13+} fraction channeled along a $\langle 112 \rangle$ direction in Si showing that $\Delta n = 1, 2,$ and 3 transitions can be induced by RCE when the ionization cross section for the excited ions is relatively small (adapted from Forster *et al.*, 1996, reprinted with permission).

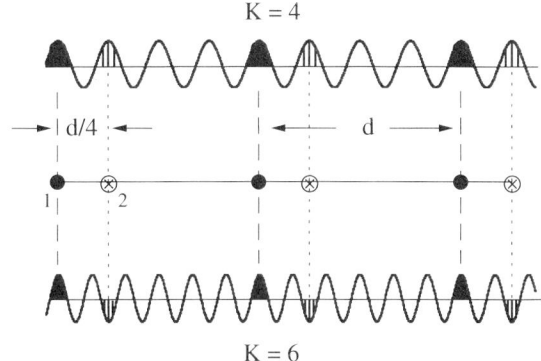

FIG. 14. An atomic string along a ⟨111⟩ direction of a diamond lattice that consists of substrings labeled 1 and 2. The sinusoidal integrand of the Fourier amplitude for the electric field shows positive and negative reinforcements in the $K = 4$ and 6 harmonics, respectively. The reinforcements are due to interferences in the $n = 1$ to 2 electric dipole transition amplitude.

Because of this displacement, the frequencies generated in the harmonic sequence $K = 4, 8, 12$ interfere constructively, whereas those in the sequence $2, 6, 10, \ldots$ interfere destructively. This effect is dramatically demonstrated in Fig. 15, where for the N^{6+} resonances in Si ⟨111⟩, the $K = 6$ resonance has completely disappeared. Significant cancellation in the $K = 6$ harmonic excitation does not occur in an Si ⟨112⟩ channel (Fig. 13), because atoms "1" and "2" do not lie in a straight line in a ⟨112⟩ direction.

The aforementioned studies of RCE identified the coherently excited ions indirectly by observing the emergent H-like ion charge fraction only. Higher-Z ions have smaller ionization cross sections, so that it is possible for a fraction of ions in higher n states to remain un-ionized when passing through a thin crystal. In this case, radiative relaxation of the excited state produced can be compared with collisional ionization. The first successful attempt involved H-like Ne^{9+} ions channeled along a ⟨111⟩ axis in a 0.2-μm Au crystal (Fujimoto et al., 1988). They observed an enhancement of Ne^{9+} $K\alpha$ X rays in the vicinity of the $K = 6$ resonance at the projectile speed where the transmitted H-like ion fraction showed a decrease.

A more detailed X-ray study of RCE involved Mg^{11+} excited via the seventh harmonic in a 0.15 μm thick Au crystal aligned along a ⟨111⟩ direction (Datz et al., 1991). Exit charge-state distributions and X-ray yields were measured as a function of the Mg^{11+} beam energy (145–170 MeV). These results are shown in Figs. 16 and 17, respectively. The seventh harmonic of the $n = 1$ to 2 resonance ($\Delta E_{ij} = 1472$ eV) should

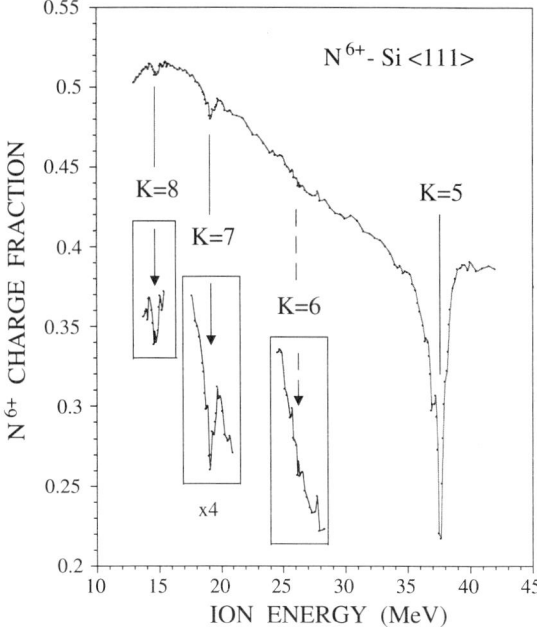

FIG. 15. Surviving N^{6+} fraction for N^{6+} ions channeled along an Si $\langle 111 \rangle$ axis of a 0.16-μm-thick crystal showing destructive interference for $K = 6$. The $K = 6$ resonance would normally be about one-half the size of the large $K = 5$ resonance if destructive interference in the $n = 1$ to 2 electric dipole transition amplitude did not occur. The ordinate scale of each inset is expanded by a factor of four.

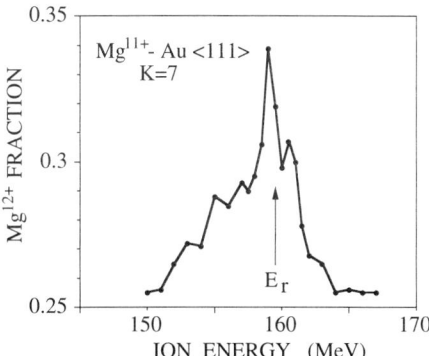

FIG. 16. Charge fraction of Mg^{12+} emerging from a $\langle 111 \rangle$ channel of a 0.2-μm-thick Au crystal as a function of entering Mg^{11+} ion energy. The arrow indicates where the RCE resonance should appear if the ion were in vacuum.

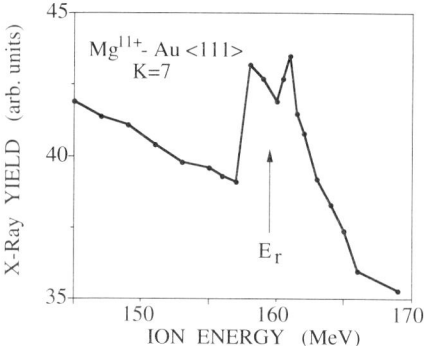

FIG. 17. Yield of Mg^{11+} Ly α X rays from Mg^{11+} ions channeled in Au $\langle 111 \rangle$ (0.2-μm-thick) as a function of energy. The arrow indicates where the RCE resonance should appear if the ion were in vacuum.

appear at 159.3 MeV if the transition were to occur in vacuum. The crystal field effects shift the $2p_x$ and $2p_y$ peaks to lower energies and the $2p_z$ state to a higher energy. Thus, the upper of the two peaks corresponds to excitation into a $2p_z$ state, and the lower to a $2p_x$ or $2p_y$ state. In the X-ray yield, the two peaks are of equal height, yet an analysis of the relative field strengths indicates that the $2p_x$ state should be stronger. A perusual of Figs. 16 and 17 shows that the excess $2p_x$ has been preferentially ionized, i.e., a comparison of the two curves shows that the probability of escape from the crystal without ionization (X-ray curve) is greater for ions in the $2p_x$ state than those in the $2p_{x,y}$ state.

One implication of the foregoing result is that the orbital alignment is not destroyed by electron collisions on a time scale commensurate with that required for ionization. An alternative explanation is that the $2p_{x,y}$ state is formed in regions of higher electron density. To further investigate this phenomena and to simplify the interpretation, experiments were carried out using RCE in planar channeling (Datz et al., 1995). Here, instead of varying the ion velocity, the distance between exciting centers can be varied by changing the tilt angle of the plane between axes. For example the condition for a resonance in a {100} plane between $\langle 100 \rangle$ and $\langle 110 \rangle$ is given by

$$E \text{ (MeV/nucleon)} = 3.03 \frac{2}{\gamma + 1} \left[\frac{\Delta E_{ij} \text{ (keV)} \times d \text{ (Å)}}{l \cos \theta + k \sin \theta} \right]^2 \quad (17)$$

where the planar indices l and k must either be both odd or even integers and θ is the tilt angle measured from a $\langle 100 \rangle$ axis (Datz et al., 1980; Moak et al., 1982).

Using the TASCC Tandem Cyclotron Facility at Chalk River, a Mg^{11+} beam at 25 MeV/u was passed through an Ni crystal (0.40 μm thick) aligned so that it could be rotated about a ⟨100⟩ axis in a {100} plane. A Q3D magnetic analyzer recorded the exit charge-state distribution, and two Si(Li) X-ray detectors aimed at 90° and 45° with respect to the beam recorded the emission of Mg^{11+} Ly α ($n = 2 \rightarrow 1$) X-rays.

As before, crystal field effects are expected to remove the degeneracy and to shift the states from vacuum levels. A schematic drawing of the wavefunctions for the $n = 2$ substates, as a function of distance from the channel midplane, is shown in Fig. 18 (O. H. Crawford, unpublished calculations, private communication, 1995; Datz et al., 1995; Garcia de Abajo and Echenique, 1996). A plot of the calculated energy shifts as a function of distance from the planar channel center is shown in Fig. 19, and a plot of the transition strengths for each of these states is shown in Fig. 20 (O. H. Crawford, unpublished calculations, private communication, 1995).

A resonance at a given angle can be associated with a given transition energy and a given position in the channel. For the $2s2p_x(w)$, an intense

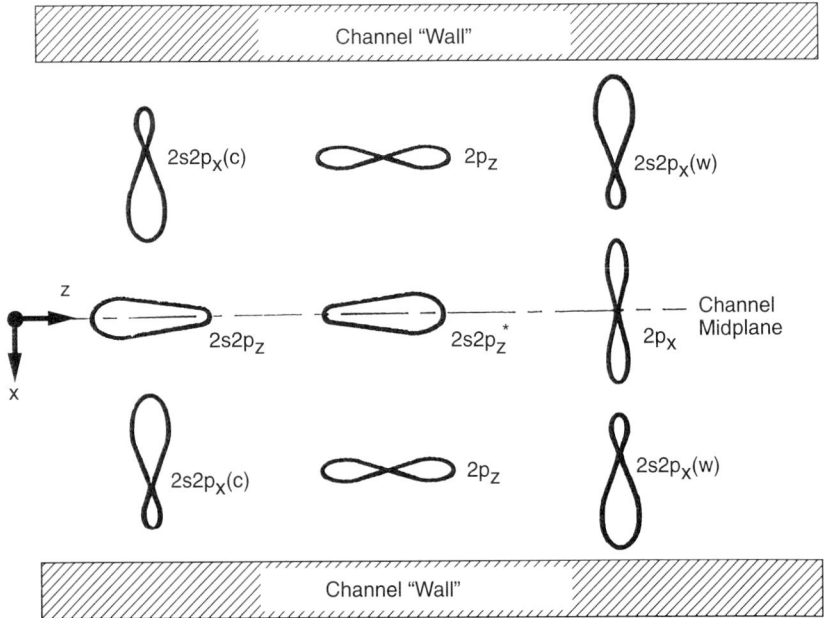

FIG. 18. Wavefunction shapes of the substates $n = 2$ as a function of distance from the channel midplane. The right-handed Cartesian directions are indicated (the y direction is pointing into the drawing), and the velocity of the projectile ion is parallel to the z direction.

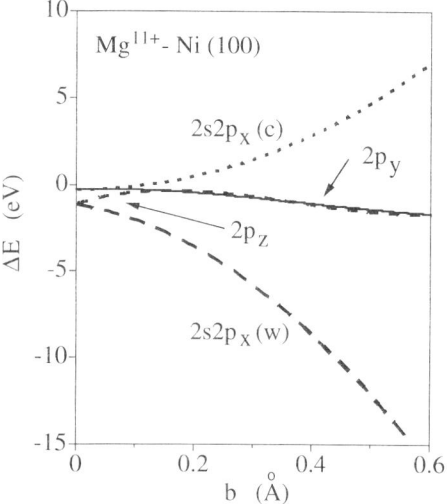

FIG. 19. Energy shifts of the various $n = 2$ substates from vacuum level, as a function of distance from the planar channel midplane. The calculation here is for 25-MeV/u Mg^{11+} moving 10° from a $\langle 100 \rangle$ axis in the (100) planar channel of Ni. Each substate is indicated by the off-midplane character shown in Fig. 18.

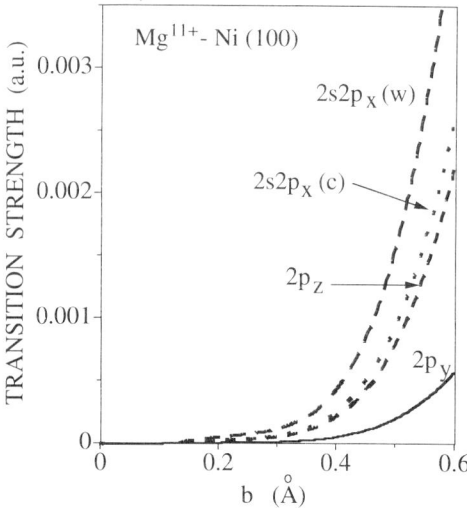

FIG. 20. Transition strengths of the substates calculated here for 25-MeV/u Mg^{11+} moving 10° from a $\langle 100 \rangle$ axis in the (100) planar channel of Ni. Each substate is indicated by the off-midplane character shown in Fig. 18.

but broad feature extending to -20 eV below the vacuum level and the $2s2p_x(c)$ state is expected to appear as a shoulder extending up to $\sim 7-8$ eV on the high side of the vacuum level. The w and c designations for "wall" or "center" indicate the direction of the large lobe for electron probability shown in Fig. 18. All of these expectations are borne out in the measured charge-state fraction shown in Fig. 21. The tilt angle in the (100) plane is shown at the top of the figure, and the corresponding transition energy obtained from Eq. (17) is shown at the bottom.

Effects can be observed in ions that have been excited and survive to exit the crystal and then relax radiatively by X-ray emission. The results of the X-ray measurements on Mg^{11+} are shown in Fig. 21 (broken lines) and compared with the ionization measurements (solid line). The first observation is that the X rays attributable to states excited to the $2s2p_x(w)$ are strongly suppressed when compared with emission from ions excited to the $2p_z$ state. The relative probabilities of excitation to $2s2p_x(w)$ and $2p_z$ are almost independent of position (see the transition amplitude in Fig. 20). Hence, the difference in ionization probability is attributable to the shape of the wavefunction. This difference is not an unexpected result because the bulk of the square of the wavefunction for the $2s2p_x(w)$ state lies perpendicular to the direction of the bombarding electron flux and extends

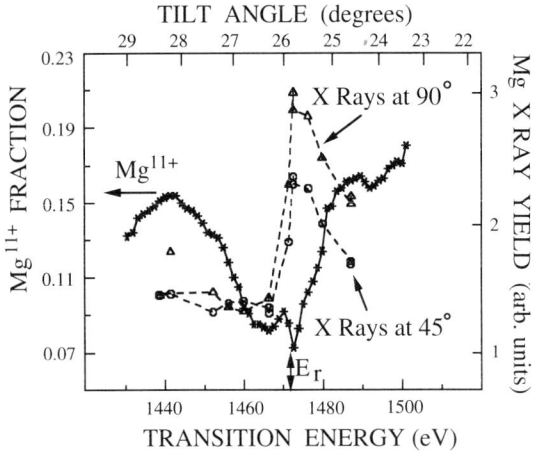

FIG. 21. Surviving charge fraction of Mg^{11+} (25 MeV/u) in the (100) planar channel of Ni as a function of tilt angle from $\langle 100 \rangle$ (top scale), and X-ray yields in detectors placed at angles of 45° and 90° to the beam as a function of transition energy (tilt angle) as compared with charge-state fraction. The angle can be related to transition energy through Eq. (17). The region from 1440 to 1470 eV can be attributed to $2s2p_x(w)$ state, the peaks at 1472 eV are due to the $2p_x$ state, and the shoulder extending up to ~ 1485 eV is due to the $2s2p_x(c)$ state.

into regions of higher electron density near the planes of atomic cores. Although the shapes of the wavefunctions for these two states [i.e., $2p_z$ and $2s2p_x(w)$] vary with position, their identities are reasonably clear. A $2p_z$ mixed with $2s$ on axis becomes almost pure $2p_z$ far off axis. The $2s2p_x(w)$ state is a pure $2p_x$ on axis but mixes with $2s$ as it moves away from the axis. The case for the $2s2p_x(c)$ state is not clear. The wavefunction for this state changes character drastically as it moves through its oscillation in the channel. Its on-axis manifestation is a $2s$ mixed with $2p_z$. In its extreme form far off axis, it is a $2s$ mixed with a $2p_x$. In between, it has s, p_z, and p_x character.

When the excited ions emerge from the crystal into vacuum, they must assume discrete states, $2p_x$, $2p_z$, or $2s$. The fraction appearing in a given state will depend on the vertical displacement in the planar channel at the exit point. The radiative lifetime for the $2s \rightarrow 1s$ transition in Mg^{11+} is 3×10^{-7} sec, so that very little $2s$ radiation will occur within the field of view of the Si(Li) detectors. The radiation pattern of the $2p_x$ state is orthogonal to the x direction. It should be isotropic in the y–z plane, and no differences (after correction for relativistic headlight effects) in intensity should be observed between the 90° and 45° detectors. The radiation pattern for the $2p_z$ state is orthogonal to the z direction, isotropic in the x–y plane, and so there should be an increased intensity on the 90° detector compared with the 45° detector. Such an effect is visible in Fig. 21. Here again, the $2s2p_x(c)$ state is more complex. Although the state lies perpendicular to the electron flux, it always extends into regions of lower electron density. Its state depends on vertical displacement at the exit point. From the X-ray yield, it is evident that there is a chance for escape from the crystal without ionization and that it shows evidence for alignment.

From the foregoing work, there is strong evidence that once a given ionic state is excited, a large fraction of the ions remain in that state and are either ionized or radiate from the state. The result is counter to the expectation that electron–ion collisions in the channel would have large enough cross sections to destroy the alignment before either of these two events could take place.

Resonant coherently excited ions have been used to study convoy electron (CE) production in crystal channels (Kimura et al., 1991). The production of CEs ejected in ion–solid collisions with electron velocity very close to the ion velocity has been extensively studied. It is now widely accepted that either electron loss to the continuum (ELC) (Breinig et al., 1982) or electron capture to the continuum (ECC) (Dettmann et al., 1974) contribute to the production of CEs, depending on the experimental conditions (Yamazaki, 1990; Kemmler, 1990). Under channeling conditions, ELC is dominant because the electron capture probability, and thus the ECC probability, is drastically reduced. In the ELC process, the important role of the excited states has been suggested. Kimura et al. demonstrated the role of the excited states in CE production by changing the population of the excited states using RCE. They channeled a 20.4-MeV beam of C^{5+} ions along a $\langle 100 \rangle$ axial direction of a 0.16-μm Au crystal; this energy corresponds to the $K = 2$ harmonic for RCE. The crystal was placed at the entrance focus of an electrostatic spherical sector spectrometer ($\Delta E/E = 0.01$). The electrons emitted into the forward cone of half-angle $\theta = 4.05°$ were energy analyzed and detected by a microchannel plate. The best channeled ions emerging from the crystal were resolved by charge state and detected. Coincidences between the CEs and the emergent C^{5+} and C^{6+} ions were measured.

The RCE effect was clearly shown for each exiting charge state when the crystal was aligned to a $\langle 100 \rangle$ direction, and the effect disappeared when the crystal was oriented in a random direction. The velocity distributions of CEs produced in channeling were almost independent of either the ion energy or the exit charge state. The peak in the C^{6+} distribution at the RCE resonant energy showed an enhancement of the electron-loss cross section of C^{5+} ions due to RCE. Also measured was structure on the RCE peak observed by varying the projectile energy as a result of Stark-shifted energy levels of the moving ions caused by the wake potential and the crystal field. Overall, the modeled results suggest that electrons lost from excited states are strongly forward-peaked and, therefore, have a large probability of becoming CEs as compared with those lost from the ground state.

Acknowledgments

This research is supported by the U.S. Department of Energy, Office of Basic Energy Sciences, Division of Chemical Sciences, under Contract Number DE-AC05-96OR22464 with Lockheed Martin Energy Research Corp.

References

Anderson, J. U., Chevallier, J., Ball, G. C., Davies, W. G., Forster, J. S., Geiger, J. S., Davies, J. A., and Geissel, H. (1993). *Phys. Rev. Lett.* **70**, 750.

Andriamonje, S., Chevallier, M., Cohen, C., Dural, J., Gaillard, M. J., Genre, R., Hage-Ali, M., Kirsh, R., L'Hoir, A., Mazuy, B., Mory, J., Moulin, J., Poizat, J. C., Remillieux, J., Schmaus, D., and Toulemonde, M. (1987). *Phys. Rev. Lett.* **59**, 2271.

Andriamonje, S., Anne, R., de Castro Faria, N. V., Chevallier, M., Cohen, C., Dural, J., Gaillard, M. J., Genre, R., Hage-Ali, M., Kirsch, R., L'Hoir, A., Farizon-Mazuy, B.,

Mory, J., Moulin, J., Poizat, J. C., Quéré, Y., Remillieux, J., Schmaus, D., and Toulemonde, M. (1989). *Phys. Rev. Lett.* **63**, 1930.

Andriamonje, S., Chevallier, M., Cohen, C., Cue, N., Dauvergne, D., Dural, J., Genre, R., Girad, Y., Kirsh, R., L'Hoir, A., Poizat, J. C., Quéré, Y., Remillieux, J., Schmaus, D., and Toulemonde, M. (1992). *Phys. Lett.* **164**, 184.

Anholt, R., Andriamonje, S. A., Morenzoni, E., Stoller, C. Molitoris, J. D., Meyerhof, W. E., Bowman, H., Xu, J.-S., Xu, Z.-Z., Rasmussen, J. O., and Hoffman, D. H. H. (1984). *Phys. Rev. Lett.* **53**, 234.

Anholt, R., Meyerhof, W. E., Xu, X.-Y., Gould, H., Feinberg, B., McDonald, R. J., Wegner, H. E., and Thieberger, P. (1987). *Phys. Rev. A* **36**, 1586.

Appleton, B. R., Datz, S., Moak, C. D., and Robinson, M. T. (1971). *Phys. Rev. B: Solid State* [3] **4**, 1452.

Appleton, B. R., Moak, C. D., Noggle, T. S., and Barrett, J. H. (1972). *Phys. Rev. Lett.* **28**, 1307.

Appleton, B. R., Richie, R. H., Biggerstaff, J. A., Noggle, T. S., Datz, S., Moak, C. D., Verbeek, H., and Neelavathi, V. N. (1979). *Phys. Rev. B: Condens. Matter* [3] **19**, 4347.

Barnett, C. F. (1989). *In* "Physics Vade Mecum" (H. L., Anderson, Ed.), P. 84. Am. Inst. Phys. New York.

Barrett, J. H. (1971). *Phys. Rev. B: Solid State* [3] **3**, 1527.

Barrett, J. H. (1990). *Nucl. Instrum. Methods Phys. Res. Sect. B* **44**, 367.

Belkacem, A., Kanter, E. P., Rehm, K. E., Bernstein, E. M., Clark, M. W., Ferguson, S. M., Tanis, J. A., Berkner, K. H., and Schneider, D. (1990). *Phys. Rev. Lett.* **64**, 380.

Bethe, H. A., and Salpeter, E. E. (1957). "Quantum Mechanics of One- and Two-Electron Atoms," p. 320. Academic Press, New York.

Breinig, M., Elston, S. B., Huldt, S., Liljeby, L., Vane, C. R., Berry, S. D., Glass, G. A., Schauer, M., Sellin, I. A., Alton, G. D., Datz, S., Overbury, S., Laubert, R., and Suter, M. (1982). *Phys. Rev. A* **25**, 3015.

Claytor, N., Feinberg, B., Gould, H., Bemis, C. E., Gomez del Campo, J., Ludemann, C. A., and Vane, C. R. (1988). *Phys. Rev. Lett.* **61**, 2081.

Crawford, O. H. (1980). *Nucl. Instrum. Methods* **170**, 21.

Crawford, O. H., and Ritchie, R. H. (1979). *Phys. Rev. A* **20**, 1848.

Datz, S., and Moak, C. D. (1983). *In* "Heavy Ion Science" (D. A. Bromley, ed.), p. 369. Plenum, New York.

Datz, S., Moak, C. D., Noggle, T. S., Appleton, B. R., and Lutz, H. O. (1969). *Phys. Rev.* **179**, 315.

Datz, S., Martin, F. W., Moak, C. D., Appleton, B. R., and Bridwell, L. B. (1972). *Radiat. Eff.* **12**, 163.

Datz, S., Moak, C. D., Crawford, O. H., Krause, H. F., Dittner, P. F., Gomez del Campo, J., Biggerstaff, J. A., Miller, P. D., Hvelplund, P., and Knudsen, H. (1978). *Phys. Rev. Lett.* **40**, 843.

Datz, S., Moak, C. D., Crawford, O. H., Krause, H. F., Miller, P. D., Dittner, P. F., Gomez del Campo, J., Biggerstaff, J. A., Knudsen, H., and Hvelplund, P. (1980). *Nucl. Instrum. Methods Phys. Res.* **170**, 15.

Datz, S., Vane, C. R., Dittner, P. F., Giese, J. P., Gomez del Campo, J., Jones, N. L., Krause, H. F., Miller, P. D., Schulz, M., Schöne, H., and Rosseel, T. M. (1989). *Phys. Rev. Lett.* **63**, 742.

Datz, S., Dittner, P. F., Gomez del Campo, J., Kimura, K., Krause, H. F., Rosseel, T. M., Vane, C. R., Iwata, Y., Komaki, K., Yamazaki, Y., Fujimoto, F., and Honda, Y. (1991). *Radiat. Eff. Defect Solids* **117**, 73.

Datz, S., Vane, C. R., Krause, H. F., and Dittner, P. F. (1992). *In* "Recombination of Atomic Ions" (W. G. Graham, ed.), p. 311. Plenum, New York.

Datz, S., Dittner, P. F., Krause, H. F., Vane, C. R., Crawford, O. H., Forster, J. S., Ball, G. S., Davies, W. G., and Geiger, J. S. (1995). *Nucl. Instrum. Methods Phys. Res. Sect. B* **100**, 272.

Dettmann, K., Harrison, K. G., and Lucas, M. W. (1974). *J. Phys. B: At. Mol. Phys.* **7**, 269.

Dittner, P. F., Vane, C. R., Krause, H. F., Gomez del Campo, J., Jones, N. L., Zeijlmans van Emmichoven, P. A., Bechthold, U., and Datz, S. (1992). *Phys. Rev. A* **45**, 2935.

Donets, E. D. (1983). *Phys. Scr.* **T3**, 11.

Dygo, A., and Turos, A. (1987). *Nucl. Instrum. Methods Phys. Res. Sect. B* **18**, 115.

Feagin, J. M., and Wanser, K. (1991). *Phys. Rev. A* **44**, 4228.

Feagin, J. M. Briggs, J. S., and Reeves, T. M. (1984). *J. Phys. B: At. Mol. Phys.* **17**, 1057.

Fontes, C. J., Sampson, D. H., and Zhang, H. L. (1995). *Phys. Rev. A* **51**, R12.

Forster, J. S., Alexander, T. K., Anderson, J. U., Ball, G. C., Davies, J. A., Davies, W. G., Geiger, J. S., Geissel, H., and Kanter, E. P. (1993). *Radiat. Eff. Defect Solids* **126**, 299.

Forster, J. S., Ball, G. C., Davies, W. G., Geiger, J. S., Anderson, J. U., Chevallier, J., Davies, J. A., and Geissel, H. (1994). *Nucl. Instrum. Methods Phys. Res. Sect. B* **90**, 172.

Forster, J. S., Ball, G. C., Davies, W. G., Geiger, J. S., Anderson, J. U., Davies, J. A., Geisel, H., and Nickel, F. (1996). *Nucl. Instrum. Methods Phys. Res. Sect. B* **107**, 27.

Fujimoto, F., Komaki, K., Ootuka, A., Vilalta, E., Iwata, Y., Hirao, Y., Hasegawa, T., Sekiguchi, M., Mizobuchi, A., Hattori, T., and Kimura, K. (1988). *Nucl. Instrum. Methods Phys. Res. Sect. B* **33**, 354.

Garcia de Abajo, F. J., and Echenique, P. (1996). *Phys. Rev. Lett.* **76**, 1856.

Gemmell, D. S. (1974). *Rev. Mod. Phys.* **46**, 129.

Gibson, W. M., Rasmussen, J. B., Ambrosius-Olesen, P., and Andreen, C. J. (1968). *Can. J. Phys.* **46**, 551.

Gould, H., Greiner, D., Lindstrom, P., Symons, T. J. M., and Crawford, H. (1984). *Phys. Rev. Lett.* **52**, 180.

Kemmler, J. (1990). *Nucl. Instrum. Methods Phys. Res. Sect. B* **48**, 612.

Kimura, K., Gibbons, J. P., Elston, S. B., Biedermann, C., DeSerio, R., Keller, N., Levin, J. C., Breinig, M., Burgdörfer, J., and Sellin, I. A. (1991). *Phys. Rev. Lett.* **66**, 25.

Krause, H. F., Datz, S., Dittner, P. F., Gomez del Campo, J., Miller, P. D., Moak, C. D., Nešković, N., and Pepmiller, P. L. (1986a). *Nucl. Instrum. Methods Phys. Res. Sect. B* **13**, 51.

Krause, H. F., Datz, S., Dittner, P. F., Gomez del Campo, J., Miller, P. D., Moak, C. D., Nešković, N., and Pepmiller, P. L. (1986b). *Phys. Rev. B: Condens. Matter* [3] **33**, 6036.

Krause, H. F., Datz, S., Dittner, P. F., Jones, N. L., and Vane, C. R. (1993). *Phys. Rev. Lett.* **71**, 348.

Krause, H. F., Barrett, J. H., Datz, S., Dittner, P. F., Jones, N. L., Gomez del Campo, J., and Vane, C. R. (1994). *Phys. Rev. A* **49**, 283.

L'Hoir, A., Andriamonje, S., Anne, R., De Castro Faria, N. V., Chevallier, M., Cohen, C., Dural, J., Gaillard, M. J., Genre, R., Hage-Ali, M., Kirsch, R., Farizon-Mazuy, B., Mory, J., Moulin, J., Poizat, J. C., Quéré, Y., Remillieux, J., Schmaus, D., and Toulemonde, M. (1990). *Nucl. Instrum. Methods Phys. Res. Sect. B* **48**, 145.

Lindhard, J., Scharff, M., and Schiott, H. E. (1963). *Mat.-Fys. Medd.—K. Dan. Vidensk. Selsk.* **33**, 14.

Lotz, W. (1967). *Z. Phys.* **206**, 205.

Lutz, H. O., Datz, S., Moak, C. D., and Noggle, T. S. (1966). *Phys. Rev. Lett.* **17**, 285.

Marrs, R. E. Elliott, S. R., and Knapp, D. A. (1994). *Phys. Rev. Lett.* **72**, 4082.

Meyerhof, W. E., Anholt, R., Eichler, J., Gould, H., Munger, C., Alonso, J., Thieberger, P., and Wegner, H. E. (1985). *Phys. Rev. A* **32**, 3291.
Miller, P. D., Krause, H. F., Biggerstaff, J. A., Crawford, O. H., Datz, S., Dittner, P. F., Gomez del Campo, J., Moak, C. D., Nešković, N., Pepmiller, P. L., and Brown, M. D. (1986). *Nucl. Instrum. Methods Phys. Res. Sect. B* **13**, 56.
Moak, C. D., Datz, S., Crawford, O. H., Krause, H. F., Dittner, P. F., Gomez del Campo, J., Biggerstaff, J. A., Miller, P. D., Hvelplund, P., and Knudsen, H. (1979). *Phys. Rev. A* **19**, 977.
Moak, C. D., Biggerstaff, J. A., Crawford, O. H., Dittner, P. F., Datz, S., Gomez, del Campo, J., Hvelplund, P., Knudsen, H., Krause, H. F., Miller, P. D., and Overbury, S. H. (1982). *Nucl. Instrum. Methods* **194**, 327.
Moores, D. L., and Reed, K. J. (1995). *Phys. Rev. A.* **51**, R9.
Müller, A., Schennach, S., Spies, W., Uwira, O., Haselbauer, J., Wagner, M., Frank, A., Becker, R., Jennewein, E., Kleinod, M., Pröbstel, U., Angert, N., Klabunde, J., Mokler, P. H., Spädtke, P., and Wolf, B. (1991). In "Cooler Rings and Their Applications" (T. Katayama and A. Noda, eds.), p. 248. World Scientific, Singapore.
Nešković, N. (1986a). *Phys. Rev. B: Condens. Matter* [3] **33**, 6030.
Nešković, N. (1986b). *Phys. Rev. B: Condens. Matter* [3] **33**, 7488.
Nešković, N., and Perović, B. (1987). *Phys. Rev. Lett.* **59**, 308.
Nešković, N., Kapetanović, G., Petrović, S., and Perović, B. (1993). *Phys. Lett. A* **179**, 343.
Nešković, N., Petrović, S., Kapetanović, G., Perović, B., and Lennard, W. N. (1994). *Nucl. Instrum. Methods Phys. Res. Sect. B* **93**, 249.
Pindzola, M. S., Moores, D. L., and Griffin, D. C. (1989). *Phys. Rev. A* **40**, 4941.
Pitarke, J. M., Ritchie, R. H., and Echenique, P. M. (1991). *Phys. Rev. B: Condens Matter* [3] **43**, 62.
Reusch, S., Mokler, P. H., Schuch, R., Justiniano, E., Schulz, M., Müller, A., and Stachura, Z. (1987). *Nucl. Instrum Methods Phys. Res. Sect. B* **23**, 137.
Robinson, M. T. (1969). *Phys. Rev.* **179**, 327.
Robinson, M. T. (1971). *Phys. Rev. B: Solid State* [3] **4**, 1461.
Robinson, M. T., and Oen, O. S. (1963a). *Appl. Phys. Lett.* **2**, 30.
Robinson, M. T., and Oen, O. S. (1963b). *Phys. Rev.* **132**, 2385.
Schnopper, H. W., Betz, H. D., Delvaille, J. P., Kalata, K., Sohval, A. R., Jones, K. W., and Wegner, H. E. (1972). *Phys. Rev. Lett.* **29**, 898.
Schnopper, H. W., Delvaille, J. P., Kalata, K., Sohval, A. R., Abdulwahab, M., Jones, K. W., and Wegner, H. E. (1974). *Phys. Rev. Lett.* **47**, 61.
Scofield, J. H. (1978). *Phys. Rev. A* **18**, 963.
Smulders, P. J. M., Dygo, A., and Boerma, D. O. (1992). *Nucl. Instrum. Methods. Phys. Res. Sect. B* **67**, 185, and references contained therein.
Stark, J. (1912). *Phys. Z.* **13**, 973.
Stark, J., and Wendt, G. (1912). *Ann. Phys. (Leipzig)* **38**, 921.
Stobbe, M. (1930). *Ann. Phys. (Leipzig)* **7**, 661.
Stöhlker, T., Kozuharov, C., Livingston, A. E., Mokler, P. H., Stachura, Z., and Warczak, A. (1992). *Z. Phys. D: At. Mol. Clusters* **23**, 121.
Tanis, J. A., Bernstein, E. M., Clark, M. W., Graham, W. G., McFarland, R. H., Morgan, T. J., Mowat, J. R., Mueller, D. W., Müller, A., Stockli, M. P., Berkner, K. H., Gohil, P., McDonald, R. J., Schlachter, A. S., and Sterns, J. W. (1986). *Phys. Rev. A* **34**, 2543.
Tawara, H., Richard, P., and Kawatsura, K. (1982). *Phys. Rev. A* **26**, 154.
Tribedi, L. C., Nanal, V., Press, M. R., Kurup, M. B., Prasad, K. G., and Tandon, P. N. (1994). *Phys. Rev. A* **49**, 374.

Vane, C. R., Dittner, P. F., Krause, H. F., Gomez del Campo, J., Jones, N. L., Zeijlmans van Emmichoven, P. A., Bechthold, U., and Datz, S. (1992). *Nucl. Instrum. Methods Phys. Res. Sect. B* **67**, 256.

Vane, C. R., Datz, S., Dittner, P. F., Giese, J., Jones, N. L., Krause, H. F., Rosseel, T. M., and Peterson, R. S. (1994). *Phys. Rev. A* **49**, 1847.

Wanser, K., and Feagin, J. M. (1991). *Nucl. Instrum. Methods Phys. Res. Sect. B* **56**, 145.

Yamazaki, Y. (1990). *Nucl. Instrum. Methods Phys. Res. Sect. B* **48**, 97.

Younger, S. M. (1980). *Phys. Rev. A* **22**, 111, 1425.

Younger, S. M. (1981). *Phys. Rev. A* **24**, 1278.

Zhang, H. L., and Sampson, D. H. (1990). *Phys. Rev. A* **42**, 5378.

Ziegler, J. F., Biersack, J. P., and Littmark, U. (1985). "The Stopping and Range of Ions in Solids," Vol. 1, p. 41. Pergamon, New York.

EVAPORATIVE COOLING OF TRAPPED ATOMS

WOLFGANG KETTERLE AND N. J. VAN DRUTEN

Department of Physics and Research Laboratory of Electronics
Massachusetts Institute of Technology, Cambridge, Massachusetts

I. Introduction 181
II. Theoretical Models for Evaporative Cooling 184
    A. General Scaling Laws 184
    B. The Speed of Evaporation and Loss Processes 186
    C. Models 193
    D. The Dimension of Evaporation 197
    E. The Number of Collisions for Thermalization 199
    F. The Number of Collisions to Bose–Einstein Condensation 200
    G. Desirable Extensions of the Models 200
III. The Role of Collisions for Real Atoms 201
    A. Elastic and Inelastic Collisions 201
    B. Atoms for Evaporative Cooling 203
    C. Enhanced Relaxation in the Condensate 206
    D. Special Topics 207
IV. Experimental Techniques 209
    A. Precooling 209
    B. Conservative Traps 212
    C. Laser Cooling inside Conservative Traps 216
    D. Adiabatic Compression 217
    E. Evaporation Techniques 218
V. Summary of Evaporative Cooling Experiments 227
    A. Hydrogen 227
    B. Alkali Atoms 228
    C. Comparison 229
VI. Outlook 229
    References 231

I. Introduction

Evaporation is a well-known phenomenon in everyday life. It describes the conversion of liquid to the gaseous state. It either happens as a slow process, e.g., when a wet surface dries, or can be forced by means of heat or ventilation. In a more abstract sense, evaporation describes the process of energetic particles leaving a system with a finite binding energy. This happens naturally since there are always high energy particles in the tail of the Maxwell–Boltzmann distribution. This process results in cooling: Since the evaporating particles carry away more than their share of thermal energy, the temperature of the system decreases. Due to the lower temperature, the evaporation process slows down, unless evaporation is forced by modifying the system in such a way that less energetic particles can escape from the system.

Evaporative cooling cools a cup of coffee, is used to cool apples by overtree sprinkling (Evans et al., 1995), and is employed in technical water coolers (Berman, 1961). Stars evaporate from globular clusters (Spitzer, 1987). Evaporating neutrons carry away mega–electron volts of excitation energies of compound nuclei (Nolan and Twin, 1988). On the other end of the scale, at energies of pico–electron volts, evaporative cooling has led to the coldest temperatures ever observed in the universe: submicro-Kelvin temperatures generated in atom traps.

The last example is the subject of this chapter: evaporative cooling of trapped neutral atoms. The history of this technique is only a little bit shorter than the history of neutral atom trapping. In 1985, the year of the first realization of a trap for neutral atoms (Migdall et al., 1985), Harald Hess (1985, 1986) suggested evaporative cooling as an efficient way to cool trapped atoms with the goal of achieving Bose–Einstein condensation. The original suggestion focused on trapped atomic hydrogen, but in 1994 this technique was extended to alkali atoms by combining evaporative cooling with laser cooling (Davis et al., 1994; Petrich et al., 1994b). It was the key technique to achieve Bose–Einstein condensation (BEC), a long-sought goal in atomic physics (Anderson et al., 1995; Bradley et al., 1995; Davis et al., 1995b).

The recent observations of Bose–Einstein condensation showed dramatically the potential of evaporative cooling. Through evaporative cooling, phase-space density could be increased by six orders of magnitude in these experiments. Evaporative cooling was used to reach temperature and densities that were unprecedented for trapped atoms, and greatly exceeded what had been reached before by laser cooling. It should be noted that laser cooling has recently broken the recoil limit in three dimensions (3D) (Lawall et al., 1995, Lee et al., 1996), and reached extremely cold temperatures of 3 nK in 1D (Reichel et al., 1995). However, none of these optical sub-recoil techniques have been realized so far at high densities ($>10^{12}$ cm^{-3}). The current density limitations are caused by absorption of light, radiation trapping, and excited state collisions. They might be avoided in dark optical traps which are being pursued theoretically (Dum et al., 1994; Pellizzari and Ritsch, 1995) and experimentally (Hemmerich et al., 1995; Davidson et al., 1995).

What makes evaporative cooling so attractive is not only its simplicity, but also that it works over a wide range of temperatures and densities. The major requirement for the application of evaporative cooling is that the thermalization time be short compared with the lifetime of the sample. The lifetime might be technically determined by loss or heating processes of the atom trap, or it could be intrinsically limited by inelastic collisions ("bad collisions"). Eventually, since elastic collisions ("good collisions") are necessary for evaporative cooling, it is the ratio of good to bad collisions that sets the limit to evaporative cooling. This limit depends on the atomic species, as will be discussed in Section III, but for favorable species such as alkali atoms, the fundamental temperature limit is in the pico-Kelvin range. To overcome all technical limitations and reach this fundamental limit of evaporative cooling remains a formidable challenge.

An often-mentioned disadvantage of evaporative cooling is the loss of atoms. However, as discussed in Sections II and V, the efficiency of evaporative cooling is quite high. In recent experiments, phase-space density increases of six orders of magnitude were accomplished by losing only a factor of about 1000 in the number of atoms. Although the latter reduction is substantial, it can be compensated for by relatively straightforward improvements in loading the trap that can result in orders of magnitude higher numbers of atoms. Evaporative cooling is now an established cooling technique, and this will spur major efforts to get larger samples of trapped atoms—evaporative cooling allows transformation of higher initial numbers into higher final densities and lower temperatures.

Experiments on evaporative cooling have gone through rapid recent development. In the last few years, novel traps have been developed for the purpose of evaporative cooling, and novel ways of evaporating atoms were realized. We summarize these advances in Section IV. Equally important was the development of novel techniques to diagnose dense and cold samples of trapped atoms, but the description of these is beyond the scope of this chapter.

The process of evaporative cooling has been the subject of several theoretical studies, which we summarize in Section II. Evaporative cooling is a relatively simple classical process. Therefore, its theoretical foundation is well established. Some future research may address optimization of the cooling process and verification of models. However, we expect evaporative cooling to be mainly a tool to obtain colder and denser samples of trapped atoms. There is considerable hope of discovering new physics at ultralow temperatures, and the observation of Bose−Einstein condensation in dilute atomic gases might be just the first step. However, in this chapter, we will not discuss experiments that become possible through evaporative cooling; rather, we concentrate on the cooling process itself.

Previous review papers on evaporative cooling have focused on spin polarized hydrogen, the only gas that was cooled by this technique prior to 1994. The older reviews (Greytak and Kleppner, 1984; Silvera and Walraven, 1986) describe properties of spin polarized hydrogen in depth, and treat relaxation/recombination processes exhaustively. The latest reviews of the leading groups focus on wall-free confinement, evaporative cooling, and optical diagnostics (Walraven and Hijmans, 1994; Greytak, 1995; Walraven, 1996; Silvera and Reynolds, 1992; Silvera, 1995a,b).

We have attempted to give a complete guide to the relevant topics, and we hope that our work contributes to the successful cross-fertilization between the spin-polarized-hydrogen and laser-cooling communities. Most parts of this chapter are meant as an introduction to further study of the many references given. There are three topics where we feel that we have provided additional details or aspects not yet covered in the literature: a semiquantitative, comprehensive discussion of evaporative cooling both in hydrogen and alkalis (Sections II.A and B), a detailed discussion of rf-induced evaporation (Section IV.E), and a comparison among all experiments done so far on evaporative cooling (Section V).

II. Theoretical Models for Evaporative Cooling

A. GENERAL SCALING LAWS

Several models describing the evaporative cooling process have been published. However, as we discuss here and in Section II.B, most of the dynamics of evaporative cooling is model independent and follows from simple considerations.

First, evaporative cooling happens on an exponential scale: Within a certain time interval (naturally measured in units of collision times or relaxation times), all relevant parameters (number of atoms, temperature, density) change by a certain factor. The characteristic quantities for the evaporation process are therefore logarithmic derivatives such as

$$\alpha = \frac{d(\ln T)}{d(\ln N)} = \frac{\dot{T}/T}{\dot{N}/N} \qquad (1)$$

or, if evaporative cooling is described as a process with finite steps (Davis et al., 1995c),

$$\alpha = \frac{\ln(T'/T)}{\ln(N'/N)} \qquad (2)$$

where $T' = T + \Delta T$ and $N' = N + \Delta N$. (We are mainly following the notation of Walraven, 1996, which is the most comprehensive previous review article on the theory of evaporative cooling). If α is constant during the evaporation process, the temperature drops at $T(t)/T(0) = [N(t)/N(0)]^\alpha$. In a power law potential in d dimensions, $U(r) \propto r^{d/\delta}$, all relevant quantities scale as $[N(t)/N(0)]^x$ during evaporative cooling, where x depends only on δ and α. δ is defined in such a way that the volume scales as T^δ. All other quantities are products of powers of temperature, number, and volume, and their scaling is listed in Table I. Note that $\delta = 3$ for a 3D linear potential, $\frac{3}{2}$ for a 3D harmonic potential (generally anisotropic), and 0 for a box potential. As long as we are interested in scaling laws, it does not matter whether density is peak density or some form of average density. Phase space density D is defined as $n\lambda_{dB}^3$ with the thermal de Broglie wavelength $\lambda_{dB} = \sqrt{2\pi\hbar^2/mkT}$ for an atom with mass m. D is identical to the quantum occupancy number of the lowest state, as long as $D \ll 1$. Bose–Einstein condensation occurs for bosonic atoms when $D > 2.612$ (Huang, 1987).

The key parameter of the whole cooling process is α, which expresses the temperature decrease per particle lost. Technically, evaporation is controlled by limiting the depth of the potential to ηkT. The average energy of the escaping atoms is $(\eta + \kappa)kT$, where κ is a small number usually between 0 and 1, depending both on η and on the dimension of the evaporation (discussed later). For a qualitative discussion, we can ignore κ or approximate it by 1, especially in the case of large η.

For large η, the energy distribution of the trapped atoms is close to a Boltzmann distribution (to be discussed) with an average energy of $(\delta + \frac{3}{2})kT$ (assuming 3D motion), and there is a simple relation between

TABLE I
SCALING LAWS FOR EVAPORATIVE COOLING IN
A d-DIMENSIONAL POTENTIAL $U(r) \propto r^{d/\delta}$

Quantity	Exponent[a], x
Number of atoms, N	1
Temperature, T	α
Volume, V	$\delta\alpha$
Density, n	$1 - \delta\alpha$
Phase-space density, D	$1 - \alpha(\delta + 3/2)$
Elastic collision rate, $n\sigma v$	$1 - \alpha(\delta - 1/2)$

[a] Each quantity scales as N^x, where N is the number of trapped particles.

the average energy of an escaping atom and α:

$$\alpha = \frac{\eta + \kappa}{\delta + 3/2} - 1 \qquad (3)$$

This expression has an obvious meaning: α is a dimensionless quantity, characterizing how much more than the average energy $(\delta + \frac{3}{2})kT$ is removed by an evaporating atoms. These considerations show that in principle there is no upper bound for α or the efficiency of evaporative cooling. This is demonstrated by the following extreme example: With an extremely large η, one just has to wait for the event that one particle has all the energy of the system. Evaporating a single particle then cools the whole system to zero temperature. Unfortunately, this cooling strategy would take an almost infinite amount of time.

Therefore, efficiency of evaporation, and comparison between different trap geometries, can only be made if the trade-off between efficiency and cooling speed is considered. This is done by specifying loss mechanisms that are unavoidable in practice.

B. The Speed of Evaporation and Loss Processes

The considerations in the previous subsection focused on temperature decrease in evaporation as a function of the number of particles left. We now want to extend the discussion and consider the speed of evaporation, i.e., introduce time as a parameter. The situation we have in mind is forced evaporation at a constant η parameter; i.e., the threshold for evaporation is lowered in proportion to the decreasing temperature. Constant η ensures that the energy distribution is only rescaled during the cooling and does not change its shape. This assumption is reasonably well fulfilled in experiments.

We can easily obtain an analytical expression for the rate of evaporation in the following situation: We consider particles at density n_0 in a box potential, and we assume that η is large. The rate of evaporating atoms can then be obtained as follows: In an untruncated Maxwell–Boltzmann distribution, almost every collision involving an atom in the high energy tail removes the atom from the high energy tail. By detailed balance, elastic collisions produce atoms with energy larger than ηkT at a rate that is simply the number of atoms with energy larger than ηkT divided by their collision time. For a large value of η, the rate of evaporation in a truncated Boltzmann distribution is identical to the production rate of atoms with energy larger than ηkT in the untruncated distribution.

The velocity of atoms with energy ηkT is $\sqrt{2\eta kT/m} = \sqrt{\pi\eta}\,\bar{v}/2$ where $\bar{v}$ denotes the average thermal velocity. For large η, the fraction of atoms with energy larger than ηkT approaches $2e^{-\eta}\sqrt{\eta/\pi}$. The rate of evaporating atoms is thus

$$\dot{N} = -Nn_0\sigma\bar{v}\eta e^{-\eta} = \frac{-N}{\tau_{ev}} \quad (4)$$

where we have introduced the elastic collision cross section σ and the time constant for evaporation τ_{ev}. Its ratio to the elastic collision time τ_{el} is expressed by $\lambda = \tau_{ev}/\tau_{el}$. Using $1/\tau_{el} = n_0\sigma\bar{v}\sqrt{2}$, where $\bar{v}\sqrt{2}$ is the average relative velocity between two atoms, we obtain, in the limit of large η,

$$\lambda = \frac{\sqrt{2}\,e^\eta}{\eta} \quad (5)$$

This expression is also valid for any power law potential in the limit of large η (Walraven, 1996), where n_0 is now the peak density. This last point is easily explained: Evaporation is a local phenomenon where each volume element can be regarded as a square well (Luiten *et al.*, 1996). For large η, the behavior is independent of the potential because the smaller threshold for evaporation for volume elements further outside in the cloud is, at least in leading order, compensated for by the lower local density. For arbitrary η, λ can be expressed analytically by generalized gamma functions, which are obtained from a collision integral for the truncated Boltzmann distribution (Luiten, 1993; Walraven, 1996; Luiten *et al.*, 1996).

The quantity $1/\tau_{el}$ is the elastic collision rate in the center of the trap. The cloud-averaged collision rate of one atom is smaller by a factor of $2\sqrt{2}$ in a parabolic trap, 8 in a linear trap. The event rate of collisions is smaller by another factor of 2 because each collision involves two atoms.

1. Runaway Evaporation

For alkali atoms, where the dominant loss mechanism is background gas collisions, an important criterion for sustained evaporation is to maintain or increase the elastic collision rate, $n\sigma v$. From Table I and Eq. (4), it follows that the elastic collision rate varies as

$$\frac{d(n\sigma v)}{dt}\bigg/n\sigma v = \frac{1}{\tau_{el}}\left(\frac{\alpha(\delta - 1/2) - 1}{\lambda} - \frac{1}{R}\right) \quad (6)$$

In the temperature range of interest, σ is the s-wave cross section and is independent of temperature. R is the number of elastic collisions per trapping time (also called the ratio of good to bad collisions): $R = \tau_{loss}/\tau_{el}$, where τ_{loss} is the time constant for trap loss due to background gas collisions. Evaporation at constant or increasing collision rate ("runaway evaporation") requires

$$R \geq R_{\min} = \frac{\lambda}{\alpha(\delta - 1/2) - 1} \qquad (7)$$

Figure 1 shows this minimum number of elastic collisions per trapping time. The figure includes plots based on the simple expressions for α (Eq. 3) and λ (Eq. 5), which are only valid for large η, and plots using the results of the more elaborate treatment of the Amsterdam group (Walraven, 1996; Luiten et al., 1996).

The advantage of the linear confinement is clearly visible in Fig. 1. For the optimum choice of η, runaway evaporation in the linear trap occurs at a three to four times smaller ratio of good to bad collisions than in the parabolic trap. Qualitatively, one can understand this effect as a compression effect of the linear potential: In a potential $U(r) = U'r$, a cloud of temperature T has a size kT/U'. The same size is obtained by a harmonic confinement $U(r) = U''r^2/2$ with $U'' = U'^2/kT$. This means that evaporation in a linear potential is similar to evaporation in a harmonic potential with a force constant that is varied inversely proportional to the temperature. This compression greatly increases the density and elastic collision rate.

In the absence of any loss process ($R = \infty$), the minimum η for runaway evaporation is determined by

$$\alpha > \frac{1}{\delta - 1/2} \qquad (8)$$

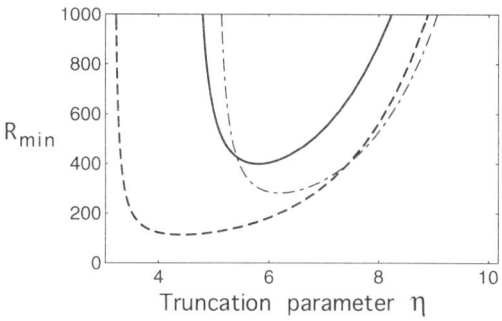

FIG. 1. The minimum ratio $R_{\min}$ of good to bad collisions needed for runaway evaporation, versus the truncation parameter η, Eq. (7). Plots are shown for parabolic (solid line) and linear (dashed line) potentials. The curves were calculated using the Amsterdam model. The dash–dotted line is obtained for a parabolic potential by using the approximations in Eqs. (3) and (5) for α and λ.

Equation (3) is not valid for such small η, and using a more accurate formula (Eq. (145) of Walraven, 1996) one obtains $\eta = 3.2$ and 4.6 for $\delta = 3$ and $\frac{3}{2}$, respectively.

The increase of phase space density with time is given by

$$\beta = 100\tau_{\text{el}} \frac{d}{dt}(\log_{10} D) = \frac{100}{\ln 10} \left(\frac{\alpha(\delta + 3/2) - 1}{\lambda} - \frac{1}{R} \right) \quad (9)$$

β was normalized so that it gives the phase-space density increase (in orders of magnitude) per 100 elastic collision times. In Fig. 2, β is plotted versus the truncation parameter η for various ratios R of good to bad collisions. β becomes negative for small R and large η because slow evaporative cooling cannot compensate for fast losses due to background gas collisions. β reaches a maximum for low values of η: By rapid evaporation of atoms, a very large time derivative of D is possible; however, the density decreases, R becomes small, and the cooling process slows down and becomes inefficient.

2. *Maximizing Phase-Space Density*

Walraven (1996) has compared the efficiency of evaporation in different power law potentials by comparing the α parameter for certain evaporation strategies. However, α describes only the change in temperature. If we regard evaporation in a linear potential as analogous to evaporation in a harmonic potential with continuous adiabatic compression, we realize that α does not provide the most meaningful comparison, because adia-

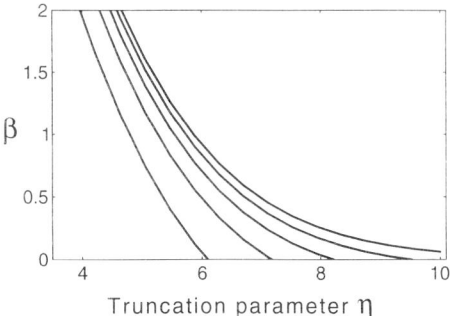

FIG. 2. Logarithmic phase-space density increase β per 100 elastic collision times, Eq. (9), versus the truncation parameter η. The ratio of good to bad collisions R is infinite (upper line), 500, 200, 100, and 50 (lower line), respectively. The potential is a 3D parabolic potential, and α and λ were obtained from the Amsterdam model.

batic compression trades in temperature against density. We therefore now focus on phase-space density D, which is invariant with respect to adiabatic changes of the potential.

By analogy with Eq. (6) we find the relative increase in phase space density with decreasing number N:

$$\gamma = -\frac{d(\ln D)}{d(\ln N)} = \frac{\alpha(\delta + 3/2)}{1 + \lambda/R} - 1 \tag{10}$$

Figure 3 plots γ versus η for different values of R for a linear and a parabolic potential. As expected, for long trapping times (large R), the optimum γ is realized with large truncation parameters η. The efficiency γ of the evaporation process is higher in a parabolic trap than in a linear trap. However, this reflects only the situation at the same value of R. For the example shown in Fig. 3 with $R = 200$, the linear trap operates in the runaway regime, whereas in the harmonic case the elastic collision rate is decreasing. This mean that the process in the linear trap works with increasing R and therefore increasing γ, whereas the trend is opposite in the harmonic trap (see Section II.B.3).

The most important figure of merit of evaporative cooling in atom traps is to achieve the maximum increase in phase-space density with the smallest loss in the number ("to reach BEC with the largest number of atoms possible"). This would mean that the goal is to achieve the largest value of the global parameter

$$\gamma_{\text{tot}} = \frac{\ln(D_{\text{final}}/D_{\text{initial}})}{\ln(N_{\text{final}}/N_{\text{initial}})} \tag{11}$$

This global parameter is optimized by optimizing γ (Eq. 10) at any moment. (This was first brought to our attention by R. Hulet). To explain this, we first note that the elastic collision rate at any moment can be expressed by the phase-space density and the number of atoms:

$$n\sigma v \propto D^{(\delta-1/2)/(\delta+3/2)} N^{2/(\delta+3/2)} \tag{12}$$

Then we divide the increase in phase-space density in many small steps of size ΔD. The efficiency γ of the second step is optimized by maximizing R (see Eq. (10)). According to Eq. (12), this is achieved by maximizing the number of atoms left after the first step. As a result, following an evaporation path that maximizes γ at any given point maximizes γ_{tot} and, thus, the number of atoms left at the final phase-space density.

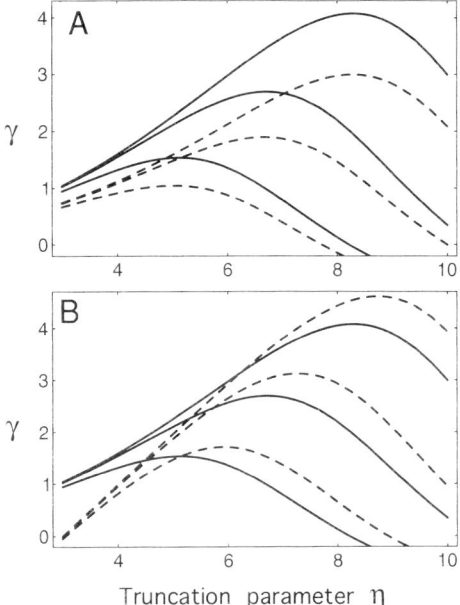

FIG. 3. The efficiency parameter of evaporative cooling, $\gamma = d(\ln D)/d(\ln N)$, versus the truncation parameter η. (A) Comparison between a 3D parabolic potential ($\delta = \frac{3}{2}$, solid lines) with a linear potential ($\delta = 3$, dashed lines). γ was calculated using Eq. (10), where α and λ were obtained from the Amsterdam model. (B) γ for a 3D parabolic potential; the solid lines are identical to those in (A), and the dashed lines are obtained by using the approximations in Eqs. (3) and (5) for α and λ. In each case, three different lines are given, for ratio of good to bad collisions R of 5000 (upper lines), 1000, and 200 (lower lines), respectively.

The same conclusion applies for cooling of atomic hydrogen, where the dominant loss process is inelastic binary collision (to be discussed). R is then independent of density and proportional to $\sqrt{T}$, resulting in

$$R \propto D^{-1/[2(\delta+3/2)]} N^{1/[2(\delta+3/2)]} \tag{13}$$

Again, optimizing N in one step maximizes R and, therefore, γ for the following step.

There is one major difference between the loss due to background gas collisions and that due to inelastic collisions. Inelastic collisions heat up the sample because they happen most frequently in regions of high density, where the potential energy is small. As a result, atoms lost due to inelastic collisions carry away less than their share in total energy, thus increasing the average energy of the trapped atoms. This effect is quantitatively discussed by Walraven (1996) and decreases the effective α, which expresses the average energy of a lost particle. Actually, α even changes sign as a function of temperature when the heating due to dipolar relaxation dominates over the evaporative cooling. The criterion $\alpha = 0$ determines the lowest temperature that can be achieved in evaporative cooling due to relaxation heating, whereas a model considering only background gas collisions allows cooling to arbitrarily small temperatures.

3. Strategies for Evaporation Cooling

It might appear that the two previous subsections on phase-space density increase and runaway evaporation describe different strategies or even conflicting goals. We now want to discuss qualitatively how they depend on each other and emphasize the major difference between evaporative cooling in alkalis and in atomic hydrogen.

The ultimate goal is the achievement of high phase-space densities. As was pointed out, maximizing γ is the optimum strategy for this goal. However, for small R, γ becomes negative; i.e., no increase in phase-space density is possible. Therefore, a phase-space density increase can only be sustained as long as R is larger than a critical value.

For loss due to background gas collisions (the dominant limitation for alkalis), R changes according to Eq. (6), or equivalently,

$$-\frac{d(\ln R)}{d(\ln N)} = \frac{\alpha(\delta - 1/2)}{1 + \lambda/R} - 1 \tag{14}$$

R varies exponentially with $1/N$, with the exponent given by the right hand side of Eq. (14). If the exponent is negative at some point during the cooling (even for optimized η), R decreases, resulting in an even more negative exponent later on. The consequence is an accelerated decrease of R and an accelerated reduction of the efficiency parameter γ, until eventually γ reaches zero and no further increase in phase-space density is possible. Work in alkali systems concentrated on realizing an initial situation with a sufficiently large R, so that R increased during evaporation. In principle, cooling with a (slowly) decreasing R is possible, as long as the desired phase-space density is reached before γ becomes zero. However, when the threshold of runaway evaporation is reached, R stays constant or increases. This guarantees both fast and efficient cooling to very high phase-space densities. The phase-space density increase is only limited by the onset of other loss processes such as three-body recombination and dipolar relaxation, which inevitably reduce R and throttle the cooling process.

In atomic hydrogen, the situation is quite different. All experiments in atomic hydrogen were done in a cryogenic environment where background gas collisions were negligible, and the dominant loss mechanism was inelastic binary collisions. R is independent of density and decreases proportionally to $\sqrt{T}$. Runaway or increasing collision rates speed up the cooling process, but in contrast to alkali atoms, they do not increase R nor improve the efficiency of evaporative cooling. In all experiments, the initial temperature, and therefore R, was large enough to allow efficient cooling in the beginning. However, during the cooling process the efficiency decreased, and the highest phase-space densities were reached when either γ became small and/or the number of atoms in the sample reached the detection limit.

We want to add one additional comment with regard to adiabatic compression. Adiabatic compression increases both temperature and density and therefore always increases R. As a result, the increase in phase-space density after adiabatic compression is always superior, and the optimum strategy is to use the tightest confinement possible. This strategy only finds its limitations when three-body recombination sets in, or when the collision time for an evaporating atom becomes comparable with its escape time from the trap. In the latter case, λ in Eq. (10) increases with density.

C. MODELS

In the previous subsections, we have sketched a semiquantitative picture of the evaporation process, which was partly a simplification and partly an extension of models by other authors. It was mainly stimulated by work of the Amsterdam group (Walraven, 1996; Luiten *et al.*, 1996) and our own previous work (Davis *et al.*, 1995c). In this section, we will summarize the different efforts aimed at modeling evaporative cooling.

1. *Amsterdam*

The Amsterdam group modeled evaporative cooling within a kinetic theory involving a numerical solution of the Boltzmann equation. These results confirmed that the energy distribution can be very well approximated by a truncated Boltzmann distribution (Luiten *et al.*, 1996). This assumption of quasi-equilibrium has been often used in previous work, but it had not been thoroughly checked before. Subsequently, a closed set of differential equations for evaporative cooling was derived under the assumption of a truncated Boltzmann distribution, i.e., that all particles with energy larger than $\eta k T$ leave the trap.

In particular, the Amsterdam group has exactly calculated the collision integral for the number of particles produced with energy larger than ηkT. For example, Eq. (140) of Walraven (1996) for $d(\ln N)/d(\ln T)$ replaces our Eq. (3) for α. Our κ is expressed by their term $(1 - X_{ev}/V_{ev})$. The average energy $\delta + \frac{3}{2}$ is generalized to their expressions $\tilde{\gamma} + \frac{3}{2}$ and C_i/Nk, which differ only when the truncation of the Boltzmann distribution becomes important. Furthermore, when the threshold energy for evaporation is ramped down (forced evaporation), the average energy of the evaporating atoms is reduced from $(\eta + \kappa)kT$, because atoms are now "shaved" away at an energy ηkT. This results in a "spilling" correction that is proportional to κ, the difference in energy between a "truly evaporated" and a "spilled" atom (see Walraven, 1996, for details). In the language of that paper, our λ parameter would be expressed as $\sqrt{2}(1 - \tilde{\xi}\tilde{\alpha})e^{\eta}(V_e/V_{ev})$. The nice feature of the Amsterdam treatment is that most of the final expressions, although seemingly complicated, have been solved analytically in terms of generalized gamma functions. Analytical expressions were not only derived for power law potentials, but also for a potential that is a very good approximation to the Ioffe–Pritchard configuration, which is frequently used for trapping hydrogen and alkalis (Luiten et al., 1996).

This model included dipolar relaxation as a loss and heating mechanism, and it was used to discuss the prospects of cooling trapped atomic hydrogen into BEC. They concluded that BEC can be obtained in magnetically trapped atomic hydrogen, and that phase-space density can be increased down to temperatures of 30 nK (Luiten et al., 1996). However, below 2 μK, the density decreases due to the strong truncation of the Boltzmann distribution, which is necessary to cool against the loss and heating due to dipolar relaxation.

2. *Davis and Coworkers (MIT)*

An analytical model for evaporation was presented earlier by our group (Davis et al., 1995c). It approximated the evaporation process as a discrete series of truncation and rethermalization processes, and it arrived at simple analytical results. These were used to discuss the threshold for runaway evaporation for different potentials.

This model described a situation that is not directly realized in the experiments. It assumed thermalization of the distribution (both in velocity and position space) before the next truncation step, and it estimated that the time for rethermalization is several collision times. This situation would be realized in radiofrequency (rf)-induced evaporation when the rf is pulsed on and off with an off time that is longer than the thermalization time.

In real experiments, the truncation is continuous. Still, the number of evaporated atoms can be approximated by an integral over the tail of the velocity distribution using detailed balance (see Section II.B). This argument shows the connection between the Amsterdam model and that of Davis *et al.*, which have the same asymptotic behavior in a box potential. For any other potential, however, there are discrepancies, because Davis *et al.* assumed replenishment of the wings of both the velocity and spatial distribution before the next truncation was performed. In the Amsterdam model, collisions fill up the empty states only within the spatially truncated cloud; the spatial wings are not involved in the evaporation process. In that sense, one might say that normal evaporation involves only velocity space, whereas the pulsed evaporation discussed by Davis *et al.* occurs in both velocity and position space. Furthermore, as was discussed, the time to replenish the tail of the velocity distribution depends on η and is faster than a collision time (by a factor $\sqrt{\pi\eta/8}$), whereas replenishment of the spatial tail takes longer. The time step in the model of Davis *et al.* was assumed to be several thermalization times and therefore corresponds to evaporation that is not forced at the maximum possible speed.

The major purpose of this discrete model was to obtain fully analytical expressions for the evaporation dynamics in the simple case of alkali atoms, where background gas collisions are the dominant loss mechanism. In Section II.B, we have extended the Amsterdam model to this situation and discussed runaway evaporation and the efficiency of evaporative cooling in the presence of background gas collisions. We feel that the present discussion is more realistic than our earlier treatment, and that the discrete model is mainly of pedagogical value because of its simplicity.

3. *Doyle and Coworkers (MIT)*

John Doyle and collaborators derived a set of coupled differential equations that described various cooling, heating, and loss processes and included adiabatic changes of the potential during the cooling process (Doyle *et al.*, 1991, 1994; Doyle, 1991). This was the first comprehensive model of evaporative cooling. The numerical solution of these equations agreed very well with the evaporative cooling experiments done at MIT. Furthermore, optimized trajectories in phase space were determined aiming at the highest final phase-space density for a given initial temperature and density combination. Those calculations predicted that it should be possible to reach BEC in hydrogen with initial conditions realized by the MIT group (Doyle, 1991).

The equations of Doyle et al. (1991, 1994) and Doyle (1991) neglected the spilling of particles due to forced evaporation, which was unimportant in the MIT experiments but is essential for the alkali experiments. It could easily be added to their formalism. The spirits of this model and of the Amsterdam model are very similar. The Amsterdam model can be regarded as a major extension of the work by Doyle et al. (1991, 1994) and Doyle (1991), especially because the predominately analytical treatment provides additional insight.

4. Earlier Work

For completeness and historical interest, we mention earlier work on modeling evaporative cooling. The concept of wall-free confinement and evaporative cooling of atomic hydrogen was conceived by Hess around 1983 or 1984, quite some time before it was published, and was discussed and refined by the MIT group. The original suggestion by Hess (1985, 1986) already discussed the interplay between cooling and dipolar heating and derived some of the scaling laws presented in Section II.A. He proposed that such a technique could cool and compress atomic hydrogen to attain Bose–Einstein condensation at a temperature of 30 μK.

Similar ideas were, probably independently, conceived by Lovelace et al. (1985) and Tommila (1986). They refer to Hess's abstract of his presentation at a meeting of the American Physical Society (APS) (Hess, 1985). Lovelace et al. proposed a dynamic trap in which dipolar relaxation would be absent. Tommila investigated the dynamics of evaporative cooling in linear, quadratic, and square potentials.

5. Monte Carlo Simulations

Very recently, two groups developed an elegant Monte Carlo trajectory technique that allowed an efficient solution of the classical kinetic equation (Holland et al., 1996b; Wu and Foot, 1996). This technique solves the Boltzman equation directly without any assumptions such as partial equilibrium or the presence of a truncated Boltzmann distribution.

This method, similar to the "Monte Carlo wavefunction" approach in quantum optics, is claimed to be computationally much less intensive than the usual collision integral treatment. These calculations show good agreement with the calculations based on the truncated Boltzmann distribution (Luiten et al., 1996). There are small differences, however, which are attributed to the deviation from the truncated Boltzmann distribution. Wu and Foot (1996) showed that the axial and radial temperature during two-dimensional (see Section II.D) forced evaporation differ by up to 25%.

6. *Quantum Mechanical Models*

Several authors have studied a quantum mechanical three-level model that strips evaporative cooling to its bare bones (Holland *et al.*, 1996a; Wiseman *et al.*, 1996). These models were developed to describe a possible realization of an "atom laser" pumped by elastic collisions. The ground state is the single state that collects the cold atoms; collisions between atoms in the first excited state result in the necessary energy exchange; and the second excited state is sufficiently damped so that the high energy atoms are lost. The goal was to obtain a microscopic model for the accumulation of particles in the ground state, including quantum statistical effects. The influence of interactions on fluctuations of the occupation numbers in this simple model of evaporative cooling was studied by Quadt *et al.* (1996). The ideal Bose gas shows anomalous fluctuations of the ground state population in the framework of the grand-canonical ensemble (Greytak and Kleppner, 1984). Those fluctuations were shown to be suppressed in the presence of interactions.

D. THE DIMENSION OF EVAPORATION

Evaporative cooling relies on the selection of energetic particles that leave the trap. This selection can be based on the total energy E or on the energy for the motion in a particular direction, e.g., E_z. We call the selection one-dimensional if it is based on $E_z > \eta kT$, two-dimensional for $E_x + E_y > \eta kT$, and three-dimensional for $E > \eta kT$. This distinction is rigorous only for separable potentials. In the context of evaporative cooling the relevant timescale is the time between elastic collisions. Thus, if the ergodic mixing time is longer than the elastic collision time, the potential can be regarded as separable. This is the situation in many atom traps (Helmerson *et al.*, 1992b; Monroe *et al.*, 1993; Davis *et al.*, 1995b), where ergodic mixing of many seconds was observed, much longer than typical collision times of milliseconds to a second.

In the case of negligible ergodic mixing, the depletion of the high-energy tail of the distribution depends on the selection method. In the opposite case of fast ergodic mixing the distribution is depleted for total energy $E > \eta kT$, independent of the dimensionality of the selection. Since the dynamics of the evaporation are determined by the dimension of the depletion, we will call this the *dimension of evaporation*.

Most evaporative cooling experiments on hydrogen used evaporation over a saddle point of the potential which is a 1D selection scheme. Radiative evaporation using rf is a three-dimensional selection scheme in a dc magnetic trap. However, in the TOP trap, it is only two-dimensional, and if gravitational forces become essential, it is only one-dimensional (Section IV.E.4).

The problem with evaporation in lower dimensions is the dramatic reduction in efficiency, as was first discussed by Surkov *et al.* (1996) for an Ioffe–Pritchard (IP) trap. The reason is that a major fraction of the atoms that have enough total energy to escape collide with other atoms and lose the high excitation energy. Therefore, several "attempts" are necessary before an atom leaves the trap.

As long as there are no loss processes, this affects only the time for evaporation. However, if a time limitation is set by inelastic collision processes and background gas collisions, one- or two-dimensional evaporation strongly decreases the efficiency of evaporative cooling. This can be readily seen with Eq. (10). Since loss processes enter in the ratio λ/R, an increase in λ has an effect similar to a decrease in the ratio of good to bad collisions.

Surkov *et al.* (1996) compared axial and three-dimensional evaporation in the Ioffe–Pritchard trap. The rate of axial evaporation was smaller by a factor of 4η (η is typically between 5 and 10). Wu and Foot compared evaporative cooling in three and two dimensions and showed a scenario in which the trap depth was ramped down by a factor of 1000. The phase-space density increase in 3D evaporation was 20 times high than in 2D, and the fraction of the remaining atoms was three times higher (Wu and Foot, 1996).

The much lower efficiency of evaporation in lower dimensions follows from a simple geometric argument: In velocity space, the evaporating atoms are mainly from a shell with velocities between $\sqrt{2\eta kT/m}$ and $\sqrt{2(\eta+1)kT/m}$. This reflects that κ (Eq. 3) is about 1. The volume of this shell which has a z component of the velocity larger than $\sqrt{2\eta kT/m}$, is a factor of 4η smaller than the whole shell. Thus, λ is reduced by approximately this factor. In 2D evaporation, λ is reduced by a factor of $3\sqrt{\eta}/2$.

Keeping λ, the number of collisions per evaporating atom, as small as possible is particularly important for atomic hydrogen, where the evaporation is limited by the decrease of R with decreasing temperature. However, the only demonstrated method for 3D selection is rf-induced evaporation. This technique has not been demonstrated yet in a cryogenic environment.

We want to add a rather technical remark: If one compares 3D with 1D evaporation, one has to consider that κ, which determined the energy $(\eta + \kappa)kT$ of an evaporating atom, is no longer between 0 and 1, but larger. This is because atoms with a larger excess energy are less affected by the reduced dimension of evaporation. If the transverse and axial motion are separable, κ is about 1 plus the average energy along the transverse directions. The larger value of κ improves the α parameter. As a result, one should not compare 1D and 3D evaporation at the same η; rather, one should evaluate Eq. (10) for γ with the appropriately modified λ and α. This follows the spirit of our earlier suggestion: Comparison between different potentials and/or evaporation methods should use the efficiency of evaporation as defined in Eq. (10) as the figure of merit.

E. THE NUMBER OF COLLISIONS FOR THERMALIZATION

Several authors have discussed the number of elastic collisions necessary for thermalization (Snoke and Wolfe, 1989; Snoke et al., 1992; Monroe et al., 1993; Davis et al., 1995a; Coakley, 1996). This concept is important for evaporative cooling if one regards the evaporation process as a truncation–replenishment cycle (Davis et al., 1995c). However, most models of evaporative cooling avoid the notion of thermalization by calculating the flux of evaporating atoms from the collision term of a Boltzmann equation. In any case, evaporative cooling creates a truncated and, therefore, nonthermal energy distribution, and the relaxation of this distribution to an equilibrium distribution is important, e.g., when the final temperature is determined at the end of the cooling process.

Snoke and Wolfe (1989) performed numerical quantum mechanical calculations on the relaxation of a nonthermal distribution and found that the distribution was indistinguishable from a thermal distribution after five collisions, independent of the nature of the nonthermal distribution.

Experimentally, elastic collision cross sections were determined by preparing a sample with different temperatures along the axial and radial directions and observing the equilibrium. In this way, the low temperature, elastic collision cross sections for cesium, sodium, and rubidium were determined (Monroe et al., 1993; Davis et al., 1995a; Newbury et al., 1995). The number of elastic collisions for such a cross-sectional relaxation was numerically determined to be 2.7 (Monroe et al., 1993). This result was later confirmed by an analytic solution of the Boltzmann transport equation (Newbury et al., 1995) and also by a Monte Carlo trajectory calculation (Wu and Foot, 1996).

We have shown in Section II.A, using detailed balance arguments, that the high energy tail of the velocity distribution is replenished in one elastic collision. Since the collision time for fast atoms enters, we arrive at the counterintuitive result that the higher the energy of the truncated tail is, the shorter is this replenishment time. This argument only applies to replenishment in velocity space; the thermalization of a spatially truncated cloud requires at least one oscillation period for the energetic atoms to reach the outer parts of the cloud, and it will eventually depend on the ergodicity of the trapping potential.

F. THE NUMBER OF COLLISIONS TO BOSE–EINSTEIN CONDENSATION

All the models discussed here show that evaporative cooling is a fast and efficient way to reduce the temperature of trapped atoms and increase the phase-space density. A possible scenario is as follows (see Figs. 2 and 3): harmonic potential, η parameter of 6, removal of 0.7% of the atoms per collision time, increase of phase-space density by a factor of 10^6 in 600 collision times, and a loss in the number by a factor of 100. Reduction of the η parameter to 5 gives the same phase-space density increase in half the number of collisions, but with 5 times fewer atoms left.

Since the phase-space density of laser-cooled atoms is typically 10^{-6} lower than is required for Bose–Einstein condensation (BEC), one can summarize the potential of evaporative cooling in the following way: About 500 collision times suffice to achieve BEC with 1% of the atoms remaining!

Real experiments have performed somewhat worse (Section V), probably due to nonoptimized cooling, lower dimension of the evaporation, and additional loss and heating processes. However, the few experiments done so far have already impressively confirmed the potential of evaporative cooling.

G. DESIRABLE EXTENSIONS OF THE MODELS

The theoretical foundations of evaporative cooling are well established. The different models we have discussed describe the salient features of the evaporation process. Extensions in the following directions appear worthwhile:

1. The calculation of optimized trajectories in phase space. There are obviously trade-offs between efficiency and cooling time, and some detailed treatment is desirable.

2. So far, all models assume that all particles with a certain (1D or 3D) energy will escape. However, all methods of evaporation rely on spatial selection of atoms. Energetic atoms that are on circular orbits have a smaller outer turning point than atoms undergoing radial oscillations. Such

nonergodic effects reduce the efficiency of evaporative cooling. So far, only one theoretical paper has addressed this problem (Surkov *et al.*, 1996).

3. All models discuss evaporation in the dilute regime. Dilute means here collisionally thin, or that the mean free path is longer than the size of the sample. One consequence is that the products of inelastic collisions leave the sample without transferring their energy to other atoms. It is obvious that the dense limit is not advantageous for evaporative cooling and should be avoided by adiabatic expansion, but there is probably some trade-off involving efficiency, speed, and collisional mixing that ensures ergodicity.

4. Most models have treated the atomic motion classically, ignoring quantum statistical effects. For the description of the cooling process, this is justified because evaporative cooling has been used to increase the phase-space density by six orders of magnitude, whereas quantum statistical effects are only pronounced in the vicinity of BEC. However, the last stage of the cooling and the actual formation of a Bose condensate are a very interesting regime to study both theoretically (Stoof, 1995; Kagan, 1995) and experimentally. So far, only simplified three-level models have been formulated, which combine evaporative cooling with quantum effects (Holland *et al.*, 1996a; Wiseman *et al.*, 1996) (see Section II.C.6).

III. The Role of Collisions for Real Atoms

A. ELASTIC AND INELASTIC COLLISIONS

Evaporative cooling is driven by elastic collisions. High density and the presence of collisions are therefore prerequisites for evaporative cooling. However, this means that inelastic collisions are unavoidable. We have already mentioned background gas collisions and dipolar relaxation as loss processes, and we wish to give a more systematic account of collision processes in this section. For a more comprehensive treatment of cold collisions, we refer to several review articles on this subject (Silvera and Walraven, 1986; Verhaar, 1995; Heinzen, 1995; Walker and Feng, 1994; Lett *et al.*, 1995).

In a magnetic trap, three kinds of inelastic collisions are relevant for evaporative cooling experiments: dipolar relaxation and spin relaxation, which are binary collisions, and three-body recombination (or dimerization). In an optical dipole trap (to be discussed), atoms can be trapped in the lowest hyperfine state, eliminating binary collisions as trap loss process. If atoms are trapped in the upper hyperfine state, hyperfine state changing collisions are important (Walker and Feng, 1994), and the loss processes are similar to the situation in a magnetic trap. In this section, we focus on the most common situation of magnetically trapped samples.

An overview of the relevant collisional parameters for hydrogen and the alkalis is given in Table II. Spin relaxation has a rather large rate coefficient, typically 10^{-12} cm^3/sec. It involves an exchange of angular momentum between the electron spin and the nuclear spin. This process is usually important only after the initial loading of the trap and leads to a "self-purification" of the sample in the sense that there are only atoms left in a hyperfine state, which does not undergo spin relaxation. This usually happens in the first few seconds after loading a magnetic trap with atomic hydrogen (Greytak, 1995).

Spin relaxation is forbidden for a doubly spin polarized sample due to angular momentum conservation, and also in the upper (weak-field-seeking) sublevel of the lower hyperfine level (as long as kT is small compared with the ground state hyperfine splitting). However, spin-flips can still occur by coupling the spin angular momentum to the orbital angular momentum. This process, called dipolar relaxation, happens with a rate coefficient G_{dip} of typically 10^{-15} cm^3/sec, 1000 times smaller than spin relaxation. The rate coefficients for dipolar relaxation are similar for hydrogen and alkalis.

As for all inelastic processes, the rate coefficient G_{dip} is constant for temperatures approaching zero, whereas for elastic collisions the cross section σ approaches a constant. This means that in the limit of low temperature and low density (so that three-body recombination can be neglected) the ratio of good to bad collisions is given by $\sqrt{2}\,\sigma\bar{v}/G_{\text{dip}}$.

This ratio reaches unity at a characteristic temperature T_*:

$$kT_* = \frac{\pi m G_{\text{dip}}^2}{16\sigma^2} \qquad (15)$$

This temperature represents a theoretical lower bound to the minimum temperatures attainable by forced evaporation. As shown by Luiten *et al.* (1996), the minimum temperature in a harmonic trap is about three times larger than T_*. The smallest temperature that can be reached by evaporative cooling with increasing density is approximately 2000 times larger. One might thus regard $1000 T_*$ as the practical temperature limit of evaporative cooling.

The rate of three-body recombination is given by Ln^2, where L is the rate coefficient. In a simple picture, one can estimate the rate for three-body recombination as a $(2 + 1)$ process, a collisional encounter between two atoms with a third one joining. This suggests a correlation between the elastic cross section and the rate for three-body recombination. Recent

TABLE II
Collision Parameters[a]

Atom	$a_<$ (a_0)	$a_>$ (a_0)	G_{ex}[b] (cm³/sec)	$G_{dip,>}$ (cm³/sec)	$L_>$ (cm⁶/sec)
¹H	absent	1.4[c]	2×10^{-13}[d]	10^{-15}[e]	10^{-38}[f]
⁷Li	10[g]	-27.3 ± 0.8[h]		2×10^{-14}[q]	2.6×10^{-28}[i]
²³Na	92 ± 25[j]	45 to 185[g]	10^{-11}[k]	3×10^{-15}[k]	2.0×10^{-28}[i]
⁸⁵Rb	?[l]	-1000 to -60[m]		3×10^{-14}[r]	
⁸⁷Rb	?[l]	85 to 140[m]		10^{-15}[r]	4×10^{-30}[i]
¹³³Cs	?[n]	-1100 to -200[n]		10^{-16}–10^{-15}[p]	5×10^{-29}[p]

[a] Where available, the parameters have been taken for temperature $T \to 0$ and magnetic field $B \to 0$. The dependence on T and B is weak, except where indicated. Subscript $>$ indicates the $F = m_F = F_{max}$ low-field-seeking doubly polarized state; subscript $<$ refers to the low-field-seeking $F = -m_F = F_{max} - 1$ state.

[b] Spin exchange coefficient G_{ex}, typical value for a mixture of low-field-seeking states.

[c] Friend and Etters (1980).

[d] Stoof et al. (1988).

[e] van Roijen et al. (1988).

[f] Hess et al. (1984).

[g] Moerdijk and Verhaar (1994).

[h] Abraham et al. (1995).

[i] Moerdijk et al. (1996a).

[j] Davis et al. (1995a).

[k] Tiesinga et al. (1991). A recent, improved calculation gives $G_{dip,>} = 7 \times 10^{-15}$ cm³/sec for sodium (Moerdijk and Verhaar, 1996). For the lower hyperfine state, recent calculations yield $G_{dip,<} \approx 10^{-16}$ cm³/sec, at magnetic fields of a few gauss (Moerdijk and Verhaar 1996).

[l] Verhaar (1995).

[m] Gardner et al. (1995).

[n] Verhaar et al. (1993).

[p] Tiesinga et al. (1992).

[q] Moerdijk and Verhaar (1996).

[r] Boesten et al. (1996).

calculations show that this correlation exists (Moerdijk et al., 1996a; Fedichev et al., 1996b).

Three-body recombination was the limiting process in the attempt to reach quantum degeneracy in spin polarized hydrogen at temperatures of 500 mK (Greytak, 1995). Evaporative cooling at much lower temperature and density turned out to be the solution to avoid this process.

B. Atoms for Evaporative Cooling

Three kinds of atoms have been used or discussed for evaporative cooling: atomic hydrogen, alkali atoms, and metastable helium. They show major differences in their collisional properties.

Atomic hydrogen is distinguished by an extremely small s-wave scattering length $a = 0.072$ nm (Friend and Etters, 1980) resulting in an elastic cross section $\sigma = 8\pi a^2 = 1.3 \times 10^{-15}$ cm^2. With a rate coefficient for dipolar relaxation of $G_{\text{dip}} = 1.0 \times 10^{-15}$ cm^3/sec (van Roijen et al., 1988), the ratio of good to bad collisions at $T = 1$ mK is 800. Since trapping experiments for atomic hydrogen are always carried out in a cryogenic environment (the trap is loaded by cryogenic precooling of the atoms), the background gas pressure is extremely low, and the trapping time is usually determined by inelastic collisions among the trapped atoms themselves rather than by background gas collisions.

The rate for three-body recombination is extremely small for spin polarized atomic hydrogen. This reflects the fact that the lowest triplet state of the hydrogen molecule is repulsive, and three-body recombination is only possible through the weak interaction of the electron magnetic dipole moments. As a result, for most realistic situations, the major loss and heating process during evaporative cooling is dipolar relaxation (Fig. 4a).

In alkali vapors, the situation is very different. The s-wave scattering length is on the order of 5 nm, resulting (e.g., for Na) in an elastic cross section of 6×10^{-12} cm^{-2}, more than 1000 times larger than for hydrogen. The rate constant for dipolar relaxation is similar to that for atomic hydrogen (see Table II). As a result, the ratio of good and bad (binary) collisions is extremely high, e.g., for Na at 10 μK it is 24,000. Most evaporative cooling experiments on alkalis work at densities of at most 10^{12} to 10^{13} cm^{-3}. The time constant for loss due to dipolar relaxation is more than one minute, and it plays only a minor role for the evaporative cooling. However, the Na experiment at MIT realized densities of 10^{14} cm^{-3}, where inelastic collisions should be important. The rate constant L for three-body recombination is typically 10 orders of magnitude larger than for atomic hydrogen and is on the order of 10^{-28} cm^3/sec (see Moerdijk et al., 1996a, for recent calculations for different atoms). Trap loss due to three-body recombination dominates for densities $n > G_{\text{dip}}/L$, which is typically 10^{13} cm^{-3}. At this density, the time constant for inelastic collisions is about 100 sec. That is why we expect that in most situations the trapping time is either limited by background gas collisions at low density, or by three-body recombination at high density (Fig. 4b). It is possible that dipolar relaxation will limit evaporative cooling of alkali atoms only if one attempts to cool samples at low density and extremely low background pressure.

Since dipolar relaxation is the dominant limitation for atomic hydrogen, early discussions on evaporative cooling in alkalis overly focused on this loss mechanism. One suggestion how to avoid this loss process was trap-

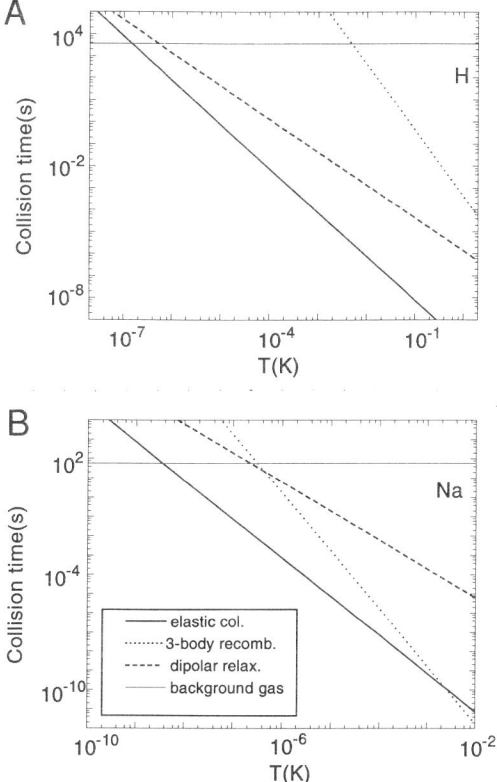

FIG. 4. Collision times for several elastic and inelastic processes, as a function of temperature, at the density needed for Bose–Einstein condensation. Shown are elastic collisions ("good" collisions) and the "bad" collisions due to background gas (for typical experimental conditions), elastic collisions, three-body recombination, and dipolar relaxation. (a) Atomic hydrogen; (b) sodium. For hydrogen, dipolar relaxation is the dominant source of bad collisions, whereas for sodium the dominant source is background gas collisions (at low densities) or three-body recombination (at high densities).

ping of the weak-field-seeking lower hyperfine state (which exists only for atoms with nuclear spin $I > \frac{1}{2}$). In this case, the energy release in dipolar relaxation approaches zero for small magnetic fields. This suppresses dipolar relaxation because of the centrifugal barrier in the exit channel (Tiesinga *et al.*, 1992) (or to state it more simply: for zero energy release, the process becomes elastic and has a vanishing rate at zero temperature). Note that reduction of G_{dip} would decrease T_* and thus decrease the lowest temperature achievable with evaporative cooling. In this sense, there is no fundamental limit for the lowest temperature achievable by evaporative cooling. Some experiments have indeed magnetically trapped rubidium (Petrich et al., 1995) and sodium (Davis et al., 1995b) in the lower hyperfine state, but because this state was easily prepared rather than for the sake of suppressing dipolar relaxation.

Our experiment on sodium was the only alkali experiment that entered the region where inelastic collisions became important. At densities higher than 10^{14} cm^{-3}, the predicted trapping time due to three-body recombination is less than 1 sec (Moerdijk et al., 1996a). The advantage of high density is the high speed of evaporation—the whole cooling process took only 7 sec. It appears that this experiment has already reached the maximum density at which evaporative cooling is still efficient.

Shlyapnikov et al. (1994) proposed another atom for evaporative cooling: spin polarized metastable helium (Fedichev et al., 1996). The attractive feature of this system is a probably large "alkali-like" elastic collision cross section of 10^{-12} cm^2. Metastable helium has an additional loss process—Penning ionization, which severely limits the densities achievable in magneto-optical traps (Bardou et al., 1992). However, this process is suppressed by five orders of magnitude in the spin polarized gas (Shlyapnikov et al., 1994; Fedichev et al., 1996). To obtain a sample of metastable helium with a high number of atoms and high density is still an experimental challenge, but efforts are under way (Rooijakkers et al., 1995).

C. Enhanced Relaxation in the Condensate

Our discussion of collisions has focused on the situation before BEC occurs. Bose condensation in a trap is accompanied by a dramatic increase of the density, and as a result, by an increase of the rates for inelastic collisions. Due to Bose symmetry, the rate coefficients for dipolar relaxation and three-body recombination are smaller than for a normal gas, by factors of 2! and 3!, respectively (Kagan et al., 1985; Hijmans et al., 1993). Thus, the total loss rate for a pure condensate with density n is given by $G_{\text{dip}} n/2 + Ln^2/6$.

The density increase during Bose condensation is limited by the repulsive interaction between the atoms. The mean field energy is given by $4\pi\hbar^2 na/m$, where a is the s-wave scattering length. At the BEC transition point with temperature T_c, the ratio r of thermal energy to mean field energy is $r = kT_c/(4\pi\hbar^2 n_c a/m)$, which is 50 for Na at 100 nK and 500 for hydrogen at 100 μK. This ratio r determines the maximum density increase of a condensate in a harmonic trap (assuming 100% condensate fraction): $n_0/n_c = 1.185 r^{3/5}$, which is about 10 for Na and 50 for hydrogen.

In the recent BEC experiments on alkalis, condensate fractions close to 100% have been observed with lifetimes of 1 sec or more (Anderson *et al.*, 1995; Davis *et al.*, 1995b). A fully condensed sample of hydrogen will decay within 1 sec (Greytak, 1995). However, if evaporative cooling is slow, the increase of the inelastic collision rate might limit the buildup of the condensate fraction to a few percent (Hijmans *et al.*, 1993; Greytak, 1995).

D. SPECIAL TOPICS

In this section we want to briefly mention two more special aspects of evaporative cooling and refer to the relevant literature.

1. *Tuning the Cross Section*

The efficiency of evaporative cooling depends on the elastic collision cross section. The success of evaporative cooling in alkalis is mainly due to the large cross section for these systems. Usually, one regards the cross section as given and fixed by the choice of atom, but there are a few intriguing developments that might allow one to change or even tailor collisional properties.

One such development is the theoretical prediction that the scattering length for atoms in the lower hyperfine state can be tuned by an external magnetic field (Moerdijk *et al.*, 1995; Tiesinga *et al.*, 1992, 1993). This tuning exploits the fact that, for a hyperfine state that is not doubly polarized, there is a different hyperfine coupling for the separated atoms and the dimer. The difference in hyperfine and Zeeman interaction energies is used to vary the binding energy of the last bound state of the molecule and, therefore, the scattering length, which is uniquely determined by this binding energy.

However, large tuning will only be possible around so-called Feshbach resonances. No such resonance occurs for lithium and sodium in the range of magnetic fields where the lower hyperfine state is weak-field-seeking (Moerdijk *et al.*, 1995). For cesium and rubidium, the prospects of finding such a resonance at sufficiently low fields is larger, although the first experimental search was not successful (Newbury *et al.*, 1995).

Another way of changing collisional properties is by dressing atoms with radiofrequency, which can create resonances in collisions. Rf radiation might also increase inelastic collision rates. Fortunately, rf-induced exchange relaxation is negligible for the experimental parameters of rf-induced evaporation (Moerdijk *et al.*, 1996b).

Finally, the cross section of elastic collisions can be increased by optical shielding (see Heinzen, 1995, for a recent review). A blue-detuned laser can be used to cause an avoided crossing between the potential curves of two colliding atoms. This creates a repulsive potential at large interatomic distances corresponding to a large value of the elastic cross section. The basic effect has been demonstrated by several groups (Sanchez-Villicana et al., 1995; Walhout et al., 1995; Marcassa et al., 1994; Katori and Shimizu, 1994); however, it is not clear if adverse effects (heating, optical pumping) of the laser light can be avoided.

2. Cooling Fermions

So far, all efforts of evaporative cooling have concentrated on bosonic atoms. The major reason is their abundance in nature and ease of manipulation. In the quantum degeneracy regime at ultralow temperatures, fermions will show a behavior radically different from bosons. The only abundant alkali isotope that is fermionic is lithium-6. So far, only laser cooling of this isotope has been done. Deuterium, the fermionic hydrogen isotope, sticks much better than hydrogen to liquid helium–coated walls, and attempts to load deuterium in magnetic traps built for hydrogen were unsuccessful (Hijmans et al., 1989; Doyle, 1996). Once fermions are successfully loaded, evaporative cooling would pose the challenge that elastic collisions can freeze out at very low temperatures (Koelman et al., 1987). This is because s-wave collisions are forbidden for spin polarized fermions, and the cross section for p-wave collisions varies with temperature as T^2. For bosons, s-wave collisions dominate below temperatures of the order $\hbar^3/km^{3/2}C_6^{1/2}$, where C_6 is the coefficient characterizing the ground state van der Waals potential. For fermions, one might regard this temperature as an estimate for the temperature below which the elastic cross section freezes out. For alkalis, these temperatures are in the micro- to milli-Kelvin range and are easily reached by laser cooling. Further evaporative cooling of laser-cooled fermions is therefore not promising.

Several solutions can be envisioned. One is trapping of fermions in different hyperfine states. Fermions in different states would still undergo s-wave collisions. The trapping of different hyperfine states in an optical trap is straightforward, whereas in a magnetic trap, spin relaxation would limit the stability of the sample already at very low densities. It has been shown that this process can be sufficiently suppressed at bias fields of 10 T (Stoof et al., 1996). Another route is to simultaneously trap two isotopes, where the thermalization happens through collisions between the different isotopes. The first experimental step toward such schemes has been done with the recent trapping of two isotopes in optical traps (Santos et al., 1996; Süptitz et al., 1994).

Finally, we want to mention that evaporative cooling is not limited to the atomic species mentioned in Section III.B. John Doyle at Harvard is planning such experiments with magnetically trapped molecules and other atomic species (Doyle *et al.*, 1995) (see Section IV.A.1).

IV. Experimental Techniques

A description of the experimental techniques for evaporative cooling is naturally divided in techniques for precooling (Section IV.A), trapping (Section IV.B), and evaporation (Section IV.E). The experimental efforts have developed along two rather separate lines, determined by the different properties of atomic hydrogen on the one hand, and the alkali-metal atoms on the other hand.

Nevertheless, there has been a substantial amount of cross-fertilization between the two fields: Magnetic trapping was first demonstrated for alkalis using laser cooling (Migdall *et al.*, 1985). Subsequently, it was used for hydrogen, in the first experiments on evaporative cooling (Hess *et al.*, 1987; Masuhara *et al.*, 1988; van Roijen *et al.*, 1988). Next, while work on laser cooling of alkalis developed toward high densities and long trapping times, bringing evaporative cooling in reach for the alkalis, optical cooling and detection techniques for hydrogen were developed (Luiten *et al.*, 1993; Setija *et al.*, 1993). The ultimate goal of evaporative cooling, reaching the quantum degenerate regime (Anderson *et al.*, 1995; Bradley *et al.*, 1995; Davis *et al.*, 1995b), was only achieved by a combination of techniques developed in once separated subfields.

A. Precooling

1. *Cryogenic Cooling*

The attainment of ultracold temperatures requires wall-free confinement; this is related to the fact that the lowest binding energy of any atom to walls is about 1 K, realized with hydrogen atoms and helium-coated walls. As a consequence, hydrogen cannot be cooled to below 75 mK in cells (Silvera, 1995a). The deepest traps that have been built with electromagnetic fields have a trap depth of about 1 K, thus requiring precooling of the atoms below this temperature as part of the loading process.

In the case of atomic hydrogen, there is a narrow temperature region in which hydrogen can already be magnetically trapped and where it does not stick to helium-covered walls. This technique was used at MIT (Hess *et al.*, 1987) and in Amsterdam (van Roijen *et al.*, 1988) to load magnetic traps.

Atomic hydrogen is generated in a low temperature rf discharge. The atoms migrate to the magnetic trapping region and thermalize with the helium-covered walls, resulting in a sample of magnetically trapped low-field seekers. Initially, both the c and d hyperfine states are present, but spin exchange collisions (predominately between the c-state atoms at these temperatures) lead to a doubly polarized sample of only d-state atoms within a few seconds.

Attempts with deuterium were unsuccessful so far, attributed to the 2.5 times larger binding energy of deuterium atoms with the walls (Hijmans et al., 1989). Note that laser cooling, the method of choice for alkalis, is difficult for atomic hydrogen because of the prohibitively short wavelength (121.6 nm) of the 1s–2p transition; only one experiment with pulsed radiation has been reported (Setija et al., 1993).

The idea of precooling by thermalization with a cryogenic environment was recently extended to a large class of paramagnetic atoms and molecules in a proposal by Doyle et al. (1995). Those species would stick to the cryogenic walls, but precooling can be achieved by a buffer gas of ^{3}He in the trapping region. The gas is kept at a temperature of 240 mK and a number density of 5×10^{15} cm^{-3}. This temperature is sufficiently low for magnetic confinement of the precooled particles. After the loading process, the ^{3}He is pumped out by lowering the cell temperature to 80 mK, thereby reducing the vapor pressure of ^{3}He to below 10^{-15} Torr. Subsequent evaporative cooling could then be used to reduce the temperature of the trapped gas.

A recently proposed scheme uses collisions to produce cold metastable helium (S. Yurgenson and J. A. Northby, personal communication, 1995). Helium atoms in a supersonic beam with a small velocity spread are excited to a metastable state by electron impact. The electron energy is chosen in such a way that the momentum transfer of the excitation process stops the helium atoms inside the trapping region of a magnetic trap.

2. Laser Cooling

Many atomic species can be cooled to submilli-Kelvin temperatures using laser cooling techniques. For evaporative cooling experiments, this has major advantages: (1) Precooling to temperatures around 100 μK allows much simpler construction of the magnetic trap using ordinary water-cooled conductors. (2) Atomic species can now be evaporatively cooled that have a much more favorable elastic collision cross section that atomic hydrogen. The ratio of good to bad collisions of alkali atoms can be 100 times higher than for hydrogen atoms (see Fig. 4). (3) The large elastic cross section of alkali atoms allows evaporative cooling times as short as several seconds, which makes a cryogenic vacuum unnecessary (note that for hydrogen the cryogenic system was not only necessary for the excellent vacuum, but also for the superconducting magnets and the wall precooling). The major challenge for the laser coolers was to create laser-cooled samples with sufficiently long trapping times and high densities to start evaporative cooling.

The standard method for laser cooling begins with loading a magneto-optical trap (MOT) (Raab *et al.*, 1987) from a vapor cell (Monroe *et al.*, 1990; Cable *et al.*, 1990), or from a cold atomic beam cooled by either chirped slowing (Ertmer *et al.*, 1985) or Zeeman slowing (Phillips *et al.*, 1985). After switching off the weak magnetic field of the MOT, the temperature is lowered to approximately 10 recoil energies, by polarization gradient cooling (Lett *et al.*, 1989). Typically, this results in 10^7 to 10^{10} trapped atoms at densities of 10^{10} cm^{-3} to 10^{11} cm^{-3}. Higher densities of about 10^{12} cm^{-3} have been achieved in smaller clouds (Drewsen *et al.*, 1994; Townsend *et al.*, 1995). The temperature is limited by the recoil limit of polarization gradient cooling, whereas the density is limited by absorption and radiation trapping of the cooling light (Walker *et al.*, 1990) and by excited state collisions (Walker and Feng, 1994), and eventually also by level shifts due to the resonant dipole interaction (Castin *et al.*, 1995). The maximum phase space density achieved in this way is 10^{-4} to 10^{-5} (Drewsen *et al.*, 1994; Townsend *et al.*, 1995).

A major step forward was the development of the dark SPOT (S̲pontaneous Force O̲ptical T̲rap) and dark polarization gradient cooling (Ketterle *et al.*, 1993). These techniques avoided the density limiting processes by keeping the atoms in a dark hyperfine state for most of the time. This allowed the confinement of more than 10^{10} atoms at densities of close to 10^{12} cm^{-3}. The dark SPOT technique closed the gap between the maximum densities usually achieved in laser cooling and the minimum densities required for evaporative cooling. This method was extended to rubidium (Anderson *et al.*, 1994) and cesium (Townsend *et al.*, 1996), and it was used in all realizations of evaporative cooling in alkali atoms (Adams *et al.*, 1995; Petrich *et al.*, 1995; Davis *et al.*, 1995a) with the exception of the Rice experiment, in which the lower density was compensated for by long cooling times (Bradley *et al.*, 1995).

Sub-recoil cooling techniques such as Raman cooling (Kasevich, 1995; Lee *et al.*, 1996; Kasevich and Chu, 1992; Davidson *et al.*, 1994; Reichel *et al.*, 1995), velocity-selective coherent population trapping (VSCPT) (Lawall *et al.*, 1994, 1995; Aspect *et al.*, 1988; Esslinger *et al.*, 1996), and adiabatic expansion of optical lattices (Kastberg *et al.*, 1995; Chen *et al.*, 1992) have been developed. These techniques might allow for even lower temperatures before loading the atoms into a magnetic trap, but this has yet to be demonstrated experimentally. Only Raman cooling was used at densities larger than 10^{11} cm^{-3} (Kasevich, 1995; Lee et al., 1996).

B. CONSERVATIVE TRAPS

Evaporative cooling of atomic gases requires an environment for the atoms that insulates them from the "hot world." Any "thermal contact" with the cooled sample or any form of residual heating would require additional evaporative cooling and greatly reduce the cooling efficiency. Wall-free confinement of cold atoms provides excellent insulation. It can be achieved by various electromagnetic traps. Since evaporative cooling is used to cool below the recoil limit, none of the variants of the magneto-optical trap can be used. Conservative traps for neutral atoms have been realized mainly by two principles: static magnetic fields and far-off-resonant optical fields. The resource letter of Newbury and Wieman (1996) contains many references on atom traps.

1. *Magnetic Traps*

Maxwell's equations do not allow for a maximum of a static magnetic field in free space (Wing, 1984; Ketterle and Pritchard, 1992). Atoms are therefore trapped in the weak-field-seeking state at a minimum of the magnetic field. There are two classes of static magnetic traps (Bergeman et al., 1987). In one case the minimum is a zero crossing of the magnetic field, characterized by the derivative B' of the magnetic field strength. The trapping potential is proportional to the absolute value of B, resulting in a linear, V-shaped potential. It is usually realized with the spherical-quadrupole field of two anti-Helmholtz coils. The other trapping topology has a parabolic minimum around a finite bias field. The parabolic trap was first suggested by Pritchard (1983) for atom trapping, and it is similar to the Ioffe configuration discussed earlier for plasma confinement by Gott et al. (1962). We refer to any configuration that is characterized by a nonzero bias field B_0 as an Ioffe–Pritchard (IP) trap. The lowest order expansion parameters for any $|B|_{min} > 0$ trap are B_0, the axial field curvature B'', and the radial field gradient B' (Bergeman et al., 1987). Several variants of this configuration have been discussed, such as the baseball trap or yin-yang trap. We regard them as alternative ways of winding coils to achieve the IP configuration, which in its most straightforward implementation consists of two pinch coils and four Ioffe bars (Bergeman et al., 1987).

Bergeman et al. (1987) discuss the field geometry of an IP trap. The trap has two different regimes. For temperatures $kT < \mu B_0$ (where μ is the atom's magnetic moment), the cloud experiences the potential of a 3D anisotropic harmonic oscillator with axial curvature $\mu B''$ and transverse curvature $\mu[(B'^2/B_0) - B''/2]$. In the case of $kT > \mu B_0$, the potential is predominantly linear along the radial direction (with a gradient of $\mu B'$) and harmonic along the axial direction. As a result, the δ parameter, characterizing the potential, changes from $\delta = \frac{3}{2}$ to $\delta = \frac{5}{2}$ for decreasing bias field. A low-bias-field IP trap is therefore advantageous for entering runaway evaporation. It features long cigar-shaped clouds. Such traps are in use in the hydrogen experiments at MIT (Greytak, 1995) and in Amsterdam (Walraven and Hijmans, 1994).

The efficiency and speed of evaporation are improved by strong confinement and adiabatic compression (see Section IV.D). Therefore, all groups pursuing evaporative cooling have developed tightly confining traps. In the hydrogen experiments this was achieved with superconducting magnets. Typical values in Amsterdam are $B'' = 440 \text{ G}/\text{cm}^2$, $B' = 2 \times 10^4$ G/cm (Luiten, 1993). Only two groups in the alkali community have built superconducting magnetic traps: The historic experiment in Pritchard's group, which was the second magnetic atom trap ever built (Bagnato et al., 1987), and a recent effort at Caltech (Willems and Libbrecht, 1995). Most alkali trappers try to avoid the added complexity of a cryogenic system; furthermore, in a cryostat the optical access to the trapping region would be limited.

Permanent magnets offer a simple alternative. The Rice group has combined efficient loading with tight field curvatures of about 1000 G/cm² (Tollet et al., 1995). Similar efforts were pursued by two groups in Germany (Ricci et al., 1994; Frerichs et al., 1992). An additional advantage compared with conventional magnets is the absence of noise on the trapping fields, which might otherwise cause heating. However, permanent magnets have the disadvantage that the magnetic fields cannot be changed or switched off, e.g., for adiabatic compression or time of flight diagnostics.

The linear trap offers superior confinement compared with a parabolic trap. This follows from the simple argument that coils that are a distance R_{coil} away from the trapped cloud and generate a field B_{coil} at the coil, produce a field gradient of $B' \approx B_{coil}/R_{coil}$ and a curvature of $B'' \approx B_{coil}/R_{coil}^2$. A cloud of size r in a linear potential with gradient B' would be confined to the same size by a curvature that is equal to B'/r. The "effective curvature" of linear confinement is therefore B_{coil}/rR_{coil}. This exceeds the curvature of a parabolic trap by R_{coil}/r, which is usually an order of magnitude or more.

A linear potential was used for the first magnitude trap (Migdall et al., 1985). When it was employed for the first demonstrations of evaporative cooling with alkali atoms (Davis et al., 1994; Petrich et al., 1994b), trap loss due to Majorana flops (Schwinger, 1937; Migdall et al., 1985; Phillips et al., 1985; Bergeman et al., 1989) near the zero of the magnetic field was encountered. These spin-flips to the high-field-seeking state occur due to a violation of the adiabatic condition that requires the Larmor frequency to be larger than the angular frequency at which the moving atoms perceive a rotating magnetic field. This region of low magnetic field has a size of about $\sqrt{2\hbar v/\pi\mu B'}$, which is about 1 μm for a velocity $v = 1$ m/sec and $B' = 1000$ G/cm, and was dubbed "the hole in the trap." As long as the hole is small compared with the cloud diameter, the trapping time is long (even longer than a minute) and evaporative cooling can increase the phase-space density by more than two orders of magnitude (Davis et al., 1995a). However, as the temperature drops, the trap loss due to the hole becomes prohibitive for further cooling.

Two methods have been demonstrated to plug the hole. One solution is to add a rotating magnetic field to the anti-Helmholtz trap. The rotating frequency is much higher than the orbiting frequency of the atoms, and much lower than the Larmor frequency. The resulting time-averaged, orbiting potential (TOP) trap is harmonic, but much tighter than what could be obtained by dc magnets of the same size (Petrich et al., 1995) (see Note added in proof). Typical field curvatures in the TOP trap are $B'^2/B_{\rm rot}$, where $B_{\rm rot}$ is the (constant) magnitude of the rotating field (about 10 G).

Another solution is to plug the hole by using electric dipole forces from a tightly focused blue-detuned laser beam (see also Section IV.B.2), which repels atoms from the center of the trap (optically plugged magnetic trap) (Davis et al., 1995b). The optically plugged trap achieved very tight confinement corresponding to a curvature of about B'/x_0, where x_0, the separation of the potential minimum from the zero of the magnetic field, was about 50 μm. The geometric average of the three curvatures exceeded the value for the TOP trap by a factor of 12. The disadvantage is the need for precision (micrometer) alignment of the laser beam with respect to the trap center.

To summarize, sufficiently tight confinement in a magnetic trap has so far only been achieved with one of the following experimental "nuisances": superconducting coils, permanent magnets, rotating magnetic fields, or the optical plug (see Note added in proof).

2. Optical Traps

Optical dipole traps (Chu et al., 1986) rely on the principle that an off-resonant laser beam attracts or repels atoms, depending on whether it is red- or blue-detuned. The trap depth depends on the laser intensity divided by the detuning, whereas the spontaneous rate of light scattering scales as the intensity divided by the square of the detuning, i.e., as the trap depth divided by the detuning. Heating due to spontaneous scattering can therefore be avoided if intense, far-detuned laser light is used (Phillips, 1992; Miller et al., 1993).

The only evaporative cooling experiment in a dipole trap was performed at Stanford by Adams et al. (1995). The trap consisted of two crossed red-detuned Nd:YAG laser beams (1.06 μm) far detuned from the 589-nm sodium resonance. Up to 8 W of total power in the two beams provided up to 900 μK trapping potential with an estimated heating rate of 1 photon/sec or 0.5 μK/sec. Such a heating rate requires fast evaporative cooling, which is possible due to the high atomic densities in dipole traps. The heating should be negligible if a CO_2 laser at 10.6 μm is used. Such a quasi-electrostatic trap has been demonstrated by Takekoshi and Knize (1996).

If only a single beam is used, the dipole trap has weak confinement along the axial direction. A recently demonstrated electro-optical hybrid trap used electrostatic confinement along the axial direction by surrounding the beam waist with a high voltage electrode (Lemonde et al., 1996).

A blue-detuned dipole trap has the advantage that atoms are repelled by the laser beams and trapped in the dark; this greatly reduces residual spontaneous scattering and, therefore, the heating rate. In such a trap, the atoms are surrounded by repulsive walls of blue-detuned light. Gravity can replace one of the walls. Such a trap was demonstrated by Davidson et al. (1995). A very flexible trap geometry can be achieved by rotating a laser beam with an acousto-optical modulator (Thompson et al., 1995). Larger trapping volumes can be realized within an enhancement cavity (Han et al., 1995). Another related trapping geometry is an atom cavity employing evanescent-wave mirrors and gravity for confinement (Aminoff et al., 1993).

A problem needing further study is heating due to beam jitter, especially if more than one laser beam is used to form the trapping potential. In comparison with magnetic traps, optical traps have the advantage that the trapping potential can be switched in less than 1 μsec, whereas magnetic traps have typical (inductive) time constants of 1 msec. Furthermore, some precision experiments (e.g., search for an electron electric–dipole moment (EDM), frequency standards) require the absence of inhomogeneous magnetic fields. Blue-defined traps are harder to load than red-detuned traps because they are missing the "sucking" action of red-detuned light.

An interesting future direction could be a hybrid approach using dipole and magnetic forces. A first realization of his concept was the optically plugged magnetic trap (Davis et al., 1995b), but one can envision further variants such as a red-detuned laser focused into a magnetic trap. The magnetic field would act as an enormous funnel to collect atoms and evaporatively cool them into the "deep dimple" of the potential created by the optical dipole force. Another possibility is to generate a Bose condensate in a magnetic trap and then (adiabatically or suddenly) transfer the atoms into some form of dipole trap—maybe by rotating a blue-detuned laser beam around the atoms similar to the trap just mentioned above (Thompson et al., 1995). In any case, dipole forces can by applied with extremely high spatial and time resolution, and we expect them to play an important role in experiments with evaporatively cooled atoms.

3. *Special Traps*

The limit for evaporative cooling of hydrogen in a conventional magnetic trap is dipolar relaxation. This is avoided if atomic hydrogen can be trapped in the ground hyperfine state (the *a* state). This can be achieved in a microwave trap (Agosta et al., 1989), which is the magnetic dipole analog of the electric-dipole-force laser trap mentioned in the previous section. The ultimate goal is to combine such a trap with a conventional magnetic trap as a hybrid trap (Silvera and Reynolds, 1991; Silvera, 1995b). Since the trap depth of the microwave trap is only on the order of several milli-Kelvin, hydrogen would be first trapped in the *d* state, evaporatively cooled to a few hundred micro-Kelvin, then transferred into the microwave trap, and further cooled by evaporation.

A microwave trap was recently demonstrated with laser-cooled cesium (Spreeuw et al., 1994). This trap was only 0.1 mK deep, mainly due to the strong influence of gravity on Cs. The gravitational pull on the Cs atoms was compensated for by a magnetic field gradient, which affected the trapping strength significantly (due to the resulting gradient in the detuning of the microwave field). For hydrogen, the situation would be much better.

Another way to trap strong-field-seeking atoms is an ac magnetic trap. It was suggested in 1985 for atomic hydrogen (Lovelace et al., 1985), and realized in 1991 with Cs (Cornell et al., 1991). This trap is much weaker than dc magnetic traps, and it has not been pursued beyond the first demonstration. Finally, ac electric fields offer another possibility to trap strong-field seekers (Riis and Barnett, 1993; Shimizu and Morinaga, 1992). However, the forces are very weak unless extremely small separations between the electrodes are chosen.

C. LASER COOLING INSIDE CONSERVATIVE TRAPS

Laser cooling the atoms after they have been loaded into a conservative trap has a double advantage: As the temperature is reduced, the density increases proportional to $T^{-\delta}$, resulting in an increase in phase space of

$T^{-\delta-3/2}$ without loss in the number. However, this is only possible until limitations of laser cooling are reached, which are due to absorption of the cooling light or to inelastic collisions.

So far, Doppler cooling inside a magnetic trap has been demonstrated by Pritchard's group at MIT for sodium (Helmerson *et al.*, 1992b), by the Rice group for lithium (Tollet *et al.*, 1995), and by the Amsterdam group for hydrogen (Setija *et al.*, 1993). In all these experiments, the atoms were loaded directly into a magnetic trap, not transferred from another trap (e.g, a MOT). Atoms released from a MOT are usually already cooled to or below the ultimate temperature of Doppler cooling, and cooling after the transfer would only compensate for the temperature increase during the transfer—which is usually only a small factor.

The gain in phase-space density is larger when sub-Doppler or even sub-recoil schemes are employed. The group in Boulder has recently cooled magnetically trapped atoms with gravitational Sisyphus cooling; this cooling scheme involves both rf transitions and optical pumping in such a way that the atom experiences a steeper potential when moving outward from the trap center than when moving inward (Newbury *et al.*, 1994). This scheme realized the earlier proposal of Pritchard (1983) and Pritchard and Ketterle (1992), then called cyclic cooling.

Raman cooling down to the recoil limit was applied to sodium atoms in an optical dipole trap, resulting in a final density of 4×10^{11} cm^{-3} and temperature of 1.0 μK (Lee *et al.*, 1996), the highest phase-space density which has been achieved so far by optical cooling. The limitation was probably related to inelastic hyperfine changing collisions.

Cyclic cooling can in principle reach sub-recoil temperatures. However, the schemes which have reached sub-recoil temperatures thus far are not compatible with magnetic trapping, but they are compatible (at least in principle) with optical dipole traps (Dum *et al.*, 1994; Pellizzari and Ritsch, 1995). With the recent progress in Raman cooling (Lee *et al.*, 1996; Reichel *et al.*, 1995) and velocity-selective coherent population trapping (Lawall *et al.*, 1995; Esslinger *et al.*, 1996), there is hope to reach extremely high phase-space densities in far-off-resonant dipole traps, requiring only a modest step of evaporative cooling to reach quantum degeneracy.

D. ADIABATIC COMPRESSION

Adiabatic compression is a simple technique, but we want to emphasize the importance of such a compression of the atom cloud prior to evaporative cooling. If, after the loading, the potential is adiabatically increased by a factor a, the temperature rises by a factor $a^{2\delta/(2\delta+3)}$ and the density by $a^{3\delta/(2\delta+3)}$. Phase-space density is, of course, constant, but the elastic collision rate and, therefore, the speed of evaporative cooling has been increased by a factor $a^{4\delta/(2\delta+3)}$. The ratio of good to bad collisions goes up, either in proportion to the speed (when background gas collisions are the dominant loss mechanism) or in proportion to $\sqrt{T}$ in the case of dipolar relaxation. In the evaporative cooling experiments in Boulder and by our group, adiabatic compression was essential. Davis et al. (1995a) increased the collision rate by a factor of 20 by this technique. Generally, it is always favorable to work at the tightest confinement until three-body recombination sets in or the sample becomes collisionally thick. Those limits were reached in the MIT sodium experiment (Davis et al., 1995b). In hydrogen experiments, the density to reach collisional thickness is relatively low because the sample is cigar-shaped and the evaporation is one-dimensional along the long axis (see Section IV.E.3). In these experiments, the dense regime is avoided by lowering the radial confinement (Graytak, 1995; Doyle et al., 1994).

E. EVAPORATION TECHNIQUES

Evaporative cooling is most efficient if the high energy atoms, produced in collisions, leave the trap without experiencing additional collisions. All techniques of evaporation select the most energetic particles spatially; i.e., it is a selection based on potential energy. It is conceivable to perform the selection based on velocity using the Doppler shift of a narrow resonance line and thus select particles according to their kinetic energy, but since the kinetic and potential energy distributions are related by the virial theorem, there is no obvious advantage in velocity selection. The different evaporation techniques to be discussed here differ in the manner of spatial selection. As will be discussed, the evaporation mechanism and the trap geometry play important roles in the efficiency of the cooling.

1. *Direct Contact with Walls*

The simplest conceivable scheme for evaporation is to have a sticky wall close to the atom cloud, which absorbs the high energy tail of the distribution. Such a method was already used in the first realization of electromagnetic confinement of neutral particles, the neutron storage ring by Paul and collaborators (Kügler et al., 1985). A movable beam scraper was used to remove neutrons on outer orbits. Absorption to the walls provided the initial evaporative cooling in the MIT hydrogen experiment (Hess et al., 1987). During the loading, the walls are kept at a higher temperature (about 250 mK), at which the desorption rate is higher than the rate of surface recombination. Subsequently, the wall temperature is lowered, and it acts as an absorbing surface providing evaporative cooling. As the cloud shrinks, the evaporation process slows down, and it is subsequently forced by ramping down the axial trap depth (to be discussed) or by moving the sample closer to the walls (D. G. Fried, personal communication, 1996).

In Amsterdam, the surface was kept at a constant temperature of 200 mK, and the evaporation was triggered by boiling off the helium film from the surface of a bolometer, thus creating a sticky surface for atomic hydrogen (where the atoms recombine) (Luiten *et al.*, 1993). Various aspects of whether a wall is sticky (i.e., it removes atoms) or provides thermalization are exhaustively discussed by Setija (1995).

Evaporation to the walls is initially a two- (or even three-) dimensional process, depending on the geometry of the cell. Forcing evaporation by moving the sample toward one wall may result in 1D evaporation, which is much less effective.

2. *Lowering the Total Trapping Potential*

Evaporative cooling can be forced by lowering the total trapping potential. This was done in the crossed dipole trap in Stanford, wherein the total laser power was ramped down (Adams *et al.*, 1995). This is essentially an adiabatic decompression; however, after decompression there are fewer bound states in the potential, and evaporation occurs because states that were bound in the tight potential correspond to continuum states after decompression. The method is very simple, but it has the major disadvantage that the collision rate becomes very small during the decompression. One might have to repeat several compression/decompression cycles, where intermediate compression speeds up the thermalization. A potential disadvantage of this repetitive scheme is that the cold atoms are not left alone; rather, they are continuously compressed and decompressed. Unless this is done fully adiabatically, some heating might occur.

3. *Saddle Point of the Potential*

The disadvantage of the previous method is due to the proportionality between trap depth and confinement. This was avoided in the MIT hydrogen experiments, where evaporation was forced by lowering only the axial trapping potential (Masuhara *et al.*, 1988; Hess, 1986). The hottest atoms escaped over the saddle point along the axial direction. The trapping volume, and therefore the collision rate, could be controlled independently through the radial potential. This decoupling between confinement and trap depth allowed continuous evaporation. Eventually, when the mean free path became comparable with the axial length of the sample, the radial confinement was lowered to avoid collision times shorter than the axial oscillation period.

The saddle-point method is a 1D selection method because it selects the atoms due to their axial energy. If the ergodic mixing time is longer than the time between elastic collisions, this decreases both the speed and efficiency of evaporation (Surkov *et al.*, 1994, 1996). In some of the MIT experiments on hydrogen ergodic mixing was fast enough to ensure 3D evaporation (Doyle, 1996).

4. *Radiative Evaporation*

Pritchard and collaborators were the first to perform rf spectroscopy of magnetically trapped atoms. They used the rf lineshape to determine the energy distribution of the trapped atoms (Martin *et al.*, 1988; Helmerson *et al.*, 1992a). In 1989, they suggested that this method could be used to perform evaporative cooling. More generally, they introduced the concept of *radiative evaporation*, which means that a radiation field is used to transfer atoms from a trapped to an untrapped state in an energy-selective way (Pritchard *et al.*, 1989). There are two ways of implementing this concept: (1) narrow rf or optical transitions, which are energy-selective due to the one-to-one correspondence between Zeeman shift and potential energy in a magnetic trap, or (2) a laser beam focused to the edge of a large cloud, which causes spatially selective optical pumping to an untrapped state. In the latter case, the linewidth of the transition may be broad.

The Amsterdam group independently suggested the concept of spatially selective microwave or optical transitions, and they proposed an additional way of spatial selection: A near-resonant laser beam would be absorbed in the outer layer of the cloud and automatically perform spatially selective optical pumping (Hijmans *et al.*, 1989). The cold atoms are shielded by the more energetic atoms and are unaffected. This method, called light-induced evaporation, was demonstrated in a proof-of-principle experiment in Amsterdam by Setija *et al.* (1993).

The most successful implementation of radiative evaporation in rf-induced evaporation. The original suggestion (Pritchard *et al.*, 1989) mentioned three advantages compared with the saddle-point method: (1) The magnetic trapping potential does not have to be modified for the evaporation and can thus be kept at the optimum confinement. (2) The escape rate and the potential energy of the escaping atoms is easily controlled by the amplitude and frequency of the applied radiation. (3) The space where the particles escape from the trap is given by the shell where the rf transition is tuned into resonance by the local magnetic field, and thus it is not limited to a saddle point in the trapping potential.

The use of rf-induced evaporation was first reported at the 1993 OSA meeting (Ketterle *et al.*, 1993). Most of the evaporative cooling experiments on alkalis used this technique, as will be discussed in Section V.B. In the following, we discuss two aspects of rf-induced evaporation in more depth: the process of evaporation in the adiabatic and diabatic limits and the dimension of the rf-induced evaporation for different traps.

Adiabatic and Diabatic limit of Rf-induced Evaporation. Rf-induced evaporation can be described in the dressed-atom formalism (Cohen-Tannoudji *et al.*, 1992), where the different m_F states of an atom with spin F are coupled to an rf field, which we assume to be linearly polarized: $\vec{B}_{rf}(t) = B_0 \cos(\omega t)\hat{e}_{rf}$.

The coupling matrix element between levels $|F, m_F\rangle$ and $|F, m_F \pm 1\rangle$ is $(1/4)g\mu_B B_0(\hat{e}_{rf} \times \hat{e}_z)\sqrt{F(F+1) - m_F(m_F \pm 1)}$, where g is the atomic g factor, and the trapping field points along the quantization axis $\hat{e}_z$.

The adiabatic potential curves $U(r)$ are obtained by finding the dressed-state eigenvalues for the local magnetic field $B(r)$. In the dressed-atom picture, one considers the total energy of the atom plus the field of N rf photons. Without coupling, this simply means that $N\hbar\omega$ is added to the atomic Zeeman energies, resulting in a Zeeman pattern for each N that is vertically displaced by $N\hbar\omega$. At positions where the rf field is in resonance, curves with $\Delta N = 1$ cross. The coupling transforms these crossings into avoided crossings. This determines the pattern of the adibatic energy levels (Fig. 5b).

A slowly moving atom stays on the adiabatic potential curve. As an example, we assume that an atom in the $|F, F\rangle$ hyperfine state moves outward from the center of the trap. When it comes close to resonance, the rf field mixes this state with the other m_F states, which leads to a flatter potential curve. Beyond the resonance point, the atomic state has been adiabatically transformed into a nontrapped, strong-field-seeking state, and the atom is repelled from the trap. While traversing the avoided crossing, the atom has thus emitted $2F$ rf photons in a stimulated way and reversed the direction of both electron and nuclear spin.

This picture of rf-induced evaporation emphazises the similarity with other evaporation methods because the radiofrequency leads to an adiabatic potential surface with a trap depth that is approximately $|m_F|\hbar(\omega - \omega_0)$, where ω_0 is the rf resonance frequency in the center of the trap. The evaporation process is then just the "spilling" of the most energetic atoms out of the trap.

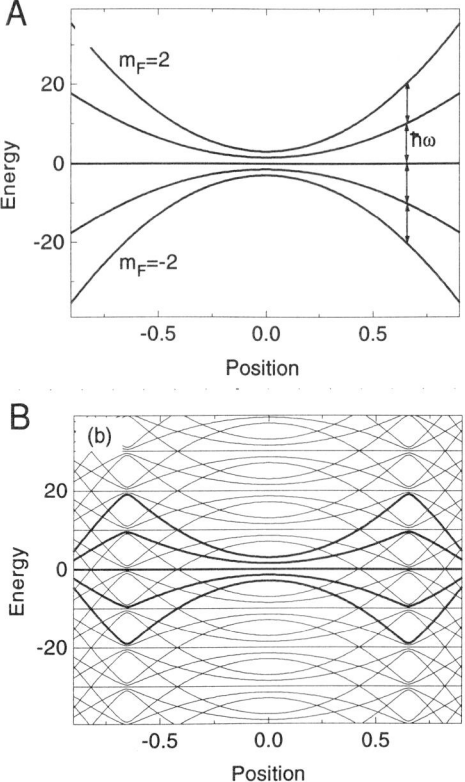

FIG. 5. Potential curves for rf-induced evaporation of an atom with spin $F = 2$, in a quadratic trap with nonzero bias field. (a) Diabatic curves, for different m_F sublevels, relevant for weak rf fields. The arrows indicate the position where the rf, with frequency ω, is resonant. (b) Adiabatic curves (dressed-atom energy levels), relevant for strong rf fields. The pattern is periodic in energy, with spacing $\hbar\omega$. One set of curves is drawn with fat lines, to show how trapped states adiabatically evolve into untrapped states, and vice versa, around the resonance position.

This adibatic picture is valid as long as an adiabaticity condition is fulfilled. This condition requires that the energy gap due to the avoided crossing be larger than the energy uncertainty related to the finite amount of time that an atom with velocity v spends in the region of resonance. For a two-level system coupled by a matrix element V_{12}, and an atom moving with velocity v along the x axis, the transition probability P between the adiabatic curves is given by the Landau–Zener formula (Rubbmark *et al.*,

1981) $P = 1 - \exp(-\xi)$, with $\xi = 2\pi |V_{12}|^2/\hbar g \mu_B B' v$. The Landau–Zener theory is strictly valid only for a two-level system, and we use it here only for a qualitative discussion of the two limiting cases.

For a weak rf field, $\xi \ll 1$, P is much smaller than unity; i.e., the atoms predominately stay on the diabatic surface shown in Fig. 5(a). The transition probability for a spin-flip is $P \approx \xi$, which describes the diabatic limit of rf-induced evaporative cooling: The atomic energy levels are almost unperturbed, the atoms slosh many times through the resonance shell, and only after $1/P$ oscillations are they spin-flipped from the $|F, F\rangle$ to the $|F, F - 1\rangle$ hyperfine state.

The adiabatic limit is clearly the ideal situation for evaporative cooling. However, the evaporation process in a trap (with oscillation time T_{osc}) saturates at a lower rf power. The condition for saturation is $P \approx T_{\text{osc}}/\tau_{\text{el}}$, which means that an energetic atom evaporates before it collides again. In the case of $F > 1$, we have the additional complication that an atom that has been spin-flipped from the $|F, F\rangle$ state to the $|F, F - 1\rangle$ state is still trapped (see Fig. 5) and can undergo spin exchange collisions with the trapped atoms, leading to heating and trap loss. The evaporating atom reaches an untrapped state after losing $F\hbar$ of spin angular momentum which due to the randomness of spin-flips up and down requires a time of $F^2 T_{\text{osc}}/P$. This has to be shorter than the spin exchange time $\tau_{\text{ex}} \approx 1/G_{\text{ex}} n_0$. To achieve the full efficiency of evaporation, the probability P for spin-flips thus has to satisfy

$$P > T_{\text{osc}} \left(\frac{1}{\tau_{\text{el}}} + \frac{F^2}{\tau_{\text{ex}}} \right) \tag{16}$$

With a typical rate coefficient for spin relaxation of $G_{\text{ex}} \approx 10^{-12}$ cm^3/sec and an elastic cross section of 10^{-12} cm^2, the second term is dominant for velocities smaller than a few centimeters per second.

Only the magnetic field component of the rf *perpendicular* to the magnetic trapping field induces spinflips. The directions of the trapping field on the surface for evaporation cover the full solid angle in certain traps, e.g., in the spherical quadrupole trap and in an IP trap with small bias field. As a result, there are two points where the transition matrix elements are zero, because the trapping field and the rf field are parallel. In an area around these points, the diabatic coupling case is realized. In practice, the rf transition can be sufficiently saturated so that this area is small and does not strongly affect the speed of evaporation.

Dimension of Rf-Induced Evaporation. Rf-induced evaporation spin-flips atoms on a resonant shell where $g\mu_B B = \hbar \omega_{\text{rf}}$. This shell surrounds the trapped atoms. Rf-induced evaporation is a 3D evaporation technique if shells with $|B|$ = const are equipotential surfaces. This is the case in a dc magnetic trap, as long as gravity can be neglected.

However, gravity pulls the cloud downward in the magnetic potential; the bottom of the cloud experiences higher magnetic field. Since evaporating atoms have an average excess energy of κkT with $\kappa \approx 1$, evaporation happens mainly at the bottom when the gravitational energy varies by more than $\pm kT$ over the equipotential surface $U = \eta kT$. For harmonic confinement along the z axis, $U = U'' z^2/2$, this equipotential surface is at $z \approx \sqrt{2\eta kT/U''}$. Evaporative cooling becomes one-dimensional for

$$kT < \frac{2\eta(mg)^2}{U''} \tag{17}$$

To express this condition in more practical units, we denote by $B'_{grav} = mg/\mu$ the magnetic field gradient that counterbalances gravity (e.g. 4.1 G/cm for Na in the $|2,2\rangle$ hyperfine state). With the magnetic field curvature B'' along the z axis, Eq. (17) is equivalent to

$$kT < \frac{2\eta\mu B'^2_{grav}}{B''} \tag{18}$$

For $B'' = 500$ G/cm^2 and $\eta = 5$, the limiting temperature is 1 μK for ^{7}Li, 10 μK for Na, and 150 μK for Rb. Below this temperature, evaporative cooling becomes less efficient. In the Li experiment at Rice (Bradley et al., 1995), this only happens at temperatures close to the onset of quantum degeneracy. The MIT Na experiment operated at a vertical field curvature of $\approx 5 \times 10^4$ G/cm^2, which was high enough to avoid 1D evaporation (Davis et al., 1995b). One-dimensional evaporation will affect the cooling most strongly for small ratios of good to bad collisions. One-dimensional evaporation might be unavoidable because a sufficiently large B'' cannot be provided for the lowest desired temperature. In this case, the experimental strategy should be to speed up evaporation in the first phase and obtain a large value for R, so that evaporation does not fall below the threshold for runaway evaporation when λ increases due to the onset of 1D evaporation (see Eq. (7)).

For a linear potential characterized by a field gradient B', the criterion for 1D evaporation is

$$B' > \eta B'_{grav} \tag{19}$$

which is independent of temperature. In the adiabatic potential picture, 1D evaporation can be visualized as a volcano crater with a slanted rim due to gravity. Atoms spill out at the lowest part of the rim, which is now a saddle point of the total adiabatic potential.

The geometry of evaporation is very different in the TOP trap (see Section IV.B.1). The surface for evaporation is always the portion of the outer shell of the cloud that is exposed to the highest magnetic field. In the TOP trap the magnetic field varies periodically, and the shell for evaporation are the points $\vec{r}$, where $\max_t |B(\vec{r}, t)| = \hbar \omega_{rf}/g\mu_B$ and the maximum is taken over one period of the rotating field (Fig. 6). This shell is approximately a cylinder, or more precisely a barrel, but the curvature is negligible. For a rotating horizontal bias field, gravity displaces the cloud only along the cylinder axis, and evaporation is always 2D. If the bias field rotates in a different way, the cylinder is tilted with respect to the vertical axis, and a displacement of the cloud by gravity leads to 1D evaporation. The critical displacement is about $1/\sqrt{8\eta}$ of the cloud diameter.

The major step toward evaporative cooling in alkalis was the attainment of a sufficiently large ratio of good to bad collisions for runaway evaporation. Our theoretical discussion in Section IV.D suggests that this ratio might be an order of magnitude larger for 1D evaporation. The three successful BEC experiments avoided 1D evaporation because of three different reasons: the magnetic field rotating around a vertical axis for Rb (Anderson *et al.*, 1995), the light mass of Li (Bradley *et al.*, 1995), and the tight confinement for Na (Davis *et al.*, 1995b). However, the important question of the dimension of evaporation was first addressed in detail by

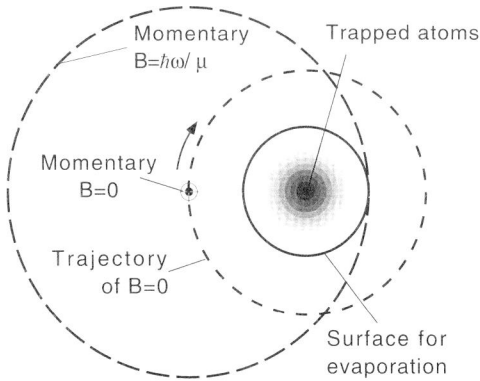

FIG. 6. The geometry of rf-induced evaporation in the TOP trap. Shown here is the plane through the trap center and the rotating bias field.

Surkov *et al.* (1996), so that probably these three solutions were found without really knowing the problem!

5. *Other Schemes*

There are other possible methods for evaporating atoms. Every loss process that happens predominately for atoms in the outer part of the cloud has a cooling effect. In the TOP trap, the rotating zero of the magnetic field leads to Majorana flops and was used for the initial cooling of the cloud (Petrich *et al.*, 1995). Similarly, in the optically plugged trap (Davis *et al.*, 1996b), the plug was initially "submerged" by the initial atom cloud, which might have resulted in some additional cooling due to Majorana flops. However, in both cases the evaporation is of lower dimension and is probably inferior to rf-induced evaporation.

6. *Relation to Other Cooling Methods*

Evaporative cooling relies on selective removal of energetic atoms and on elastic collisions. Another cooling scheme that relies on collisions is sympathetic cooling: One species is cooled by collisional thermalization with another species that is kept at a cold temperature. So far, this scheme has only been realized with trapped ions (Larson *et al.*, 1986; Imajo *et al.*, 1996).

The selection of energetic atoms is of course also an important part of many optical cooling schemes. In Raman cooling (Kasevich and Chu, 1992), velocity-selective coherent population trapping (VSCPT) (Aspect *et al.*, 1988), and cyclic cooling (Pritchard, 1983; Newbury *et al.*, 1994; Cirac and Lewenstein, 1995) [which are examples of phase-space optical pumping (Pritchard *et al.*, 1987)], energetic atoms are selectively excited using narrow optical or rf resonances. However, in contrast to evaporative cooling, these atoms are cooled and recycled. The success of evaporative cooling demonstrates that it is much easier to select atoms and waste them than to cool them!

There is one cooling method that relies on the absence of collisions: one-dimensional adiabatic cooling, which has been discussed for obtaining antihydrogen below 1 mK (Shlyapnikov *et al.*, 1993). In this case, the goal is to avoid the mixing with the motion in the other two dimensions to get the maximum cooling effect in the vertical dimension. That is just the opposite of what is desirable in evaporative cooling, but studies of ergodic behavior of trapped atoms are relevant for both adiabatic and evaporative cooling (Surkov *et al.*, 1994, 1996).

V. Summary of Evaporative Cooling Experiments

Evaporative cooling of atomic gases has so far only been demonstrated by a limited number (six) of groups. In this section, we give an overview and a comparison of the evaporative cooling experiments reported so far. Although the number of experiments is limited, and most of them were demonstration experiments rather than careful studies and optimizations of evaporative cooling, we feel that the tabular comparison of the speed and efficiency of the cooling provides insight.

A. HYDROGEN

Evaporative cooling of an atomic gas was first observed by the Greytak/Kleppner group (Hess *et al.*, 1987). The temperature of the gas was estimated to be 40 mK from the long lifetime of the trapped atoms, consistent with evaporative cooling by adsorption and recombination on the walls of the cell. Because of the limited diagnostics (only the number of trapped atoms could be measured by nuclear spin resonance), little information about the evaporation process could be gained.

This group proceeded in the next year to forced evaporative cooling (Masuhara *et al.*, 1988). As the atoms (evaporatively) cooled, the axial depth of the trapping potential was lowered exponentially, while the radial confinement was simultaneously reduced. The radial trap depth was always kept larger than the axial depth, resulting in evaporation over a saddle point of the potential. A considerable increase in phase-space density was observed. Much more information could be extracted by a more sophisticated detection scheme (Doyle *et al.*, 1989). With a bolometric detection technique, and by quickly ramping down the trapping rim, the energy distribution of the trapped particles was measured. Temperatures as low as 800 μK were reported in 1989.

The closest approach to BEC of hydrogen to date was made in 1991 (Doyle, 1991; Doyle *et al.*, 1991) when densities as high as 8×10^{13} cm^{-3} and temperatures as low as 100 μK were reported. Further progress was impeded by the limits of the bolometric detection method. The sticking probability of hydrogen on the superfluid ^{4}He surface was measured at submilli-Kelvin temperatures, and quantum reflection was demonstrated. The evaporative cooling process was in good agreement with model calculations.

Very recently, trapped atomic hydrogen was observed by two-photon spectroscopy (Cesar *et al.*, 1996). This provided not only greatly improved sensitivity, but also nondestructive monitoring of the cooling process, and should result in further improvement of the evaporative cooling.

In Amsterdam, magnetic trapping of hydrogen was reported in 1988 (van Roijen et al., 1988). In 1993, using Lyman α spectroscopy, an accurate comparison of the cooling process with a theoretical model was made (Luiten et al., 1993, 1994). These experiments were performed in a fixed field geometry, and the evaporation process was suddenly switched on by boiling off the helium coating on a bolometer, thus creating a sticky surface for hydrogen atoms.

In the same year, this group succeeded in demonstrating Doppler cooling for hydrogen, the first laser cooling experiment with hydrogen. This allowed them to reach significantly higher initial phase-space densities before evaporative cooling. Furthermore, light-induced evaporation was demonstrated (Setija et al., 1993).

B. ALKALI ATOMS

The achievement of evaporative cooling of alkali atoms was announced by the MIT and JILA groups at IQEC in May 1994 and reported at ICAP-14 (Davis et al., 1994; Petrich et al., 1994b). In the same summer, the Stanford groups also observed evaporative cooling, in an optical trap (Lee et al., 1995). The results appeared in referred journals in the first half of 1995 (Adams et al., 1995; Davis et al., 1995a; Petrich et al., 1995), and by the end of 1995, two of these groups had reported the observation of BEC (Anderson et al., 1995; Davis et al., 1995b), while a fourth group had observed evaporative cooling of alkalis and also obtained evidence for BEC (Bradley et al., 1995).

In the JILA experiment, a dark SPOT trap was loaded from a room temperature vapor cell at 10^{-11} Torr (Anderson et al., 1994). The low pressure was needed to achieve long trapping times, but it slowed down the loading process to 5 min. Approximately 10^6 atoms were trapped. A magneto-optic compression scheme (Petrich et al., 1994a), polarization gradient cooling, and optical pumping were applied before suddenly switching on the TOP trap, and starting rf-induced evaporation (Anderson et al., 1995).

In the MIT sodium experiments, a thermal beam of sodium atoms was slowed in a Zeeman slower (Joffe et al., 1993) and trapped in a dark SPOT trap (Ketterle et al., 1993). This technique yielded large numbers of atoms ($\approx 10^{10}$) at high densities (5×10^{11} cm^{-3}), within a 2-sec loading time. After polarization gradient cooling, a spherical-quadrupole trap was suddenly switched on, and rf-induced evaporation was performed. Majorana flops were prevented by an optical plug (Davis et al., 1995b).

In the Stanford experiment, a crossed-dipole trap was loaded from a magneto-optical trap. In the final stage of the loading, the (temporal) dark SPOT technique was used to enhance the density to 4×10^{12} cm^{-3}. Within two seconds, approximately 5000 atoms were loaded in the dipole trap, and evaporation was done by adiabatically ramping down the trapping potential (Adams et al., 1995).

In the Rice experiment, an atomic beam was Zeeman slowed, deflected and collimated, and passed through the center of a permanent-magnet trap (Tollet et al., 1995). Doppler cooling provided the dissipation necessary to load the atoms into the trap. This resulted in 2×10^8 atoms at a density of 7×10^{10} cm^{-3} and a temperature of 200 μK. Evaporative cooling was done by ramping down an rf field, which, due to the high bias field of the trap, was in the gigahertz range. Trapping times of longer than 10 min allowed long cooling times, which were necessary due to the low initial density (Bradley et al., 1995).

C. COMPARISON

Table III compiles the results of the experiments just mentioned. Although few careful comparisons with theory have been made so far, the table clearly shows that evaporative cooling has fulfilled its promises of efficiency and speed. In several cases, an order of magnitude decrease in the number of atoms resulted in a phase-space density increase of two to three orders of magnitude. In many experiments, this increase only required a few hundred elastic collision times. A rate of one order of magnitude phase-space density increase per second and an increase of the elastic collision rate by two orders of magnitude were also demonstrated. In agreement with our discussion in Section II, a large increase in phase-space density was only possible with constant or increasing collision rate.

VI. Outlook

The efficiency and versatility of evaporative cooling have been impressively demonstrated. Three groups have succeeded in reaching sufficient phase-space density to observe Bose–Einstein condensation of alkali atoms. In these experiments, evaporative cooling served to gain the last six orders of magnitude in phase-space density to the BEC transition line, and further evaporative cooling down to clouds with a condensate fraction of almost 100% was demonstrated (Davis et al., 1995b; Anderson et al., 1995). These results have triggered an enormous interest in evaporative cooling and

TABLE III
Overview of Evaporative Cooling Experiments

Atom	N^a	$n_0{}^a$ (cm^{-3})	T^a (K)	D^a	$\tau_{el}^{-1\,a}$ (sec^{-1})	t^b (sec)	$\gamma_{tot}{}^c$
^{1}H^d	2×10^{12}	5×10^{12}	0.05	2×10^{-6}	20	200	2.5
	3×10^{11}	8×10^{12}	3×10^{-3}	2×10^{-4}	7		
^{1}H^e	7×10^{12}	2×10^{13}	1.1×10^{-3}	2×10^{-3}	9	250	1.6
	3×10^{11}	8×10^{13}	1×10^{-4}	0.4	13		
^{1}H^f	5×10^{11}	3×10^{11}	0.2	2×10^{-8}	2	100	0.8
	4×10^{10}	4×10^{11}	0.06	1.5×10^{-7}	2		
^{1}H^g	8×10^{10}	5×10^{12}	0.011	2×10^{-5}	8	40	0.8
	7×10^{9}	4×10^{12}	3×10^{-3}	1.3×10^{-4}	4		
^{7}Li^h	2×10^{8}	7×10^{10}	2×10^{-4}	7×10^{-6}	3	300	1.7
	1×10^{5}	1.4×10^{12}	4×10^{-7}	2.6^l	2		
^{23}Na^i	5×10^{3}	4×10^{12}	1.4×10^{-4}	1.2×10^{-4}	8×10^{2}	2	1.5
	5×10^{2}	6×10^{11}	4×10^{-6}	4×10^{-3}	20		
^{23}Na^j	1×10^{9}	1×10^{11}	2×10^{-4}	2×10^{-6}	23	7	1.9
	7×10^{5}	1.5×10^{14}	2×10^{-6}	2.6^l	3×10^{3}		
^{87}Rb^k	4×10^{6}	4×10^{10}	9×10^{-5}	3×10^{-7}	5	70	3.0
	2×10^{4}	3×10^{12}	1.7×10^{-7}	2.6^l	15		

[a] Upper row: initial value, lower row: final value, for number of atoms N, peak density n_0, temperature T, phase space density D, and peak elastic collision rate τ_{el}^{-1}.

[b] Duration of forced evaporation sweep t.

[c] The overall efficiency of evaporation $\gamma_{tot} = \log(D_f/D_i)/\log(N_i/N_f)$.

[d] Masuhara et al. (1988), MIT, cryogenic Ioffe–Pritchard (IP) trap, saddle-point evaporation.

[e] Doyle et al. (1991), MIT, cryogenic IP trap, saddle-point evaporation.

[f] Luiten et al. (1993), Amsterdam, cryogenic IP trap, saddle-point evaporation.

[g] Setija et al. (1993), Amsterdam, cryogenic IP trap, light-induced evaporation.

[h] Bradley et al. (1995), Rice, permanent-magnet IP trap, rf-induced evaporation.

[i] Adams et al. (1995), Stanford, crossed-dipole trap, evaporation by lowering the trap potential.

[j] Davis et al. (1995b), MIT, optically plugged, linear magnetic trap, rf-induced evaporation.

[k] Anderson et al. (1995), JILA, time-averaged, orbiting potential trap, rf-induced evaporation.

[l] Bose–Einstein condensation was reached; the number in this row reflect the situation at the transition point.

have led to a flurry of activity. We expect several parts of this chapter to be outdated soon by the rapid pace of future developments!

Doppler cooling was the method to reach 100 μK temperatures; polarization gradient cooling gave access to a few μK. Although an optical sub-recoil scheme holds the record for the lowest 1D temperature (3 nK), we regard evaporative cooling to be the most promising technique for the nano-Kelvin regime. Nano-Kelvin atoms are (as far as we know) the coldest matter in the universe. The step from micro-Kelvin to nano-Kelvin gave us access to new physics—quantum degenerate atomic gases. In contrast to many other cooling techniques, evaporative cooling is almost entirely classical, and it is amusing that it was this technique that revealed a hidden aspect of the quantum world.

Acknowledgments

We are grateful to M. R. Andrews, D. S. Durfee, D. M. Kurn, and M.-O. Mewes for valuable discussions and to J. Doyle for helpful comments on the manuscript. This work was supported by ONR, NSF, JSEP, and the Sloan Foundation. N.J.v.D. would like to acknowledge a Talent fellowship from "Nederlandse Organisatie voor Wetenschappelijk Onderzoek (NWO)" and a Fulbright fellowship from NACEE.

Note added in proof. The discussion on tightly confining magnetic traps in Section IV.B.1 was incomplete. The comparison between different traps assumed nearly isotropic confinement. An anisotropic Ioffe–Pritchard trap with tight radial confinement achieves a larger geometric mean of the three magnetic field curvatures than the TOP trap. The mean curvature is the figure of merit of a trap for evaporative cooling. Efficient evaporation and Bose–Einstein condensation in an optimized IP trap with only d.c. electromagnets have recently been obtained in our group. Evaporative cooling began with $\approx 2 \times 10^9$ atoms and reached BEC with 1.5×10^7 atoms after 15 to 25 sec (Mewes *et al.*, 1996).

References

Abraham, E. R. I., McAlexander, W. I., Sackett, C. A., and Hulet, R. G. (1995). *Phys. Rev. Lett.* **74**, 1315.

Adams, C. S., Lee, H. J., Davidson, N., Kasevich, M., and Chu, S. (1995). *Phys. Rev. Lett.* **74**, 3577.

Agosta, C. C. Silvera, I. F., Stoof, H. T. C., and Verhaar, B. J. (1989). *Phys. Rev. Lett.* **62**, 2361.

Aminoff, C. G., Steane, A. M., Bouyer, P., Desbiolles, P., Dalibard, J., and Cohen-Tannoudji, C. (1993). *Phys. Rev. Lett.* **71**, 3083.
Anderson, M. H., Petrich, W., Ensher, J. R., and Cornell, E. A. (1994). *Phys. Rev. A* **50**, R3597.
Anderson, M. H., Ensher, J. R., Matthews, M. R., Wieman, C. E., and Cornell, E. A. (1995). *Science* **269**, 198.
Aspect, A., Arimondo, E., Kaiser, R., Vansteenkiste, N., and Cohen-Tannoudji, C. (1988). *Phys. Rev. Lett.* **61**, 826.
Bagnato, V. S., Lafyatis, G. P., Martin, A. G., Raab, E. L., Ahmad-Bitar, R. N., and Pritchard, D. E. (1987). *Phys. Rev. Lett.* **58**, 2194.
Bardou, F., Emile, O., Courty, J.-M., Westbrook, C. I., and Aspect, A. (1992). *Europhys. Lett.* **20**, 681.
Bergeman, T., Erez, G., and Metcalf, H. (1987). *Phys. Rev. A* **35**, 1535.
Bergeman, T., McNicholl, P., Kycia, J., Metcalf, H., and Balazs, N. L. (1989). *J. Opt. Soc. Am. B* **6**, 2249.
Berman, L. D. (1961). "Evaporative Cooling of Circulating Water." Pergamon, New York.
Boesten, H. M. J. M., Moerdijk, A. J., Verhaar, B. J. (1996). *Phys. Rev. A* **54**, R29.
Bradley, C. C., Sackett, C. A., Tollet, J. J., and Hulet, R. G. (1995). *Phys. Rev. Lett.* **75**, 1687.
Cable, A., Prentiss, M., and Bigelow, N. P. (1990). *Opt. Lett.* **15**, 507.
Castin, Y., Dalibard, J., and Cohen-Tannoudji, C. (1995). In "Bose-Einstein Condensation" (A. Griffin, D. W. Snoke, and S. Stringari, eds.), BEC-93, p. 173. Cambridge Univ. Press, Cambridge, UK.
Cesar, C. L., Fried, D. G., Killian, T. C., Polcyn, A. D., Sandberg, J. C., Yu, I. A., Greytak, T. J., Kleppner, D., and Doyle, J. M. (1996). *Phys. Rev. Lett.* (in press).
Chen, J., Story, J. G., Tollet, J. J., and Hulet, R. G. (1992). *Phys. Rev. Lett.* **69**, 1344.
Chu, S., Bjorkholm, J. E., Ashkin, A., and Cable, A. (1986). *Phys. Rev. Lett.* **57**, 314.
Cirac, J. I. and Lewenstein, M. (1995). *Phys. Rev. A* **52**, 4737.
Coakley, K. J. (1996). Submitted.
Cohen-Tannoudji, C., Dupont-Roc, J., and Grynberg, G. (1992). "Atom-Photon Interactions." Wiley, New York.
Cornell, E. A., Monroe, C., and Wieman, C. E. (1991). *Phys. Rev. Lett.* **67**, 2439.
Davidson, N., Lee, H.-J., Kasevich, M., and Chu, S. (1994). *Phys. Rev. Lett.* **72**, 3158.
Davidson, N., Lee, H. J., Adams, C. S., Kasevich, M., and Chu, S. (1995). *Phys. Rev. Lett.* **74**, 1311.
Davis, K. B., Mewes, M. O., Joffe, M. A., and Ketterle, W. (1994). *Int. Conf. At. Phys. 14th*, Boulder, CO, 1994, Book Abst., 1-M3.
Davis, K. B., Mewes, M.-O., Joffe, M. A., Andrews, M. R., and Ketterle, W. (1995a). *Phys. Rev. Lett.* **74**, 5202.
Davis, K. B., Mewes, M.-O., Andrews, M. R., van Druten, N. J., Durfee, D. S., Kurn, D. M., and Ketterle, W. (1995b). *Phys. Rev. Lett.* **75**, 3969.
Davis, K. B., Mewes, M.-O., and Ketterle, W. (1995c). *Appl. Phys. [Part] B* **B60**, 155.
Doyle, J. (1996). Private communication.
Doyle, J. M. (1991). Ph.D. Thesis, Massachusetts Institute of Technology, Cambridge, MA.
Doyle, J. M., Sandberg, J. C., Masuhara, N., Yu, I. A., Kleppner, D., and Greytak, T. J. (1989). *J. Opt. Soc. Am. B* **6**, 2244.
Doyle, J. M., Sandberg, J. C., Yu, I. A., Cesar, C. L., Kleppner, D., and Greytak, T. J. (1991). *Phys. Rev. Lett.* **67**, 603.
Doyle, J. M., Sandberg, J. C., Yu, I. A., Cesar, C. L., Kleppner, D., and Greytak, T. J. (1994). *Physica B (Amsterdam)* **194-196**, 13.

Doyle, J. M., Friedrich, B., Kim, J., and Patterson, D. (1995). *Phys. Rev. A* **52**, R2515.
Drewsen, M., Laurent, P., Nadir, A., Santarelli, G., Clairon, A., Castin, Y., Grison, D., and Salomon, C. (1994). *Appl. Phys.* [*Part*] *B* **B59**, 283.
Dum, R., Marte, P., Pellizzari, T., and Zoller, P. (1994). *Phys. Rev. Lett.* **73**, 2829.
Ertmer, W., Blatt, R., Hall, J. L., and Zhu, M. (1985). *Phys. Rev. Lett.* **54**, 996.
Esslinger, T., Sander, F., Weidemüller, M., Hemmerich, A., and Hänsch, T. W. (1996). *Phys. Rev. Lett.* **76**, 2432.
Evans, R. G., Kroeger, M. W., and Mahan, M. O. (1995). *App. Eng. Agric.* **11**, 93.
Fedichev, P. O., Reynolds, M. W., Rahmanov, U. M., and Shlyapnikov, G. V. (1996a). *Phys. Rev. A* **53**, 1447.
Fedichev, P. O., Reynolds, M. O., and Shlyapnikov, G. V. (1996b). Submitted.
Frerichs, V., Kaenders, W., and Meschede, D. (1992). *Appl. Phys.* [*Part*] *B* **B55**, 242.
Friend, D. G., and Etters, R. D. (1980). *J. Low Temp. Phys.* **39**, 409.
Gardner, J. R., Cline, R. A., Miller, J. D., Heinzen, D. J., Boesten, H. M. J. M., and Verhaar, B. J. (1995). *Phys. Rev. Lett.* **74**, 3764.
Gott, Y. V., Ioffe, M. S., and Tel'kovskii, V. G. (1962). *Nucl. Fusion*, *Suppl.*, Pt. 3, 1045 and 1284.
Greytak, T. J. (1995). *In* "Bose-Einstein Condensation" (A. Griffin, D. W. Snoke, and S. Stringari, eds.), BEC-93, p. 131. Cambridge Univ. Press, Cambridge, UK.
Greytak, T. J., and Kleppner, D. (1984). *In* "New Trends in Atomic Physics" (G. Grynberg and R. Stora, eds.), Les Houches Summer School, 1982, p. 1125. North-Holland Publ., Amsterdam.
Han, D. J., Mochizuki, K., Gardner, J. R., and Heinzen, D. J. (1995). *Bull. Am. Phys. Soc.* [2] **40**, 1348.
Heinzen, D. J. (1995). *At. Phys.* **14**, 369.
Helmerson, K., Martin, A., and Pritchard, D. E. (1992a). *J. Opt. Soc. Am. B* **9**, 483.
Helmerson, K., Martin, A., and Pritchard, D. E. (1992b). *J. Opt. Soc. Am. B* **9**, 1988.
Hemmerich, A., Weidemüller, M., Esslinger, T., Zimmermann, C., and Hänsch, T. W. (1995). *Phys. Rev. Lett.* **75**, 37.
Hess, H. (1985). *Bull. Am. Phys. Soc.* [2] **30**, 854.
Hess, H. F. (1986). *Phys. Rev. B: Condens. Matter* [3] **34**, 3476.
Hess, H. F., Bell, D. A., Kochanski, G. P., Kleppner, D., and Greytak, T. J. (1984). *Phys. Rev. Lett.* **52**, 1520.
Hess, H., Kochanski, G. P., Doyle, J. M., Masuhara, N., Kleppner, D., and Greytak, T. J. (1987). *Phys. Rev. Lett.* **59**, 672.
Hijmans, T. W., Luiten, O. J., Setija, I. D., and Walraven, J. T. M. (1989). *J. Opt. Soc. Am B* **6**, 2235.
Hijmans, T. W., Kagan, Y., Shlyapnikov, G. V., and Walraven, J. T. M. (1993). *Phys. Rev. B: Condens. Matter* [3] **48**, 12886.
Holland, M., Burnett, K., Gardiner, C., Cirac, J. I., and Zoller, P. (1996a). *Phys. Rev. A.*, to appear.
Holland, M., Williams, J., Coakley, K., and Cooper, J. (1996b). *Quantum Semiclass. Opt.* **8**, 571.
Huang, K. (1987). "Statistical Mechanics," 2nd ed. Wiley, New York.
Imajo, H., Hayasaka, K., Ohmukai, R., Tanaka, U., Watanabe, M., and Urabe, S. (1996). *Phys. Rev. A* **53**, 122.
Joffe, M. A., Ketterle, W., Martin, A., and Pritchard, D. E. (1993). *J. Opt. Soc. Am. B* **10**, 2257.
Kagan, Y. (1995). *In* "Bose-Einstein Condensation" (A. Griffin, D. W. Snoke, and S. Stringari, eds.), BEC-93, p. 202. Cambridge Univ. Press, Cambridge, UK.

Kagan, Y., Svistunov, B. V., and Shlyapnikov, G. V. (1985). *JETP Lett.* (*Engl. Transl.*) **42**, 209.
Kasevich, M. A. (1995). *Bull. Am. Phys. Soc.* [2] **40**, 1270.
Kasevich, M., and Chu, S. (1992). *Phys. Rev. Lett.* **69**, 1741.
Kastberg, A., Phillips, W. D., Rolston, S. L. Spreeuw, R. J. C., and Jessen, P. S. (1995). *Phys. Rev. Lett.* **74**, 1542.
Katori, H. and Shimizu, F. (1994). *Phys. Rev. Lett.* **73**, 2555.
Ketterle, W., Davis, K. B., Joffe, M. A., Martin, A., and Pritchard, D. E. (1993). Invited oral presentation at the Annual Meeting of the Optical Society of America, Toronto, Canada, October 3–8.
Ketterle, W., and Pritchard, D. E. (1992). *Appl. Phys.* [*Part*] *B* **B54**, 403.
Ketterle, W., Davis, K. B., Joffe, M. A., Martin, A., and Pritchard, D. E. (1993). *Phys. Rev. Lett.* **70**, 2253.
Koelman, J. M. V. A., Stoof, H. T. C., and Verhaar, B. J. (1987). *Phys. Rev. Lett.* **59**, 676.
Kügler, K.-J., Moritz, K., Paul, W., and Trinks, U. (1985). *Nuc. Instrum. Methods Phys. Res.* **228**, 240.
Larson, D. J., Bergquist, J. C., Bollinger, J. J., Itano, W. M., and Wineland, D. J. (1986). *Phys. Rev. Lett.* **57**, 70.
Lawall, J., Bardou, F., Saubamea, B., Shimizu, K., Leduc, M., Aspect, A., and Cohen-Tannoudji, C. (1994). *Phys. Rev. Lett.* **73**, 1915.
Lawall, J., Kulin, S., Saubamea, B., Bigelow, N., Leduc, M., and Cohen-Tannoudji, C. (1995). *Phys. Rev. Lett.* **75**, 4194.
Lee, H. J., Adams, C. S., Kasevich, M., and Chu, S. (1996). *Phys. Rev. Lett.* **76**, 2658.
Lee, H. J., Adams, C. S., Davidson, N., Young, B., Weitz, M., Kasevich, M., and Chu, S. (1995). *At. Phys.* **14**, 258.
Lemonde, P., Morice, O., Peik, E., Reichel, J., Perrin, H., Hänsel, W., and Salomon, C. (1996). *Europhys. Lett.* **32**, 555.
Lett, P. D., Julienne, P. S., and Phillips, W. D. (1995). *Annu. Rev. Phys. Chem.* **46**, 423.
Lett, P. D., Phillips, W. D., Rolston, S. L., Tanner, C. E., Watts, R. N., and Westbrook, C. I. (1989). *J. Opt. Soc. Am. B* **6**, 2084.
Lovelace, R. V. E., Mahanian, C., Tommila, T. J., and Lee, D. M. (1985). *Nature* (*London*) **318**, 30.
Luiten, O. J. (1993). Ph.D. Thesis, University of Amsterdam.
Luiten, O. J., Werij, H. G. C., Setija, I. D., Reynolds, M. W., Hijmans, T. W., and Walraven, J. T. M. (1993). *Phys. Rev. Lett.* **70**, 544.
Luiten, O. J., Werij, H. G. C., Reynolds, M. W., Setija, I. D., Hijmans, T. W., and Walraven, J. T. M. (1994). *Physica B* (*Amsterdam*) **194–196**, 897.
Luiten, O. J., Reynolds, M. W., and Walraven, J. T. M. (1996). *Phys. Rev. A* **53**, 381.
Marcassa, L., Muniz, S., de Queiroz, E., Zilio, S., Bagnato, V., Weiner, J., Julienne, P. S., and Suominen, K.-A. (1994). *Phys. Rev. Lett.* **73**, 1911.
Martin, A. G., Helmerson, K., Bagnato, V. S., Lafyatis, G. P., and Pritchard, D. E. (1988). *Phys. Rev. Lett.* **61**, 2431.
Masuhara, N., Doyle, J. M., Sandberg, J. C., Kleppner, D., Greytak, T. J., Hess, H. F., and Kochanski, G. P. (1988). *Phys. Rev. Lett.* **61**, 935.
Mewes, M.-O., Andrews, M. R., van Druten, N. J., Kurn, D. M., Durfee, D. S., and Ketterle, W. (1996). *Phys. Rev. Lett.*, in press.
Migdall, A. L., Prodan, J. V., Phillips, W. D., Bergeman, T. H., and Metcalf, H. J. (1985). *Phys. Rev. Lett.* **54**, 2596.
Miller, J. D., Cline, R. A., and Heinzen, D. J. (1993). *Phys. Rev. A* **47**, R4567.
Moerdijk, A. J., and Verhaar, B. J. (1994). *Phys. Rev. Lett.* **73**, 518.
Moerdijk, A. J., Verhaar, B. J., and Axelsson, A. (1995). *Phys. Rev. A* **51**, 4852.

Moerdijk, A. J., and Verhaar, B. J. (1996). *Phys. Rev. A* **53**, R19.
Moerdijk, A. J., Boesten, H. M. J. M., and Verhaar, B. J. (1996a). *Phys. Rev. A* **53**, 916.
Moerdijk, A. J., Verhaar, B. J., and Nagtegaal, T. M. (1996b). *Phys. Rev. A* **53**, 4343.
Monroe, C. R., Swann, W., Robinson, H., and Wieman, C. E. (1990). *Phys. Rev. Lett.* **65**, 1571.
Monroe, C. R., Cornell, E. A., Sackett, C. A., Myatt, C. J., and Wieman, C. E. (1993). *Phys. Rev. Lett.* **70**, 414.
Newbury, N. R., and Wieman, C. E. (1996). *Am. J. Phys.* **64**, 18.
Newbury, N. R., and Myatt, C. J., Cornell, E. A., and Wieman, C. E. (1994). *Phys. Rev. Lett.* **74**, 2196.
Newbury, N. R., Myatt, C. J., and Wieman, C. E. (1995). *Phys. Rev. A* **51**, R2680.
Nolan, P. J., and Twin, P. J. (1988). *Annu. Rev. Nucl., Part. Sci.* **38**, 533.
Pellizzari, T., and Ritsch, H. (1995). *Europhys. Lett.* **31**, 133.
Petrich, W., Anderson, M. H. Ensher, J. R., and Cornell, E. A. (1994a). *J. Opt. Soc. Am. B* **11**, 1332.
Petrich, W., Anderson, M. H., Ensher, J. R., and Cornell, E. A. (1994b). *Int. Conf. At. Phys., 14th,* Boulder, CO, 1994, Book Abstr. 1M-7.
Petrich, W., Anderson, M. H., Ensher, J. R., and Cornell, E. A. (1995). *Phys. Rev. Lett.* **74**, 3352.
Phillips, W. D. (1992). *Proc. Int. Sch. Phys. "Enrico Fermi"* **118**, 289.
Phillips, W. D., Prodan, J. V., and Metcalf, H. J. (1985). *J. Opt. Soc. Am. B* **2**, 1751.
Pritchard, D. E. (1983). *Phys. Rev. Lett.* **51**, 1336.
Pritchard, D. E., and Ketterle, W. (1992). *Proc. Int. Sch. Phys. "Enrico Fermi"* **118**, 473.
Pritchard, D. E., Helmerson, K., Bagnato, V. S., Lafyatis, G. P., and Martin, A. G. (1987). *Springer Ser. Opt. Sci.* **55**, 68.
Pritchard, D. E., Helmerson, K., and Martin, A. G. (1989). *At. Phys.* **11** 179.
Quadt, R., Wiseman, H. M., and Walls, D. F. (1996). *Phys. Lett. A*, to appear.
Raab, E. L., Prentiss, M., Cable, A., Chu, S., and Pritchard, D. E. (1987). *Phys. Rev. Lett.* **59**, 2631.
Reichel, J., Bardou, F., Dahan, M. B., Peik, E., Rand, S., Salomon, C., and Cohen-Tannoudji, C. (1995). *Phys. Rev. Lett.* **75**, 4575.
Ricci, L., Zimmermann, C., Vuletić, V., and Hänsch, T. W. (1994). *Appl. Phys. [Part] B* **B59**, 195.
Riis, E., and Barnett, S. M. (1993). *Europhys. Lett.* **21**, 533.
Rooijakkers, W., Hogervorst, W., and Vassen, W. (1995). *Phys. Rev. Lett.* **74**, 3348.
Rubbmark, J. R., Kash, M. M., Littman, M. G., and Kleppner, D. (1981). *Phys. Rev. A* **23**, 3107.
Sanchez-Villicana, V., Gensemer, S. D., Tan, K. Y. N., Kumarakrishnan, A., Dinneen, T. P., Süptitz, W., and Gould, P. L. (1995). *Phys. Rev. Lett.* **74**, 4619.
Santos, M. S., Nussenzveig, P., Marcassa, L. G., Helmerson, K., Flemming, J., Zilio, S. C., and Bagnato, V. S. (1996). *Phys. Rev. A* **52**, R4340.
Schwinger, J. (1937). *Phys. Rev.* **51**, 648.
Setija, I. D. (1995). Ph.D. Thesis, University of Amsterdam.
Setija, I. D., Werij, H. G. C., Luiten, O. J., Reynolds, M. W., Hijmans, T. W., and Walraven, J. T. M. (1993). *Phys. Rev. Lett.* **70**, 2257.
Shimizu, F., and Morinaga, M. (1992). *Jpn. J. Appl. Phys.* **31**, L1721.
Shlyapnikov, G. V., Walraven, J. T. M., and Surkov, E. L. (1993). *Hyperfine Interact.* **76**, 31.
Shlyapnikov, G. V., Walraven, J. T. M., Rahmanov, U. M., and Reynolds, M. W. (1994). *Phys. Rev. Lett.* **73**, 3247.
Silvera, I. F. (1995a). *J. Low Temp. Phys.* **101**, 49.

Silvera, I. F. (1995b). *In* "Bose-Einstein Condensation" (A. Griffin, D. W. Snoke, and S. Stringari, eds.), BEC-93, p. 160. Cambridge Univ. Press, Cambridge, UK.
Silvera, I. F., and Reynolds, M. (1991). *Physica B (Amsterdam)* **169**, 449.
Silvera, I. F., and Reynolds, M. (1992). *J. Low Temp. Phys.* **87**, 343.
Silvera, I. F., and Walraven, J. T. M. (1986). *Prog. Low Temp. Phys.* **10**, 139.
Snoke, D. W., and Wolfe, J. P. (1989). *Phys. Rev. B: Condens. Matter* [3] **39**, 4030.
Snoke, D. W., Rühle, W. W., Lu, Y.-C., and Bauser, E. (1992). *Phys. Rev. B: Condens. Matter* [3] **45**, 10979.
Spitzer, L., Jr. (1987). "Dynamical Evolution of Globular Clusters." Princeton Univ. Press, Princeton, NJ.
Spreeuw, R. J. C., Gerz, C., Goldner, L. S., Phillips, W. D., Rolston, S. L., Westbrook, C. I., Reynolds, M. W., and Silvera, I. F. (1994). *Phys. Rev. Lett.* **72**, 3162.
Stoof, H. T. C. (1995). *In* "Bose-Einstein Condensation" (A. Griffin, D. W. Snoke, and. S. Stringari, eds.), BEC-93, p. 226. Cambridge Univ. Press, Cambridge, UK.
Stoof, H. T. C., Koelman, J. M. V. A., and Verhaar, B. J. (1988). *Phys. Rev. B: Condens. Matter* [3] **38**, 4688.
Stoof, H. T. C., Houbiers, M., Sackett, C. A., and Hulet, R. G. (1996). *Phys. Rev. Lett.* **76**, 10.
Süptitz, W., Wokurka, G., Strauch, F., Kohns, P., and Ertmer, W. (1994). *Opt. Lett.* **19**, 1571.
Surkov, E. L., Walraven, J. T. M., and Shlyapnikov, G. V. (1994). *Phys. Rev. A* **49**, 4778.
Surkov, E. L., Walraven, J. T. M., and Shlyapnikov, G. V. (1996). *Phys. Rev. A* **53**, 3403.
Takekoshi, T., Yeh, J. R., and Knize, R. J. (1996). *Opt. Lett.* **21**, 77.
Thompson, R., Steinberg, A., Helmerson, K., and Phillips, W. D. (1995). *Res. Conf. Bose-Einstein Condens.*, Mont Ste Odile, France, *1995*, Book Abstr. p. 74.
Tiesinga, E., Kuppens, S. J. M., Verhaar, B. J., and Stoof, H. T. C. (1991). *Phys. Rev. A* **43**, 5188.
Tiesinga, E., Moerdijk, A. J., Verhaar, B. J., and Stoof, H. T. C. (1992). *Phys. Rev. A* **46**, R1167.
Tiesinga, E., Verhaar, B. J., and Stoof, H. T. C. (1993). *Phys. Rev. A* **47**, 4114.
Tollet, J. J., Bradley, C. C., Sackett, C. A., and Hulet, R. G. (1995). *Phys. Rev. A* **51**, R22.
Tommila, T. (1986). *Europhys. Lett.* **2**, 789.
Townsend, C. G., Edwards, N. H., Cooper, C. J., Zetie, K. P., Foot, C. J., Steane, A. M., Szriftgiser, P., Perrin, H., and Dalibard, J. (1995). *Phys. Rev. A* **52**, 1423.
Townsend, C. G., Edwards, N. H., Zetie, K. P., Cooper, C. J., Rink, J., and Foot, C. J. (1996). *Phys. Rev. A* **53**, 1702.
van Roijen, R., Berkhout, J. J., Jaakkola, S., and Walraven, J. T. M. (1988). *Phys. Rev. Lett.* **61**, 931.
Verhaar, B. J. (1995). *At. Phys.* **14**, 351.
Verhaar, B. J., Gibble, K., and Chu, S. (1993). *Phys. Rev. A* **48**, R3429.
Walhout, M., Sterr, U., Orzel, C., Hoogerland, M., and Rolston, S. L. (1995). *Phys. Rev. Lett.* **74**, 506.
Walker, T., and Feng, P. (1994). *Adv. At. Mol. Opt. Phys.* **34**, 125.
Walker, T., Sesko, D., and Wieman, C. (1990). *Phys. Rev. Lett.* **64**, 408.
Walraven, J. T. M. (1996). *Proc. Scott. Univ. Summer Sch. Phys.* **44**.
Walraven, J. T. M., and Hijmans, T. W. (1994). *Physica B (Amsterdam)* **197**, 417.
Willems, P. A., and Libbrecht, K. G. (1995). *Phys. Rev. A* **51**, 1403.
Wing, W. H. (1984). *Prog. Quantum Electron.* **8**, 181.
Wiseman, H., Martins, A., and Walls, D. (1996). *Quantum Semiclass. Opt.* **8**, 737.
Wu, H., and Foot, C. J. (1996). To be published.

NONCLASSICAL STATES OF MOTION IN ION TRAPS*

J. I. CIRAC
Departamento de Fisica Aplicada
Universidad de Castilla–La Mancha
Campus Universitario
Ciudad Real, Spain

A. S. PARKINS
Department of Physics
University of Waikato
Hamilton, New Zealand

R. BLATT
Institut für Experimentalphysik
Universität Innsbruck
Innsbruck, Austria

P. ZOLLER
Institut für Theoretische Physik
Universität Innsbruck
Innsbruck, Austria

I. Introduction	238
II. Model	244
A. $\|g\rangle \to \|r\rangle$ Transition	245
B. $\|g\rangle \to \|e\rangle$ Transition	250
C. $\|r\rangle \to \|e\rangle$ Transition	252
III. Sideband Cooling: Preparation in the Ground State	252
A. Laser Cooling in the Lamb–Dicke Limit	253
B. "Designing" Two-Level Atoms	255
IV. Preparation of Fock States	258
A. Fock States via Adiabatic Passage	258
B. Fock States via Observation of Quantum Jumps	260
C. Trapping States	267
V. Preparation of Squeezed States	274
A. Sudden Change of the Trap Frequency	274
B. Bichromatic Excitation	275

*A version of this chapter has been submitted to *Advances of Atomic and Molecular Physics*.

VI. Preparation of Schrödinger Cat States 278
VII. Analysis of the Nonclassical States of Motion 283
    A. Collapses and Revivals 283
    B. Distinguishing between Pure and Mixed States 290
VIII. Conclusions 292
    References....................................... 292

I. Introduction

In the field of quantum optics there is an obvious interest in schemes for the preparation of profoundly nonclassical states of quantum systems. Nonclassical states are states that cannot be described using the standard approaches of classical physics, and their existence and properties highlight some of the most intriguing features of quantum mechanics. In quantum optics, nonclassical states are usually discussed in the context of nonclassical states of light. In the present chapter, we will summarize recent theoretical work on the generation and detection of nonclassical states of motion of laser-cooled trapped ions. An ion moving in a (harmonic) trapping potential is a realization of a quantum system consisting of an atom with two or more levels strongly coupled to a harmonic oscillator (the quantized ion motion) in the limit where coherent interactions can dominant over dissipative processes, and thus provides an ideal testing ground for these fundamental questions. Furthermore, there are formal theoretical analogies between models of ions moving in a trap and cavity quantum electrodynamics (cavity QED, or CQED), so that many of the fundamental questions can be investigated in the light of a new and different physical setting. Together with this opportunity to study and test the theory of quantum mechanics at a fundamental level, nonclassical states of ion motion also offer very interesting practical possibilities in a diverse range of fields, such as precision spectroscopy (see, e.g., Wineland *et al.*, 1992, 1994) and, more recently, quantum computation (see, e.g., Ekert, 1994; Cirac and Zoller, 1995).

The study of nonclassical states has, for the most part, been centered on the quantized harmonic oscillator and perhaps the most commonly discussed nonclassical states of the harmonic oscillator have been Fock (or number) states, quadrature-squeezed states, and Schrödinger Cat states, the properties of which we will elaborate on during the course of this chapter. In the context of quantum optics, the electromagnetic field has naturally provided the most popular medium for the theoretical and experimental study of nonclassical states, with added interest due to potential applications. Squeezed states of light have had perhaps the highest profile, a consequence of the reduced quantum noise that characterizes them and that offers unique possibilities in such fields as low noise communications and high sensitivity measurements (for reviews, see *Journal of the Optical Society of America*, 1987; *Journal of Modern Optics*, 1987; *Applied Physics*, 1992).

In the simplest and most idealized case of nonclassical state preparation, one considers various operations or transformations to be performed on a single isolated mode of the electromagnetic field. A real situation that approaches this ideal case arises in the field of cavity quantum electrodynamics, where the mode in question is one of the resonant modes of a cavity, typically formed by a pair of closely spaced high reflectivity mirrors (for reviews of QED, see, for example, Haroche, 1992; Kimble, 1992; Meystre, 1992; Walther, 1992; Berman, 1994). In such a case, the manipulation of the cavity field mode is carried out via its interaction with an atom (e.g., from an atomic beam). For the case of an effective two-level atom, this atom–field (dipole) interaction is described theoretically by the so-called "Jaynes–Cummings model" (JCM) (Jaynes and Cummings, 1965).

Despite its relative simplicity, the JCM possesses a great wealth of properties, which have been studied extensively during the past 15–20 years (for a review of the properties of the JCM, see Shore and Knight, 1993). These unique properties of the JCM offer a variety of possibilities for the generation of highly nonclassical states of the cavity mode field. Sub-Poissonian (or intensity-squeezed) light and antibunched light have already been experimentally realized in cavity QED configurations (Rempe *et al.*, 1990; 1991), while numerous proposals exist for the preparation of, for example, Fock states (e.g., Krause *et al.*, 1987; Brune *et al.*, 1990, 1992; Holland *et al.*, 1991), Schrödinger Cat states (e.g., Gea-Banacloche, 1990, 1991; Meystre *et al.*, 1990; Brune *et al.*, 1992; Davidovich *et al.*, 1993; Garraway and Knight, 1994; Goetsch *et al.*, 1995), and general superposition states (Vogel *et al.*, 1993; Parkins *et al.*, 1993, 1995). Other interesting proposals for the JCM include observation of collapses and revivals of the atomic population inversion (Eberly *et al.*, 1980; Narozhny *et al.*, 1981; Yoo *et al.*, 1981; experimental evidence of collapse and revival was found by Rempe *et al.*, 1987), quantum nondemolition measurement (Brune *et al.*, 1992; Holland *et al.*, 1991), quantum tomography (Smithey *et al.*, 1993; Bardroff *et al.*, 1995), tests of Bell's inequalities (Pheonix and Barnett, 1993; Cirac and Zoller, 1994), quantum state teleportation (Davidovich *et al.*, 1994; Sleator and Weinfurter, 1994; Cirac and Parkins, 1994), and quantum computation (Sleator and Weinfurter, 1995; T. Pellizzari, S. Gardiner, J. I. Cirac, and P. Zoller, unpublished, 1995.

The great majority of these phenomena rely on operation in the "strong coupling" regime of cavity QED, whereby the dipole coupling strength between the atomic transition and the cavity mode exceeds the rates associated with dissipative processes in the system, i.e., with atomic spontaneous emission and cavity damping. While considerable experimental effort has gone into maximizing the coupling strength and attaining the strong coupling regime, dissipation is still a major limiting factor in cavity QED experiments and makes the realization of many of the preceding proposals extremely difficult, if not impossible with present experimental technology. This is a very general problem facing experiments that attempt to see coherent quantum phenomena at a level that might be regarded as macroscopic (see for example, Caldeira and Leggett, 1985; Zurek, 1991; Landauer, 1995). A further difficulty that confronts cavity QED experiments that employ (thermal) atomic beams is the unavoidable fluctuation in the number of atoms interacting with the cavity mode. A controlled and faithful realization of the single-atom, single-field-mode JCM is thus very difficult to achieve, or at least to monitor (Carmichael, 1995; Kochan, 1995).

At the same time as the field of cavity QED has been evolving, the quite distinct field of ion trapping and cooling has been advancing with similarly impressive speed (for reviews, see Wineland *et al.*, 1993; *Journal of Modern Optics*, 1992; Blatt, 1992). Although the primary goals of ion trap research may have differed from those of cavity QED, it is clear that the two fields have certain features in common, for the system comprising a "two-level" ion trapped in a harmonic potential and interacting with a laser field can be described in terms of a quantized harmonic oscillator interacting with the internal levels of a two-level atom, where, of course, the harmonic oscillator now corresponds to the motion of the ion (in contrast to the radiation field). The coupling is made possible because photon absorption and emission is accompanied by a recoil of the ion. In this chapter, we review a variety of methods for the preparation of nonclassical states of this quantized harmonic oscillator, that is, of the motion of the ion, where, in some cases, close analogies with cavity QED configurations and schemes can be made. For example, under certain special operating conditions related to the parameters of the trap and to the laser intensity and frequency, the dynamics of the system can be modeled directly in terms of the JCM. This enables many of the ideas from cavity QED to be imported to the ion trap system.

However, while direct analogies with cavity QED can be drawn, significant differences do exist that make the ion trap configuration particularly appealing and arguably unique. In particular, other regimes of operation exist for which different interaction Hamiltonians, unachievable in the context of cavity QED, provide the appropriate description and can lead to interesting dynamics and state preparation. In addition, dissipative rates in the ion trap can be essentially negligible in comparison with the coherent coupling strength. The motion of the ion is only very weakly damped due to the interaction of the charged ion with the end caps of the potential and due to collisions with the background gas (Wineland *et al.*, 1994). Typical damping times can be as long as hundreds of seconds. Spontaneous emission can be avoided or "controlled" by using electric-dipole forbidden transitions or Raman transitions between metastable ground states of the ion, leading to decay times of the order of tens of seconds (*Journal of Modern Optics*, 1992). These times are longer by several orders of magnitude than typical interaction times of the ion with the laser field (typically less than milliseconds). Furthermore, as we shall see, both the strength and the characteristics of the ion–laser interaction (i.e., of the coupling between the two-level system and the harmonic oscillator) are readily manipulated by simple changes of the laser intensity or frequency (Cirac *et al.*, 1993b; Wineland *et al.*, 1994). Thus, one has the freedom, for example, to change continuously between regimes of strong and weak coupling (Cirac *et al.*, 1994c).

Note also that the motion of trapped ions can now be routinely cooled to the absolute ground state of the harmonic trapping potential (Diedrich *et al.*, 1989; Monroe *et al.*, 1995).[1] This is important: To study coherent quantum phenomena, one would like to start from a well-defined pure state of the system. While dissipative processes are required to produce such cooling to the ground state, the separation of dissipative from coherent evolution can generally be achieved in a very elegant way for trapped ions, using "auxiliary" internal levels of the ion. For example, a closed three-level internal atomic system may be used, consisting of a dipole-allowed electronic transition (for cooling) and a dipole-forbidden electronic transition (for coherent evolution) connected via a common ground state. The dipole-allowed transition also facilitates atomic state measurements with nearly 100% efficiency via the technique of quantum jumps in fluorescence (Nagourney *et al.*, 1986; Sauter *et al.*, 1986; Bergquist *et al.*, 1986).

With regard to the theoretical analysis of the trapped ion configuration, we note that the interaction of a trapped ion with one or several laser beams is, in general, a very complicated problem. However, it can be greatly simplified under certain limiting conditions that can indeed be realized in experiments. One of these conditions is known as the *Lamb–Dicke limit* (LDL) (Stenholm, 1986), whereby the motion of the ion is restricted to a region of space much smaller than the wavelength (or effective wavelength) of the exciting laser field. In this limit, one can define

---

[1] These experiments also clearly demonstrated the quantized nature of the motion at this level via the presence of asymmetric motional sidebands in probe-field spectra.

a small parameter η (the Lamb–Dicke parameter) and expand the equations of motion in this parameter. Retaining only the lowest order terms in η allows for a considerable simplification of the analysis of the dynamics. A second limiting condition concerns the laser intensity. Here, one can consider two regimes: the *low excitation regime* and the *strong excitation regime*. In the former regime the Hamiltonian describing the evolution of the ion is typically reduced to the JCM or similar models, whereas in the second regime the situation becomes similar to that found in atom interferometry (Adams *et al.*, 1994; Weitz *et al.*, 1994). Additionally, in some of the cases in which, for instance, a dipole-allowed transition is utilized, it is possible to simplify the equations of motion given that the trap frequency is larger than the spontaneous emission linewidth of the transition, corresponding to the regime of *strong confinement* (Stenholm, 1986; Jefferts *et al.*, 1995).

In recent years, a range of proposals, typically based on one or more of these limiting conditions, have been presented for the preparation of nonclassical states of motion of a trapped ion. The nonclassical states in question have included Fock states (Cirac *et al.*, 1993b, 1994c; Blatt *et al.*, 1995; Eschner *et al.*, 1995), superpositions of Fock states (Cirac *et al.*, 1994c), states with sub-Poissonian occupation statistics (Matos Filho and Vogel, 1994), squeezed states (Heinzen and Wineland, 1990; Cirac *et al.*, 1993a; Zeng and Lin, 1995a), and Schrödinger Cat states (Poyatos *et al.*, 1995). Other interesting phenomena that have been predicted and/or observed with trapped ions include the interference fringes in the fluorescence from a pair of trapped ions (Eichmann *et al.*, 1993) and the Zeno effect (Cirac *et al.*, 1994d). The interaction of a trapped ion with a single mode of a high-Q cavity has also been analyzed; it has been shown how one might transfer back and forth the states of the two harmonic oscillators (the cavity mode and the ion motion) (Zeng and Lin, 1994), and how this interaction can be used for cooling, both in the good cavity (Cirac *et al.*, 1993c) and bad cavity (Cirac *et al.*, 1995) limits.

An outline of the chapter is as follows: In Section II we introduce a generic three-level model describing the internal structure of an ion and analyze independently the situations in which a laser (in standing wave or traveling wave configurations) excites each of the three possible transitions. We expand the corresponding Hamiltonian and Liouvillian in the LDL and specialize to the case of low excitation and strong confinement. In Section III, we review the method of sideband cooling, which is the most effective means of cooling a trapped ion to the ground state of the harmonic potential. We first analyze this problem for a two-level system and give the conditions for efficient sideband cooling; then, we show how a three-level system, under appropriate excitation, can effectively be described in terms of a two-level system also satisfying the conditions for efficient sideband cooling. In Section IV, we describe three different methods for the preparation of *Fock* states of the ion motion. The first scheme requires coherent evolution and is based on an adiabatic change of the laser frequency such that the ion follows one of the dressed states of the coupled system. This method is related to that proposed by Raimond *et al.* (1989) in the cavity QED context. The second scheme is based on the observation of quantum jumps from the manifold of dressed states of the Jaynes–Cummings Hamiltonian to a third weakly coupled atomic level. Observation of the quantum jump in the sense of continuous measurement coincides with the preparation of a Fock state of the motion. The third scheme is based on the use of two laser fields that interact with a weak and strong transition successively. Application of a well-defined pulse length on the weak (dipole-forbidden) transition leaves the population of some states outside of the oscillation manifold unchanged (trapping states). Interaction on the strong transition with second laser is used to repopulate one of these states via the third intermediate level. This scheme is related to the trapping states studied by Filipowicz *et al.* (1986) in the context of the micromaser. In contrast to the first method, these two last methods are based on the presence of the dissipation in the laser–ion interaction. In Section V we describe how *squeezed states* of the atomic motion can be prepared. We concentrate on two particular methods. The first involves a sudden change in the trap frequency, whereas the second is based on the bichromatic excitation of the trapped ion. In the latter case, the squeezed state is produced when the beat frequency between two standing wave light fields is equal to twice the trap frequency, and the preparation of a squeezed state is accompanied by a dark resonance in the fluorescence emitted by the ion. In Section VI we outline a scheme for the generation of Schrödinger Cat states of the motion of the ion using laser pulses of well-defined area in the strong excitation regime. First, a laser pulse prepares a linear combination of the internal ground and excited levels of the ion; then, a second laser pulse splits the wavefunction of the ion into two counterpropagating wavepackets; finally, a measurement of the integral state of the ion projects the state of the motion into the desired state. In Section VII, we discuss two methods for experimentally analyzing the prepared nonclassical states. The first method is based on the observation of collapses and revivals of the atomic population inversion, in direct analogy with the cavity QED phenomenon. The time dependent behavior of the atomic inversion reflects the vibrational state distribution of the ion and is intimately related to the discrete (quantized) nature of the trap levels. The second method uses quantum interference effects to distinguish between pure states and statistical mixtures, an ability that is vital to the detection of coherent quantum superposition states. Finally, Section VIII contains a summary of the results reviewed in this chapter.

II. Model

We consider a single ion confined in a harmonic trap[2] and interacting with one or several laser beams. We will assume that the lasers are directed along one of the principal axes of the harmonic potential. This assumption will simplify the problem, since it enables us to consider the ion motion in only one dimension. Hence, the Hamiltonian describing the free motion of the ion in the trap is taken to be of the form

$$H_{tp} = \frac{P^2}{2M} + \frac{1}{2} M\nu^2 R^2 \qquad (1)$$

where R and P are the position and momentum operators respectively, M is the ion mass, and ν is the oscillation frequency, Using the relations

$$R = \sqrt{\frac{1}{2M\nu}} (a + a^\dagger), \qquad P = i\sqrt{\frac{M\nu}{2}} (a^\dagger - a) \qquad (2)$$

we can rewrite the Hamiltonian (1) in the familiar form

$$H_{tp} = \nu a^\dagger a \qquad (3)$$

where $a^\dagger$ and a are the creation and annihilation operators for the harmonic oscillator, respectively, and we have taken $\hbar = 1$.

We will assume that the internal structure of the ion takes the form of a three-level system, with levels $|g\rangle$, $|e\rangle$, and $|r\rangle$ (see Fig. 1). We further assume that the transition $|g\rangle \to |e\rangle$ is a dipole-allowed transition, whereas the other two transitions ($|g\rangle \to |r\rangle$ and $|r\rangle \to |e\rangle$) are dipole-forbidden transitions. These transitions can be excited by laser beams of frequencies close the the corresponding transition frequencies. Obviously, emission or absorption of laser photons will modify the atomic motion. The analysis of this problem is, in general, extremely difficult. In most parts of this chapter, we will focus on the Lamb–Dicke limit (LDL), i.e., on the limit in which the motion of the ion is restricted to a region much smaller than the wavelength of the laser exciting a given transition (Stenholm, 1986). This will allow us to expand the Hamiltonian describing the interaction of the ion with the laser light in terms of the Lamb–Dicke parameter $\eta_i =$

---

[2] We will neglect any effects due to micromotion; such effects have been studied by, e.g., Joos and Lindner (1989a, b), Glauber (1992), Gheorghe and Vedel (1992), and Cirac et al. (1994a).

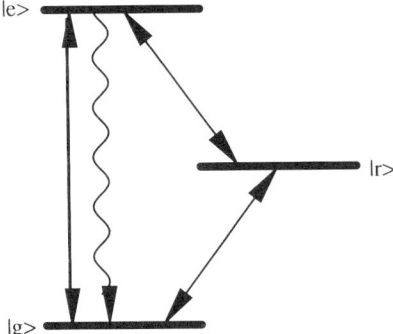

FIG. 1. Internal configuration of the ion considered in this chapter. Transition $|g\rangle \to |e\rangle$ is an electric-dipole allowed transition, whereas the other two $|g\rangle \to |r\rangle$ and $|r\rangle \to |e\rangle$ are electric-dipole forbidden transitions.

$2\pi a_0/\lambda_i$, where $a_0 = 1/(2M\nu)^{1/2}$ is the size of the ground state of the harmonic potential, and λ_i is the wavelength of the laser light exciting transition i.

In the following sections we will examine how an appropriate laser excitation can modify the atomic motion in such a way that it corresponds to a nonclassical state. To begin with, however, we will describe mathematically the interaction of the ion with the laser light in the LDL, since this is central to many of the schemes that follow. To simplify the presentation, we will consider each of the possible excitations individually.

A. $|g\rangle \to |r\rangle$ Transition

Let us consider first the situation in which only the laser driving the dipole-forbidden transition $|g\rangle \to |r\rangle$ is on. We will assume in the following that the interaction time with the laser beam is much shorter than the lifetime of level $|r\rangle$, so that we can neglect dissipation. Under these circumstances, this transition is fully described by the Hamiltonian

$$H_1 = H_1^f + H_1^i \qquad (4)$$

where $H_1^f = H_{tp} + H_g$ describes the evolution in the absence of a laser field, and H_1^i describes the interaction with the laser. In a frame rotating with the laser frequency, we have $H_g = \delta_1 |g\rangle\langle g|$ and

$$H_1^i = \frac{\Omega_1}{2} \sin[\eta_1(a + a^\dagger) + \phi_1](|r\rangle\langle g| + |g\rangle\langle r|) \qquad (5a)$$

$$H_1^{i,\pm} = \frac{\Omega_1}{2} \{|r\rangle\langle g| \exp[\pm i\eta_1(a + a^\dagger)] + h.c.\} \qquad (5b)$$

for standing wave and traveling wave configurations, respectively. Here, $\delta_1 = \omega_L - \omega_0$ is the laser detuning from the internal transition, Ω_1 is the Rabi frequency, and η_1 is the Lamb–Dicke parameter corresponding to this particular transition. The index $+$ ($-$) denotes that the laser plane wave propagates in the positive (negative) R direction, while ϕ_1 defines the position of the center of the trap in the laser standing wave.

Let us now concentrate on the LDL. In this limit, the Hamiltonian (5a) can be expanded in powers of η_1. Only the lowest orders will contribute significantly to the dynamical behavior of the ion. Thus, we can assume an effective Hamiltonian of the form

$$H_1^i = \frac{\Omega_1}{2} \{|r\rangle\langle g| [\alpha_0 + \alpha_\pm (a + a^\dagger) + O(\eta_1^2)] + h.c.\} \quad (6)$$

Here α_0 and $\alpha_\pm$ are numerical factors depending on the dimensionless Lamb–Dicke parameter and the laser configuration. For a standing wave configuration,

$$\alpha_0 = \sin(\phi_1) \quad (7a)$$

$$\alpha_\pm = \eta_1 \cos(\phi_1) \quad (7b)$$

and for a traveling wave configuration,

$$\alpha_0 = 1 \quad (8a)$$

$$\alpha_\pm = \pm i\eta_1 \quad (8b)$$

The Hamiltonian (6) can be further simplified if one considers that the laser field will only excite transitions that are close to resonance with the field frequency. To elaborate on this simplification, let us discuss the eigenvalue spectrum and dressed states of the Hamiltonian (4) as a function of the laser frequency ω_L. Let us denote by $|n, g\rangle$ ($|n, r\rangle$) the state of the ion in which the integral two-level system is in the ground (excited) state and n is the excitation number of the harmonic oscillator. The bare Hamiltonian ($\Omega_1 = 0$) shows degeneracies whenever the laser detuning is a multiple of the trap frequency, $\omega_L - \omega_0 = k\nu$ ($k = 0, \pm 1, \ldots$), i.e., whenever the laser is tuned to one of the "motional sidebands," corresponding to a degeneracy between $|n, g\rangle$ and $|n + k, r\rangle$. In the presence of the laser these degeneracies become avoided crossings, and for sufficiently weak lase excitation these avoided crossings will be isolated (nonoverlapping). For example, when the laser frequency is close to the two-level transition resonance ($|\omega_L - \omega_0| \ll \nu$, i.e., $k = 0$), transitions changing the harmonic oscillator quantum number n are off resonance and can be neglected (arrows 1 and 3 in Fig. 2). In this case the

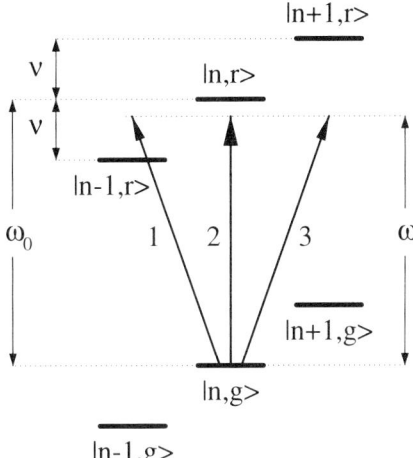

FIG. 2. Energy levels for a two-level ion moving in a harmonic potential. For $\omega \approx \omega_0$, transitions from $|n, g\rangle \to |n, e\rangle$ are close to resonance (2), whereas transitions with $|n, g\rangle \to |n \pm 1, e\rangle$ are very far off resonance (1) and (3). For $\omega \approx \omega_0 - \nu$, only transitions from $|n, g\rangle \to |n - 1, e\rangle$ are close to resonance (1), whereas all other transitions are very far off resonance. For $\omega \approx \omega_0 + \nu$, only transitions from $|n, g\rangle \to |n + 1, e\rangle$ are close to resonance (3).

Hamiltonian (4) can be approximated by

$$H^0 = H_{tp} - \frac{1}{2}\delta\sigma_z + \frac{\Omega}{2}(\alpha_0 \sigma_+ + h.c.) \qquad (9)$$

where we have dropped the subscript 1 and used the spin $\frac{1}{2}$ notation $\sigma_+ = (\sigma_-)^\dagger = |r\rangle\langle g|$, $\sigma_z = |r\rangle\langle r| - |g\rangle\langle g|$. For laser frequencies close to the lower motional sideband resonance $|\omega_L - (\omega_0 - \nu)| \ll \nu$ ($k = -1$), only transitions decreasing the quantum number n by one are important (arrow 1 in Fig. 2). Now, H_1 can be approximated by a Hamiltonian of the Jaynes–Cummings type:

$$H^{JC}_\pm = H_{tp} - \frac{1}{2}\omega\sigma_z + \frac{\Omega}{2}(\alpha_\pm \sigma_+ a + h.c.) \qquad (10)$$

Similarly, for $|\omega_L - (\omega_0 + \nu)| \ll \nu$, only transitions increasing the quantum number n by one ($k = +1$) contribute to the Hamiltonian H_{int} (arrow 3 in Fig. 2). In this case, H can be approximated by the "*anti-Jaynes–Cummings*" Hamiltonian

$$H^{AJC}_\pm = H_{tp} - \frac{1}{2}\delta\sigma_z + \frac{\Omega}{2}(\alpha_\pm \sigma_+ a^\dagger + h.c.) \qquad (11)$$

The mathematical conditions that must hold for these approximations to be valid can be derived using time dependent perturbation theory. The analysis reveals that the effective Rabi frequencies have to be much smaller than the trap frequency, $(\alpha_i \Omega/\nu)^2 \ll 1$ ($i = 0, \pm$), which we will call the *low excitation regime*. If this condition is not fulfilled, the Stark shifts produced by the laser interaction may be of the order of the energy difference between trap levels, and therefore the resonant conditions are strongly modified. For example, under laser excitation with $|\omega_L - (\omega_0 + \nu)| \ll \nu$, one can satisfy replace the total Hamiltonian (4) by H^{JC} when

$$\left[\frac{\alpha_{0,-}\Omega}{\nu}\right]^2 \ll 1 \qquad (12)$$

since in this case the transitions induced by H^0 and H^{AJC} are negligible. Note that apart from transitions, these other two Hamiltonians may give rise to phase shifts of the order of $(\alpha_{0,-}\Omega)^2/\nu$. However, they play no role in the preparation of nonclassical states of motion, and therefore we will not take them into account. Note also that at the node of the standing wave one has $\alpha_0 = 0$, and therefore the validity condition of our approach is greatly relaxed. That is, with a higher laser intensity one can satisfy condition (12) given that α_- is of the order of the Lamb–Dicke parameter η_1. It is worth mentioning that the Hamiltonians (9), (10), and (11) may be valid even in the case when the LDL does not hold, provided one is in the low excitation regime. Here, however, we will assume the LDL, since it is the typical situation in Paul traps and the formulas are simpler and more transparent.[3]

For an isolated avoided crossing, the dressed states are readily obtained by diagonalizing the 2 × 2 matrix of degenerate states. Figure 3 displays a numerical example of a dressed energy spectrum. In this figure, we have plotted the exact eigenvalues of H (calculated numerically) as a function of the laser frequency ω_L, for $\eta = 0.5$, $\Omega = 0.3\nu$, and a standing wave configuration with $\phi = \pi/4$. The dotted line represents the energies of the "bare" states. For the laser frequencies far below the two-level resonance (point "C" in Fig. 3), transitions between states $|g\rangle$ and $|r\rangle$ do not take place, since they are very far off resonance. Hence, the eigenstates of H are those of the free part $H_{tp} + H_I$, i.e., $|n, g\rangle$ and $|n, r\rangle$. For increasing values of the laser frequency, states $|n, g\rangle$ couple to states $|n-1, r\rangle$ via the Hamiltonian H^{JC}, and the eigenstates (or dressed states)

---

[3] In fact, the LDL conditions are required for sideband cooling, which is in turn required as a precursor to some of the schemes to prepare nonclassical states of motion (see Section III).

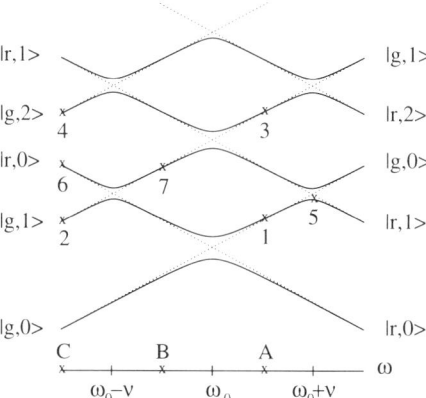

FIG. 3. Energy level diagram for a two-level ion interacting with laser light, as a function of the laser frequency ω_L. Dotted lines correspond to the bare states.

are linear combinations of these two states. In particular, for $\omega_L = \omega_0 - \nu$, it is easy to show that the eigenstates are

$$|n, \pm\rangle = \frac{1}{\sqrt{2}}(|n+1, g\rangle \pm |n, r\rangle) \quad (n = 0, 1, \ldots) \quad (13)$$

and $|0, g\rangle$, which corresponds to the first level anticrossing of Fig. 3. The energy level scheme in this case is represented in Fig. 4, which is the well-known Jaynes–Cummings ladder of states.

For increasing values of the laser frequency, but still far from resonance (point B), states $|n, g\rangle$ and $|n, r\rangle$ are again decoupled. If the laser frequency is increased to the two-level resonance, states $|n, g\rangle$ couple to states $|n, r\rangle$ via the Hamiltonian H^0, and for $\omega_L = \omega_0$, the eigenstates are

$$|n, \pm\rangle = \frac{1}{\sqrt{2}}(|n, g\rangle \pm |n, r\rangle) \quad (14)$$

which corresponds to the second level anticrossing of Fig. 3. Similarly, for increasing values of the laser frequency one can easily identify the eigenstates of H by considering transitions between states $|n, g\rangle$ and $|n+1, r\rangle$ via the Hamiltonian H^{AJC}. In particular, for $\omega_L = \omega_0 + \nu$, the eigenstates are

$$|n, \pm\rangle = \frac{1}{\sqrt{2}}(|n, g\rangle \pm |n+1, r\rangle) \quad (n = 0, 1, \ldots) \quad (15)$$

and $|0, r\rangle$.

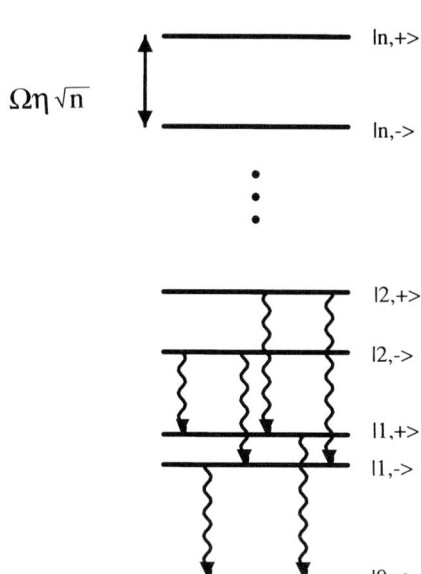

FIG. 4. Dressed states resulting from the ion–laser coupling.

B. $|g\rangle \to |e\rangle$ Transition

Let us now consider that a laser beam excites the transition $|g\rangle \to |e\rangle$. Since this is an electric-dipole allowed transition, and hence spontaneous emission is expected to play a significant role, the dynamics must be described in terms of a master equation. The master equation takes the form

$$\dot{\rho} = -i[H_2, \rho] + \mathscr{L}\rho \tag{16}$$

In a frame rotating at the laser frequency,

$$H_2 = H_2^f + H_2^i \tag{17}$$

where $H_2^f = H_{tp} + H_r$, with $H_r = \delta_2 |g\rangle\langle g|$, describes the evolution in the absence of a laser beam, and

$$H_2^i = \frac{\Omega_2}{2} \sin[\eta_2(a + a^\dagger) + \phi_2](|e\rangle\langle g| + |g\rangle\langle e|) \tag{18a}$$

$$H_2^{i,\pm} = \frac{\Omega_2}{2}\{|e\rangle\langle g| \exp[\pm i\eta_2(a + a^\dagger)] + \text{h.c.}\} \tag{18b}$$

describe the interaction with a laser beam in standing and traveling wave configurations, respectively. Here, $\delta_2 = \omega_L - \omega_0$ is the laser detuning from the internal transition, Ω_2 is the Rabi frequency, ϕ_2 gives the position of the center of the trap in the laser standing wave, and η_2 is the Lamb–Dicke parameter corresponding to this transition.

The dissipative part of the master equation (16) can be written as

$$\mathscr{L}\rho = \Gamma |g\rangle\langle g| \langle e| \tilde{\rho} |e\rangle - \tfrac{1}{2}\Gamma(|e\rangle\langle e| \rho + \rho |e\rangle\langle e|) \qquad (19)$$

where Γ is the spontaneous emission rate from level $|e\rangle$, and

$$\tilde{\rho} = \int_{-1}^{1} du N(u) e^{-i\eta_2 u(a+a^\dagger)} \rho e^{i\eta_2 u(a+a^\dagger)} \qquad (20)$$

with $N(u)$ the dipole emission pattern for spontaneous emission from $|e\rangle$ to $|g\rangle$. For example, for a $\Delta m_J = \pm 1$ transition, $N(u) = (3/8)(1 + u^2)$.

In the Lamb–Dicke limit, one can again expand the master equation (16) in terms of the Lamb–Dicke parameter η_2. The modified density operator $\tilde{\rho}$ that appears in the Liouvillian part of the master equation becomes, up to the second order in η_2,

$$\tilde{\rho} = \rho + \alpha \eta_2^2 \Big[(a^\dagger + a)\rho(a^\dagger + a) - \tfrac{1}{2}(a^\dagger + a)^2 \rho - \tfrac{1}{2}\rho(a^\dagger + a)^2 \Big] \qquad (21)$$

where the coefficient α is given by

$$\alpha = \int_{-1}^{1} du\, N(u) u^2 \qquad (22)$$

and equals $2/5$ for a $\Delta m_J = \pm 1$ transition.

The expansion of the Hamiltonian H_2^i is analogous to that for H_1^i, Eq. (6):

$$H_2^i = \frac{\Omega_2}{2}\{|e\rangle\langle g| [\alpha_0 + \alpha_\pm (a + a^\dagger) + O(\eta_2^2)] + h.c.\} \qquad (23)$$

The parameters $\alpha_{0,\pm}$ are given by (7) and (8) for standing and traveling wave configurations respectively, but with η_2 (ϕ_2) instead of η_1 (ϕ_1). The Hamiltonian H_2^i can be further simplified using arguments concerning the laser frequency. Now, however, in addition to the conditions related to the low excitation regime, $(\alpha_i \Omega/\nu)^2 \ll 1$ ($i = 0, \pm$), the system must also be in the strong confinement limit, $\Gamma \ll \nu$, in order for the structure of the energy eigenstates and coherent dynamics to manifest themselves strongly. In such a case, depending on the laser frequency, H_2 becomes either (9), (10), or (11), where now $\sigma_+ = (\sigma_-)^\dagger = |e\rangle\langle g|$ and $\sigma_z = |e\rangle\langle e| - |g\rangle\langle g|$.

C. $|r\rangle \to |e\rangle$ Transition

Let us now concentrate on the situation where only the laser driving the dipole-forbidden transition $|r\rangle \to |e\rangle$ is on. The dynamics of the atom interacting with such a laser is described by the following master equation:

$$\dot{\rho} = -i[H_3, \rho] + \mathcal{L}\rho \quad (24)$$

In a frame rotating at the laser frequency,

$$H_3 = H_3^f + H_3^i \quad (25)$$

where $H_3^f = H_{tp} + H_r$, with $H_r = \delta_3 |r\rangle\langle r|$, describes the evolution in the absence of a laser beam, and H_i describes the interaction with the laser and has a form similar to the Hamiltonians (5a) and (5b) for standing and traveling wave excitation, respectively. The dissipative part of master equation (24) is given by the Liouvillian (19).

As before, we are interested here in the LDL. In this case, the lowest order terms in the expansion of master equation (24) will suffice for our purposes. The situation is then very simple, since we only have to take into account spontaneous emissions that do not change the external (motional) state of the ion (i.e., the lowest order contribution from the LDL expansion comes from the term with $\tilde{\rho} \approx \rho$, which does not change the ion motion). Thus, after a time $t_2 \gg \alpha_i \Omega_2^{-1}, \Gamma_g^{-1}$ ($i = 0, \pm$), we can assume that the population of the state $|r, n\rangle$ has been transferred *incoherently* to the state $|g, n\rangle$. It is straightforward to show that the probability of transitions from $|r, n\rangle \to |g, m\rangle$, with $m \neq n$, is at least of order η_2^2 and, therefore, can be neglected in comparison with the probability of transitions $|r, n\rangle \to |g, n\rangle$ (of order 1), Note that this argument also applies to transitions that change the motional quantum state in directions other than that of the direction of laser propagation.

III. Sideband Cooling: Preparation in the Ground State

A prerequisite for many of the schemes for generating nonclassical states of motion is that the initial motional state of the ion be a well-defined pure state, e.g., the ground state $|0\rangle$. The best approach to preparing such a pure state of the atomic motion is sideband cooling. This laser cooling (*Journal of the Optical Society of America*, 1989; Gilbert and Wieman, 1993) mechanism has been well known for a long time (Wineland and Dehmelt, 1975; Neuhauser *et al.*, 1978) and was first demonstrated by the NIST group in 1989 (Diedrich *et al.*, 1989). More recent demonstrations have been provided by Appasamy *et al.* (1995) and Monroe *et al.* (1995).

Because of the basic importance of this cooling technique to schemes for nonclassical state preparation, we will outline briefly in this section the theory behind sideband cooling and some of the configurations used for its practical implementation.

A. LASER COOLING IN THE LAMB–DICKE LIMIT

The theoretical analysis of laser cooling of trapped ions has a strong background (Neuhauser et al., 1980; Wineland et al., 1983; Toschek, 1984; Wineland and Itano, 1987), and cooling in the Lamb–Dicke limit has been studied in detail in numerous papers for a variety of situations and conditions (Wineland and Itano, 1979; Itano and Wineland, 1982; Wineland et al., 1983, 1987; Lindberg and Stenholm, 1984; Javanainen et al., 1984; Lindberg and Javanainen, 1986; Itano et al., 1987; Wells and Cook, 1990; Cirac et al., 1992, 1993d, 1994b; Blockley and Walls, 1993). Here we will follow the approach of Cirac et al. (1992).

Consider the situation in which we excite, with a laser field, the state $|e\rangle$. In this case, using the Lamb–Dicke expansions (6) and (21), the rate at which cooling takes place is much slower than the internal dynamics of the ion, since it is proportional to η_2^2 ($\ll 1$). This enables us to adiabatically eliminate the internal degrees of freedom of the ion and to derive the following rate equation describing the dynamics of the laser cooling process (Stenholm, 1986):

$$\frac{dP_n}{dt} = (n+1)A_- P_{n+1} - [(n+1)P_+ + nP_-]p_n + nA_+ P_{n-1} \quad (26)$$

where $P_n = \text{Tr}_I \langle n| \rho |n\rangle$ is the population of the harmonic oscillator level $|n\rangle$ (Tr_I denotes the trace over the internal degrees of freedom). This equation expresses the evolution of the harmonic oscillator level populations due to transitions between the external levels. The rates determining such transitions are given by (Cirac et al., 1992)

$$A_\pm = 2\,\text{Re}[S(\mp \nu) + D] \quad (27)$$

where $S(\nu)$ is the fluctuation spectrum of the dipole force F, i.e.,

$$S(\nu) = \frac{1}{2M\nu} \int_0^\infty dt\, e^{i\nu t} \langle F(t)F(0)\rangle_{SS} \quad (28)$$

and

$$D = \alpha \Gamma \eta_2^2 \langle \rho_{ee}\rangle_{SS} \quad (29)$$

is the diffusion coefficient due to spontaneous emission. Here, ρ_{ee} is the population of the excited state, and the index SS indicates the steady state values in the case of an ion at rest at the position $R = 0$. The dipole force

$F(t)$ depends on the details of the laser excitation since it is the derivative of the interaction potential. To derive the expression for D, one has to solve the familiar Bloch equations, in the steady state, for the corresponding transition. To derive an expression for $S(\nu)$, one makes use of the quantum regression theorem, utilizing the same Bloch equations. Note that the rate equations (26) are also valid for any internal and laser configuration if D is replaced by a sum over all transitions producing spontaneous emission.

Starting from Eq. (26), one can derive the evolution equation for the mean harmonic oscillator quantum number, $\langle n \rangle = \sum_{n=0}^{\infty} n P_n$,

$$\frac{d}{dt}\langle n \rangle = -W(\langle n \rangle - \langle n \rangle_{SS}) \tag{30}$$

where

$$W = A_- - A_+ \tag{31}$$

is the cooling rate, and

$$\langle n \rangle_{SS} = \frac{A_+}{A_- - A_+} \tag{32}$$

is the steady state mean harmonic oscillator quantum number.

Let us first assume that we drive transition $|r\rangle \to |e\rangle$ with a laser beam in a standing wave configuration, such that the center of the trap coincides with the node of the standing wave (i.e., the interaction Hamiltonian is H_2^i with $\phi_2 = 0$). In this simple case, the Bloch equations are trivial, yielding $D = 0$ and

$$A_\pm = \frac{\Omega_2^2}{8} \frac{\Gamma}{(\Gamma/2)^2 + (\delta_2 \mp \nu)^2} \tag{33}$$

For the particular choice of laser detuning,

$$\delta_2 = -\nu \tag{34}$$

the steady state mean occupation number $\langle n \rangle_{SS} = (\Gamma/4\nu)^2$. Minimum temperatures occur for $\nu \gg \Gamma$, which corresponds to the strong confinement regime. For this case, $\langle n \rangle_{SS} \approx 0$, indicating that the final state of the atomic motion is essentially $|0\rangle$, i.e., a pure state.

Laser cooling in the Lamb–Dicke limit can be understood as follows: In the rest frame of the ion, the ion "sees" a laser field consisting of a carrier at frequency ω_L and small (motional) sidebands at frequencies $\omega_L \pm \nu$. Cooling occurs when absorption of laser photons from the upper laser sideband (at $\omega_L + \nu$) is stronger than from the lower sideband, since the former absorption reduces the external energy, whereas the latter increases the external energy. Hence, laser cooling is particularly efficient in the stronger confinement limit when one can tune the laser frequency so that the upper laser sideband is on resonance with the two-level transition, since in this case only the photons of the upper sideband are absorbed, and consequently the ion ends up in the ground state of the harmonic trapping potential. This is the basic mechanism of *sideband cooling*.

Finally, for a trap that is not centered around the standing wave node and for higher Rabi frequencies (but smaller than ν), one has to take into account the dynamical ac Stark shift resulting from the interaction with the carrier (which is far from resonance). Consequently, the sideband cooling condition on the laser detuning is modified to

$$\Delta = -\sqrt{\nu^2 - \Omega^2} \qquad (35)$$

B. "Designing" Two-Level Atoms

For many of the ions currently in use in Paul traps, the strong confinement condition $\nu \ll \Gamma$ is not fulfilled for dipole-allowed transitions, and hence sideband cooling is not possible (an exception is the Indium ion, for which this condition can be easily fulfilled). One might think that one could use a dipole-forbidden transition to satisfy the strong confinement condition, since for such a transition Γ is practically zero. However, in this case, the cooling rate W also tends to zero, and the cooling time becomes prohibitively long. However, auxiliary internal atomic levels and additional laser excitation can be used to circumvent this problem and facilitate efficient sideband cooling (Lindberg and Javanainen, 1986; Heinzen and Wineland, 1990; Marzoli et al., 1994). In the subsection, we will briefly describe how this may be so.

Marzoli et al. (1994) have considered schemes for efficient sideband cooling based on the use of a metastable excited atomic state. To reduce the cooling time (without changing the final energy) while using a dipole-forbidden transition, $|g\rangle \rightarrow |r\rangle$, they consider pumping from the metastable state $|r\rangle$ to a third level, $|e\rangle$. For pumping below saturation, this third level can be adiabatically eliminated in the theoretical analysis, resulting in an effective two-level system with decay rate Γ' and detuning δ'. These quantities depend on the characteristics of the pumping to the third (weakly populated) level. By choosing these characteristics appropriately, one can modify the effective decay rate and detuning so as to match the conditions required for efficient (fast) sideband cooling.

We give a more specific analysis: Following the treatment of the previous subsection, we require the steady state dipole force and the fluctuation spectrum of the dipole force for an ion at rest at the position

$R = 0$. The ion is excited by two laser fields, driving the transitions $|g\rangle \to |r\rangle$ and $|r\rangle \to |e\rangle$ respectively. The master equation for the density operator describing this situation can be found following the considerations made in Section II, setting $\eta_{1,2,3} = 0$ (ion at rest). It reads

$$\dot{\rho} = -i\delta_1[|g\rangle\langle g|, \rho] - i\frac{\Omega_1}{2}[|r\rangle\langle g| + |g\rangle\langle r|, \rho]$$

$$- i\frac{\Omega_3}{2}[|r\rangle\langle e| + |e\rangle\langle r|, \rho]$$

$$+ \frac{\Gamma}{2}(2|g\rangle\langle e|\rho|e\rangle\langle g| - |e\rangle\langle e|\rho - \rho|e\rangle\langle e|) \qquad (36)$$

In the limit in which the saturation parameter for the $|r\rangle \leftrightarrow |e\rangle$ transition is small, i.e., $s_3 = 2(\Omega_3/\Gamma)^2 \ll 1$, and for $|\delta_1| \ll \Gamma$, one can adiabatically eliminate level $|e\rangle$ since it is only very weakly occupied. The system is then reduced to an effective two-level system with ground state $|g\rangle$ and excited state $|r\rangle$, but with modified detuning, δ', and "spontaneous" emission rate, Γ'. In particular,

$$\delta' = \delta_1, \qquad \Gamma' = \frac{\Omega_3^2}{\Gamma} \qquad (37)$$

Optimum sideband cooling will take place when $\nu \gg \Gamma'$, and for detuning Δ' satisfying (34) and (35) in the low and high intensity limits, respectively. In the first experiment in which sideband cooling to the ground state was demonstrated (Diedrich et al., 1989), this was in fact the method used.

Sideband cooling in such a system can be visualized using the diagram in Fig. 5, which shows the energy levels of a three-level ion in the harmonic trap. In this figure, the state of the ion is represented by two quantum numbers, $|\alpha, n\rangle = |\alpha\rangle \otimes |n\rangle$, where $\alpha = g, r, e$ denotes the internal atomic state, and $n = 0, 1, \ldots$ is the quantum number of the harmonic oscillator. As can be seen in Fig. 5, cooling is expected to take place if the laser driving the weak transition ($|g\rangle \leftrightarrow |r\rangle$) is tuned into resonance with the transition $|g, n\rangle \leftrightarrow |r, n-1\rangle$; the second laser is tuned close to the $|r\rangle \leftrightarrow |e\rangle$ resonance. For the case of low saturation of the $|g\rangle \leftrightarrow |r\rangle$ transition and for $\Gamma' < \nu$, the cooling rate is maximal, while the final energy is minimal for detuning $\delta' = -\nu$, i.e., sideband cooling occurs. The situation is slightly more involved for the case of stronger saturation of the $|g\rangle \leftrightarrow |r\rangle$ transition. In this case, optimum cooling is achieved if the effective detuning is in resonance with the Stark shifted level (35).

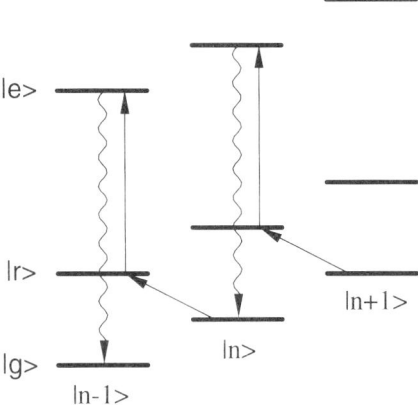

FIG. 5. Laser configuration for sideband cooling with three-level systems, employing electric-dipole forbidden transitions. The laser connecting the $|g\rangle \to |r\rangle$ transition is tuned to the lower motional sideband, and the laser connecting the $|r\rangle \to |e\rangle$ is on resonance.

This kind of treatment is easily generalized to the case in which the transition $|r\rangle \to |e\rangle$ is a dipole-allowed transition, or indeed to other three-level configurations (Marzoli et al., 1994). One particular example that is worthy of note and also very relevant to current experiments is sideband Raman cooling between two (internal) ground states of a three-level GL atomic system (Dehmelt et al., 1985; Lindberg and Javanainen, 1986; Toschek and Neuhauser, 1989; Heinzen and Wineland, 1990). In such a system, a two-photon Raman transition between the ground atomic states is driven by a pair of laser fields. The relative tuning of the laser fields is set to the lower motional sideband, so as to drive stimulated Raman transitions between the states $|i, n\rangle$ and $|j, n - 1\rangle$, where i, j denote the ground atomic states. Spontaneous Raman transitions also occur, but in the LDL these are predominantly between states of the same motional quantum number (i.e., $|j, n - 1\rangle \to |i, n - 1\rangle$). For sufficiently low laser power and sufficiently large detuning of the individual laser frequencies from the intermediate excited state, the configuration can be modeled, as before, by a two-level system (involving the two atomic ground states) with an effective spontaneous emission linewidth Γ' satisfying the condition $\Gamma' < \nu$ and thus enabling sideband cooling to occur.

The Raman technique also possesses an interesting and useful property with respect to operation in the Lamb–Dicke regime; for appropriate directions of propagation of the two laser fields, the operational Lamb–Dicke parameter is actually determined by the difference in the wavenumbers of the two light fields, rather than by the wavenumbers themselves (see e.g., Wineland et al., 1994). Provided the wavelengths of the two fields are not too different (which is the case when, for example, the internal atomic states are different hyperfine ground states), the effective Lamb–Dicke parameter can thus be very small.

We note finally that Heinzen and Wineland (1990) and Wineland et al. (1994) have analyzed the Raman system in terms of simple model Hamiltonians, as in the present chapter, showing that, in appropriate limits, the dynamics reduce to that of the Jaynes–Cummings model of cavity QED. On the experimental side, Monroe et al. (1995) have recently implemented the sideband Raman cooling technique to cool a single beryllium ion to the 3D ground state of motion.

IV. Preparation of Fock States

In this section we describe three different schemes for preparing specific Fock states of the motion of an ion. The first scheme is based on adiabatic passage along the dressed energy levels of the strongly coupled ion–laser system, facilitated by an adiabatic change in the laser frequency. The second scheme is based on the observation of quantum jumps from the manifold of dressed levels $|n, \pm \rangle$ of the Jaynes–Cummings Hamiltonian to a third weakly coupled atomic level $|r\rangle$. Observation of the quantum jump in the sense of continuous measurement coincides with the preparation of a Fock state $|n\rangle$ of the motion of the trapped ion. The third scheme is based on the application of cycles of laser pulses of well-defined "pulse area" to the transitions $|g\rangle \to |r\rangle$ and $|r\rangle \to |e\rangle$ and invokes the concept of "trapping" states.

Other schemes for the preparation of Fock states of motion have been proposed, most notably by Eschner et al. (1995), who have shown that absorption in the motional sideband produces nonclassical states of the ion motion, and that repeated null fluorescence observation leads to cooling of the ion and possibly to Fock states.

A. Fock States via Adiabatic Passage

We consider here that the electric-dipole forbidden transition $|g\rangle \to |r\rangle$ is being driven by a laser field. This situation is described by the Hamiltonian H_1, the dressed state structure of which was analyzed in Section II.A. Now, however, we consider the case in which the laser frequency is changed adiabatically. Cirac et al. (1994c) have shown that such a transformation can lead to Fock states of the motion as a result of the system adiabatically following one of the dressed states.

To illustrate this, let us assume that the initial state of the ion is $|0, g\rangle$. This corresponds to the vibrational ground state in which the ion is left after sideband cooling. The laser frequency is initially set to a certain value between ω_0 and $\omega_0 + \nu$ (denoted by A in Fig. 3), so that all transitions are off resonance. The corresponding energy level is denoted 1 in the energy diagram of Fig. 3. When the laser frequency is decreased adiabatically, the state of the ion changes according to the dressed energy diagram of Fig. 3. Once the frequency ω_L reaches point C, the state of the ion has evolved to $|1, g\rangle$ (point 2 in the figure). Now the laser is switched off and on again, but with the original frequency, point A. Thus, the state of the ion corresponds to the point 3 in the figure. By decreasing adiabatically the laser frequency again the ion ends up in the state $|2, g\rangle$ (point 4 in the figure). By repeating this cycle n times, the state of the ion becomes $|n, g\rangle$, i.e., a Fock state with n quanta is prepared. Note that after any cycle the internal state of the ion is simply the ground state $|g\rangle$, and therefore the Fock state is not modified due to spontaneous emission from the electronic transition.

The conditions for adiabatic following can be estimated straightforwardly for a laser frequency varying linearly with time. In this case, one derives the requirement $\overline{\Omega} T \gg \nu/\overline{\Omega}$, where T is the time duration of each cycle and $\overline{\Omega}$ is the effective Rabi frequency for transitions $|n, g\rangle \to |n - 1, r\rangle, |n, r\rangle$ (which, for the sale of simplicity, we assume to be similar). For example, for a trap frequency of $(2\pi)1$ MHz, and an effective Rabi frequency of $(2\pi)50$ kHz, the adiabatically condition implies interaction times $T \gg 60$ μsec. These numbers are within reach of current experiments with trapped ions. It is worth emphasizing that under low intensity conditions, and when the adiabaticity condition is satisfied, this procedure for preparing Fock states is insensitive to the specific values of the parameters involved in the process. Furthermore, it is suitable for both standing and traveling wave excitation.

General linear combinations of neighboring Fock states can also be prepared following the foregoing procedure, with the initial state of the ion in the superposition state $(\alpha |g\rangle + \beta |r\rangle)|0\rangle$. As can be deduced following the level diagram of Fig. 3, after one cycle the state of the ion becomes $(\alpha |1\rangle + \beta |0\rangle)|g\rangle$. After n cycles, the state of the ion will be $(\alpha |n\rangle + \beta |n - 1\rangle)|g\rangle$.

Fock states in two and three dimensions may be prepared as follows: We assume that the ion has been cooled via sideband cooling to its lowest state along all three dimensions; i.e., the initial state of the ion is $|0, 0, 0, g\rangle$. Applying lasers along one direction, for example, along the x axis, and with the procedure just described, the ion ends up in the state $|n_x, 0, 0, g\rangle$. Next, the lasers along the x axis are switched off, and those along the y axis switched on. By an adiabatic change of the laser frequency, the state $|n_x, n_y, 0, g\rangle$ is thus prepared. Finally, proceeding in the same manner with the z direction, one can prepare the state

$$|\Psi\rangle = |n_x, n_y, n_z\rangle \equiv |n_x\rangle_x |n_y\rangle_y |n_z\rangle_z \tag{38}$$

Obviously, by combining this procedure with that described in the previous paragraph, one can generate states of the form

$$|\Psi\rangle = \alpha |n_x, n_y, n_z\rangle + \beta |n_x, n_y + 1, n_z\rangle \tag{39}$$

As a numerical example, in Fig. 6 we have plotted the time evolution for a pulse sequence that leads to the transformation $|0,0,0\rangle \rightarrow |2,2,2\rangle$ via the adiabatic passage procedure. These curves were calculated by solving the full time dependent Schrödinger equation using a truncated basis set of oscillator eigenstates. The frequencies of the lasers along the three directions as a function of time are plotted in the upper part of Fig. 6(a). Figure 6(b) shows the populations of the trap states $|n\rangle_x$, $|n\rangle_y$, and $|n\rangle_z$ ($n = 0, 1, 2$), and Fig. 6(c) is a plot of the occupation of the internal ground state $|g\rangle$ as a function of time. As indicated in the population histogram in Fig. 6(d), the 3D Fock state $|2,2,2\rangle$ has been prepared after the completion of the three adiabatic cycles.

Finally, we note that entangled states $|\Psi\rangle = \alpha |0,1,0\rangle + \beta |1,0,0\rangle$ can also be prepared. For example, to prepare this state with $\alpha = \beta = 1/\sqrt{2}$, one can proceed as follows: In a first step, with the ion initially in its ground state, the frequency of the laser along the x direction is adiabatically increased from point A in Fig. 3 up to $\omega_0 + \nu_x$. This corresponds to moving from point 1 to 5 in Fig. 3. According to (15), the state of the ion after this first step is $(|0,0,0,g\rangle + |1,0,0,r\rangle)/\sqrt{2}$. In a second step, the frequency of the laser in the y direction is increased from C to B, which corresponds to moving the second term of the wave function from point 6 to 7 (the first part of the wavefunction remains the same, as can be deduced from the figure). Again, the internal state of the ion factorizes out in the final state, and the ion is left in its ground internal state.

B. FOCK STATES VIA OBSERVATION OF QUANTUM JUMPS

Let us consider now that the transition $|g\rangle \rightarrow |e\rangle$ is excited with a laser standing wave in such a way that the trap center coincides with one of the nodes of the standing wave (Cirac et al., 1993b). According to the discussion in Section II, this situation is described by the master equation (16).

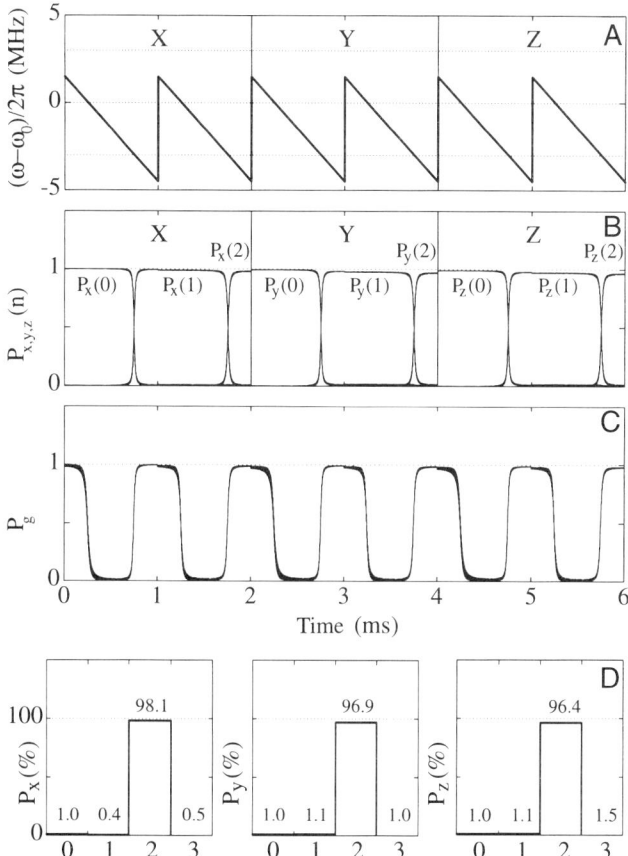

FIG. 6. Preparation of the 3D Fock state $|2\rangle_x|2\rangle_y|2\rangle_z$ by adiabatic variation of the laser detuning. (a) Variation of the laser frequency as a function of time (X, Y, and Z indicate the direction of propagation of the laser that is on during each time interval; these time intervals are separated by vertical lines in the figure. (b) Populations of the states $|n\rangle_x$, $|n\rangle_y$, and $|n\rangle_z$ ($n = 0, 1, 2$) vs time. Note that we have plotted the populations only during the time intervals in which they change due to the presence of the laser. (c) Population of the internal ground state $|g\rangle$ vs time. (d) Occupation probabilities P_x, P_y, and P_z of the Fock states of the harmonic oscillator along the directions x, y, and z, respectively, at the end of the preparation (time = 6 msec). Parameters are $\nu/2\pi = 3$ MHz, $\Omega/2\pi = 300$ kHz, $\phi = \pi/4$, and $\eta = 0.5$. Note that under these conditions we are not in the Lamb–Dicke limit.

Choosing the laser detuning $\delta_2 = -\nu$, in the LDL this master equation reduces to

$$\dot{\rho} = -i\eta_2 \frac{\Omega_2}{2}\left[|e\rangle\langle g|a + a^\dagger|g\rangle\langle e|, \rho\right]$$

$$+ \frac{\Gamma}{2}(2|g\rangle\langle e|\rho|e\rangle\langle g| - |e\rangle\langle e|\rho - \rho|e\rangle\langle e|) + O(\alpha\,\eta_2^2) \quad (40)$$

where we have retained only the terms of zero or first order in the expansion in η_2, and assumed the strong confinement condition $\nu \gg \Gamma$ in order to neglect the counterrotating terms proportional to $|e\rangle\langle g|a$ and $(|e\rangle\langle g|a)^\dagger$.

This master equation corresponds to the damped Jaynes–Cummings model. This model usually appears in cavity QED problems, where the two-level system is a two-level atom and the harmonic oscillator mode of the radiation field (a cavity mode). A significant different between the cavity QED situation and the present problem is that in cavity QED one must contend with damping of the harmonic oscillator, i.e., with cavity mode losses. Further, in the ion trap situation, the effective coupling constant between the two-level system and the harmonic oscillator, $\eta\Omega/2$, depends on the laser intensity, i.e., it can be readily changed (note that an important consequence of having the ion at the node of the standing wave is that increasing the Rabi frequency Ω does not lead to heating of the ion; this does not hold for a traveling wave light field).

According to master equation (40), in steady state the ion resides in the *pure* state $|0, g\rangle$; i.e., both the internal and external degrees of freedom of the ion end up in the ground state. This also agrees with the results of the previous section, where under the same conditions we showed that the steady state had a mean occupation number $(\Gamma/4\nu)^2$, which is negligible in the strong confinement regime. Thus, the master equation (40) describes the dynamics of sideband cooling when the trap is centered at the node of the laser standing wave. Note that since in steady state the two-level system is in its ground state, there is no fluorescence from the transition $|e\rangle$ to $|g\rangle$.

The most interesting regime of the Jaynes–Cummings model is the regime of strong coupling, in which the coupling between the two-level system and the harmonic oscillator is stronger than the spontaneous decay rate. In this case, interesting features arise as a consequence of the quantum entanglement of the two subsystems. In the trapped ion case, this regime corresponds to $\eta_2\Omega_2/2 > \Gamma$, which can be easily accomplished by suitably adjusting the Rabi frequency. As mentioned, this control of the coupling strength is in contrast with current cavity QED experiments, where the coupling is determined by atomic constants and cavity characteristics, which cannot be easily modified. In the strong coupling regime, the spectroscopy of the system is best described in terms of transitions between the "dressed states" of the Jaynes–Cummings Hamiltonian H^{JC} that was analyzed in Section II.A. The ground state is $|0, g\rangle$ and the excited dressed states take the form $|n, \pm\rangle = (|n + 1, g\rangle \pm |n, e\rangle)/\sqrt{2}$ ($n = 0, 1, \ldots$). The energies corresponding to the excited dressed states are $E_{n,\pm} = \pm \eta_2 \Omega_2 \sqrt{n}/2$, which exhibit ac Stark splitting proportional to Ω_2. The splitting between the levels $|n, \pm\rangle$ can be observed by measuring the spectrum of fluorescence corresponding, for example, to the two lowest transitions $|1, \pm\rangle \to |0, g\rangle$, which give rise to a doublet structure, the so-called vacuum Rabi splitting in cavity QED (Sanchez-Mondragon et al., 1983). Note that for conditions described here the ion does not emit light in the steady state. Hence, to observe the fluorescence spectrum, it is necessary to excite the dressed levels. This can be carried out, for example, by using broadband thermal light. Alternatively, one may simply avoid fulfilling completely the strong confinement condition that was used to neglect the rapidly oscillating terms (e.g., $|r\rangle\langle g| a^\dagger$) in (6), since these counterrotating terms actually produce excitations of the system. Figure 7 shows the spectrum computed via numerical solution of the exact master equation (16) (using a finite basis set truncation at a suitable level, and retaining all orders in the Lamb–Dicke expansion parameter) for $\nu = 10\Gamma$, $\eta_2 = 0.01$, and $\Omega_2 = 50\Gamma$, 100Γ, 150Γ, and 200Γ [(a), (b), (c), and (d), respectively]. As can be seen from this figure, the splitting between the two sidebands is proportional to Ω_2. The observation of splitting in the spectrum demonstrates the entanglement between the internal and external degrees of freedom, in analogy with what has been observed in cavity QED (Raizen et al., 1989; Zhu et al., 1990; Bernardot et al., 1992; Thompson et al., 1992). The asymmetry in the spectra is caused by the

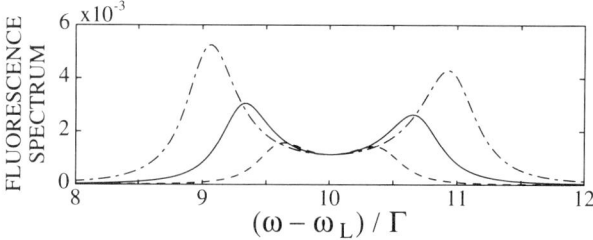

FIG. 7. Spectrum of resonance fluorescence of a trapped ion: emission doublet $|1, \pm\rangle \to |0, g\rangle$ with $\eta = 0.01$, $\nu = 10\Gamma$, and $\Omega = $ (a) 50Γ, (b) 100Γ, (c) 150Γ, and (d) 200Γ.

non-rotating-wave-approximation terms, which also lead to enhanced pumping of the levels with increasing Ω_2.

Aside from this doublet structure, additional resonances, corresponding to transitions involving higher excited states, may be observed in the fluorescence spectrum. This is shown in Fig. 8, where we have plotted the fluorescence spectrum derived from the numerical solution to master equation (16), including a broadband thermal-light driving field with mean photon number $\bar{N} = 1$. In the figure, the dressed-state level structure and the transitions giving rise to the peaks in the spectrum are also shown. Similar spectra have been derived in the context of cavity QED (see, for example, Cirac *et al.*, 1991).

An alternative approach to the observation of the level structure is a measurement of the probe field absorption spectrum. To perform such a measurement without perturbing the cooling and strong coupling, one can employ the third atomic level $|r\rangle$, which is only very weakly coupled to the $|e\rangle$–$|e\rangle$ transition in the "V" configuration. This is in the nature of atomic level schemes used for the observation of quantum jumps, where shelving

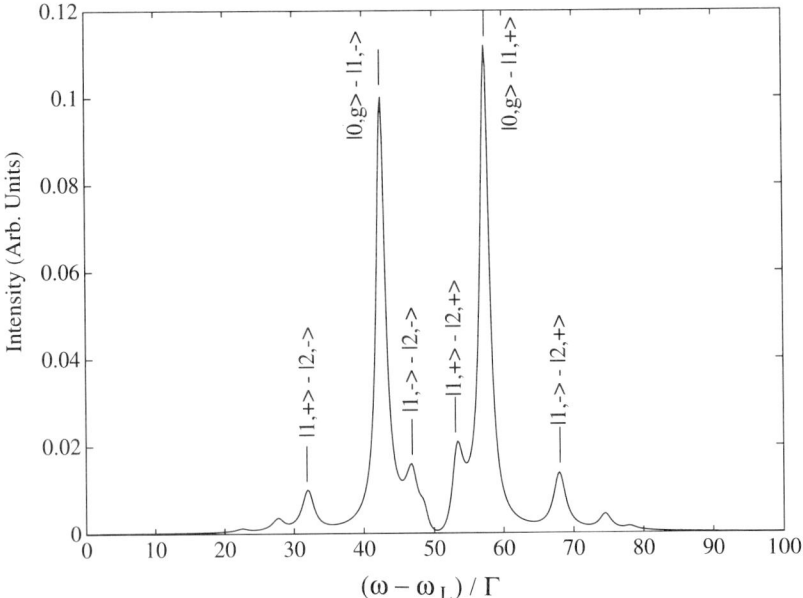

FIG. 8. Spectrum of resonance fluorescence of a trapped ion excited by a broadband thermal reservoir of mean photon number $N = 1$. Parameters are $\eta = 0.01$, $\nu = 50\Gamma$, and $\Omega = 150\Gamma$.

of the electron in the third level $|r\rangle$ produces a dark period in the observed fluorescence. The absorption spectrum can be very sensitively measured by observing quantum jumps to and from the level $|r\rangle$ as a function of the detuning of the probe laser from the $|g\rangle \to |e\rangle$ transition frequency.

Figure 9 shows an absorption spectrum for such a three-level system. To facilitate excitation of higher levels $|n, \pm \rangle$ in the JCM ladder of states, we have assumed that additional thermal light is pumping the ion (this could be readily achieved experimentally by adding a weak broadband noise to the laser). The central maximum corresponds to the transition $|0, g\rangle \to |0, r\rangle$, whereas the additional maxima corresponds to transitions between the dressed states $|n, \pm \rangle$ and the excited levels $|n, r\rangle$. A particularly interesting feature illustrated by this figure is that in the strong coupling limit, $\eta_2 \Omega_2 / 2 \gg \Gamma$, each spectral line represents a particular transition $|0, g\rangle \to |n, r\rangle$ or $|n, \pm \rangle \to |n, r\rangle$. This is a consequence of the unequal spacing of the energy levels in the JCM, which means that, at the frequency corresponding to a particular spectral line, the probe laser is to resonant with the transition frequency between a single pair of levels, and thus it will only excite the system to s single state $|n, r\rangle$.

This ability to selectivity excite a particular transition, together with the state reduction associated with the observation of quantum jumps, offers the possibility of generating Fock states of the quantized trap motion. This is because the probe laser exciting transitions to states $|n, r\rangle$ interacts only with the atomic ground state contribution to the particular dressed state $|n, \pm \rangle$ being excited. Given that we can distinguish spectroscopically between the different maxima characterizing the absorption spectrum (so that we can identify the dressed state being excited), observation of a quantum jump to the weakly coupled state $|n, r\rangle$ will tell us with certainty that the vibrational state of the ion is $|n\rangle$ and that we have produced a

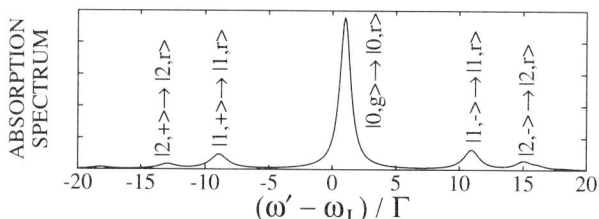

FIG. 9. Weak field absorption spectrum to level $|r\rangle$ as a function of the probe detuning for $\eta = 0.05$, $\nu = 50\Gamma$, and $\Omega = 400\Gamma$. The Rabi frequency corresponding to the $|g\rangle \leftrightarrow |r\rangle$ transition is 0.1Γ. We have taken for the spontaneous decay rate from $|r\rangle$ 0.01Γ.

Fock state of the quantized trap motion. An obvious consequence of having the freedom to choose which transition is excited is the ability to choose the Fock state that is to be produced.

Figure 10 shows the simulation of an experiment with a trapped three-level ion and the calculated fluorescence intensity in the strongly coupled two-level transition as a function of time. For this figure, we have taken the weak laser field on resonance with the transition $|2, -\rangle \to |2, r\rangle$, and we have also included a broadband thermal driving field of mean photon number $N = 2$ in order to excite the dressed levels. Quantum jumps to the state $|2, r\rangle$ are indicated when emission windows appear in the fluorescence intensity. Thus, a Fock state of the trap motion with $n = 2$ is prepared during these dark periods.

We emphasize here that in contrast to the adiabatic passage scheme of the preceding section, in which all the interactions were necessarily coherent, this second method to prepare Fock states is based on dissipation. This is so since the process of measurement (which at some stage must

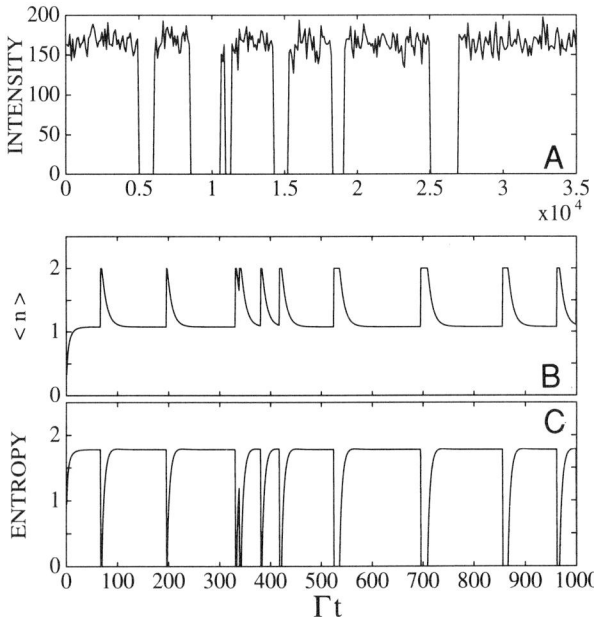

FIG. 10. Simulation of quantum jumps as a function of time. The probe laser is tuned to the $|2, -\rangle \to |2, r\rangle$ transition. (a) Fluorescence intensity on the $|g\rangle \to |e\rangle$ transition; (b) time evolution of the mean oscillator number $\langle n \rangle$; (c) time evolution of the entropy $S = -\text{Tr}[\rho \ln \rho]$ of the system.

introduce dissipation in the system) is needed here to project the motional state of the ion into the desired Fock state.

C. TRAPPING STATES

In the present section, we show that Fock states can also be prepared using a sequence of laser pulses alternating with optical pumping cycles (Blatt et al., 1995). We emphasize that, in contrast to our proposal based on adiabatic passage, the present method includes a dissipative step (here, due to optical pumping). Furthermore, in contrast to the scheme based on the observation of quantum jumps, here the Fock state is prepared in the *steady state*, at the end of the pulse sequence.

The scheme presented here is related to the so-called *trapping states* that have been discussed in the context of cavity QED (Filipowicz et al., 1986; Meystre, 1987). In this case, several two-level atoms in an atomic beam with ground state $|a\rangle$ and excited state $|b\rangle$ are sent successively through a microwave cavity with the atoms initially prepared in the excited state. The ith atom entering the cavity will undergo Rabi flopping $|b\rangle_i \otimes |n\rangle \leftrightarrow |a\rangle_i \otimes |n+1\rangle$ with $|n\rangle$, the n-photon state. For a given flight time of the atom through the cavity (i.e., for a fixed atomic velocity), the Rabi frequency will be proportional to $\sqrt{n+1}$. Thus, certain n will satisfy the *trapping condition*, whereby the atom undergoes a complete Rabi cycle and leaves the cavity again in the excited state, leaving the initial photon state unchanged. According to Filipowicz et al. (1986), a sequence of atoms will thus prepare the cavity in a Fock state determined by this trapping condition. More recently, Slosser et al. (1989) have shown that, using atoms in a well-defined superposition of ground and excited states, other nonclassical states can be produced.

For the case of a trapped ion, we will consider that the ion, initially prepared in its internal ground state $|g\rangle$, is excited by a laser pulse. We assume that the center of the trap coincides with the node of the standing wave laser field, and that the laser detuning $\delta_1 = -\nu$, so that the ion will experience transitions $|g, n\rangle \leftrightarrow |r, n-1\rangle$ (see Section II.A). This laser pulse excitation is followed by an optical pumping pulse resonant with the $|r\rangle \to |e\rangle$ transition, which, according to our analysis of Section II.C, transfers the atoms back to the ground state: $|r, n-1\rangle \leftrightarrow |e, n-1\rangle \to |g, n-1\rangle$. Note that in order to succeed in this optical pumping pulse, the duration time of this pulse must be much longer than the inverses of the corresponding Rabi frequency Ω_3 and the spontaneous emission rate Γ.

Consider now a given initial distribution of oscillator states, and a fixed pulse duration of the laser exciting transitions $|g, n\rangle \leftrightarrow |r, n-1\rangle$. As in the cavity QED case, certain n will again satisfy a trapping condition where the ion undergoes a complete Rabi oscillation, returning to the atomic ground state. Applying a sequence of laser pulses alternating with optical pumping cycles will accumulate ions in these these trapping states and thus prepare Fock states of the motion. Similar schemes can be developed based on the "anti-Jaynes–Cummings" coupling H^{AJC} given in (11); this can be realized by tuning the laser to the upper motional sideband. We note that these trapping states can be interpreted as *dark states* of the combined atom-plus-trap system, and accumulation of atoms in these trapping states is reminiscent of the sub-recoil cooling schemes for free atoms as proposed by Kasevich and Chu (1992).

After each cycle, the ion ends up in its internal ground state $|g\rangle$, whereas the state of the motion changes. Let us denote by P_n^k ($k = 0, 1, \ldots$) the probability of finding the atom in the motional state $|n\rangle$ after the kth cycle. According to the preceding discussion, the second part of the cycle simply transfers the population of the state $|r, n\rangle$ to the state $|g, n\rangle$ incoherently. Therefore, using the analysis of Section II, one can easily derive the map that transformations P_n^k into $P_{n'}^{k+1}$. We find

$$P_n^{k+1} = P_n^k \cos^2(\sqrt{n}\ \tau_1) + P_{n+1}^k \sin^2(\sqrt{n+1}\ \tau_1) \qquad (41)$$

where $\tau_1 = \eta_1 \Omega_1 / 2 t_1$, with t_1 the time duration of the first pulse. For $\tau_1 = \pi/\sqrt{n_0}$ the population of state $|n_0\rangle$ increases in each cycle, since it receives part of the population that was in the state $|n_0 + 1\rangle$, which in turn receives part of the population of $|n_0 + 2\rangle$, etc. Thus, population tends to accumulate in the states $|\Psi_m\rangle \equiv |m^2 n_0\rangle$ ($m = 0, 1, \ldots$). In particular, all the population that was initially in the states $|n\rangle$ with $m^2 n_0 \leq n < (m+1)^2 n_0$ ends up in the state $|\Psi_m\rangle$.

Let us assume that the initial state of the ion is given by a statistical mixture described by the density operator

$$\rho(0) = |g\rangle\langle g| \otimes \sum_{n=0}^{\infty} P_n^0 |n\rangle\langle n| \qquad (42)$$

where P_n is the probability of finding the ion in state $|n\rangle$. This is the case when the atom has been optically cooled, for example, by means of Doppler cooling. In this situation, the state of the ion is (42), where P_n is given by the distribution

$$P_n^0 = (1 - e^{-\hbar \nu / \kappa T}) \exp \frac{-\hbar \nu n}{\kappa T} \qquad (43)$$

and the temperature T depends on the laser cooling conditions (such as detunings and laser intensities). In this case, after several cycles the steady state density operator will be

$$\rho_{SS} = |g\rangle\langle g| \sum_{m=0}^{\infty} P_m^{SS} |\Psi_m\rangle\langle\Psi_m| \tag{44}$$

where

$$P_m^{SS} = \sum_{n=m^2 n_0}^{(m+1)^2 n_0 - 1} P_n^0 \tag{45}$$

State (44) is very similar to the one obtained in the cavity QED context using the trapping states technique. Note also that if the initial state was the one obtained after sideband cooling, i.e., $|g, 0\rangle$, there will be no change in the population. This is a special case of (43) with $T = 0$.

Now assume that we perform addition cycles, but using a laser detuning $\delta_1 = \nu$ [i.e., Hamiltonian (11)] in the first part of each cycle, and for a time $\tau_1' \pi/\sqrt{n_0 + 1}$. The second part of the cycle remains as in the previous procedure. Now, each cycle performs the mapping

$$P_n^{k+1} = P_n^k \cos^2\left(\sqrt{n+1}\ \tau_1'\right) + P_{n-1}^k \sin^2\left(\sqrt{n}\ \tau_1'\right) \tag{46}$$

After several cycles, the population tends to accumulate in the states $|\Psi_m'\rangle \equiv |m^2(n_0 + 1) - 1\rangle$ ($m = 1, 2, \ldots$). In particular, all the population that was initially in the states $|n\rangle$ with

$$m^2(n_0 + 1) \leq n + 1 < (m+1)^2(n_0 + 1)$$

ends up in the state $|\Psi_{m+1}'\rangle$.

It is then clear that if the initial state was $|g, 0\rangle$, after a few cycles it will become $|g, n_0\rangle$ and, therefore, the Fock state of motion $|n_0\rangle$ will be prepared. One can find a whole variety of distributions by combining cycles of type (a) with those of type (b). Consider, for example, an experiment where the time τ_1' cannot be matched exactly. In that case the desired Fock state would not be precisely produced. In addition, corrections to the Lamb–Dicke expansion will cause, after very many cycles, the population of the state $|n_0\rangle$ to decrease. Thus, one can combine cycles (a) and (b) to stabilize the procedure and maintain the desired Fock state.

Let us now illustrate this method with some numerical calculations for realistic experimental parameters with a trapped Ca^+ ion, which is a three-level cascade system as shown in Fig. 1. Assuming that the ion has been laser cooled (e.g., by Doppler cooling), we start our investigation with a thermal initial distribution, as shown in Fig. 11(a), of mean vibrational quantum number $\langle n \rangle = 7$. We seek to prepare a Fock state $n_0 = 7$ and correspondingly choose the appropriate Rabi frequency and interaction time for the weak transition.

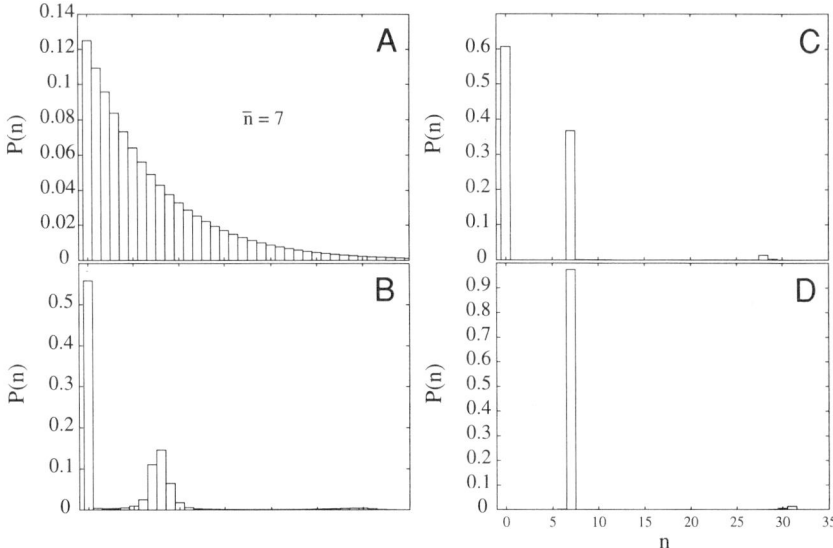

FIG. 11. Preparation of a Fock state $|7\rangle$ via trapping states. (a) Initial state distribution, e.g., after Doppler cooling with $\langle n \rangle = 7$, time $t = 0$. (b) Nonclassical state after application of several pulse cycles with $\eta\Omega_1 = 790$, $\Omega_2 = 10^5$ sec^{-1} $t_1 = t_2 = 3$ msec and detuning $\delta_1 = -\nu$ (lower sideband application) after a total time $t = 0.045$ sec. (c) As in (b), after a total time $t = 1$ sec. (d) Starting as in (c), with the detuning $\delta_1 = +\nu$ (upper sideband) and after a total time $t = 2$ sec.

Consider, then, excitation consisting first of a pulse of length $t_1 = 2\tau_1/\eta_1\Omega_1$, with $\tau_1 = \pi/\sqrt{n_0}$ ($n_0 = 7$), on the week transition $|n, g\rangle \to |n-1, r\rangle$, followed by a second pulse on the strong $|n-1, r\rangle \to |n-1, g\rangle$ transition, as previously explained. This excitation cycle is repeated a few times until a steady state is obtained. As an example, Fig. 11 shows a simulation of such a pulse sequence. The population remains trapped in the states $|\Psi_m\rangle = |7m^2\rangle$ ($m = 0, 1, \ldots$). Given the initial thermal state, most of the population will remain in the ground level of the harmonic trap $|0\rangle$, a considerable fraction will end up in the level $|7\rangle$, while only a small fraction will be left in the level $|28\rangle$, etc. Figure 11(b) shows an intermediate result after 0.045 sec and Fig. 11(c) represents the final population distribution after 1 sec.

Eventually, radiation can be applied on the upper on the upper sideband of the weak transition (i.e., $\delta_1 = +\nu$) during time $t_1 = 2\tau_1'/\eta_1\Omega_1$ with $\tau_1' = \pi/\sqrt{n_0 + 1}$ ($n_0 = 7$), followed by a pulse on the strong transition. As discussed, the population will remain trapped in the states $|\Psi_m'\rangle = |8m^2 - 1\rangle$ ($m = 1, 2, \ldots$). Now, of course, population is no longer trapped in state $|0, g\rangle$. Thus, by slightly readjusting the interaction time τ_1 and the detuning, the trapping condition for state $|7, g\rangle$ can still be fulfilled, and consequently all the population is transferred to state $|7, g\rangle$ and trapped. This is shown in Fig. 11(d), after excitation on the upper sideband for 1 sec. Note that a small amount of population has shifted to the state $|31\rangle$ (i.e., $|\Psi'_2\rangle$). We emphasize that this second step is not possible in the context of the micromaser (Slosser *et al.*, 1989), since it requires a Hamiltonian with an antiresonant Jaynes–Cummings coupling, which is not available in a micromaser.

The dynamical evolution of the nonclassical states prepared during the pulse sequence shown in Fig. 11 can be followed in Fig. 12, where the entropy and fluorescence of the system are plotted as functions of time. Both quantities are evaluated after completion of successive laser cycles (t_1, t_2). Starting from its initial value set by the thermal distribution, the entropy decays exponentially to a steady state value characterized by the population of the two trapping states $|0\rangle$ and $|7\rangle$, as shown in Fig. 11(c). Similarly, the fluorescence decays due to optical pumping to the trapping states. Switching to upper sideband excitation, i.e., $\delta_1 = +\nu$, first results in an increased entropy, since there are more oscillator levels occupied (i.e., more degrees of freedom) during intermediate pumping cycles, until eventually a Fock state $|7\rangle$ is produced with the entropy approaching zero. Accordingly, there is higher fluorescence at the beginning since now the population of the trapping state $|0\rangle$ also contributes. Note that zero entropy is not completely attained, since a weakly populated higher trap-

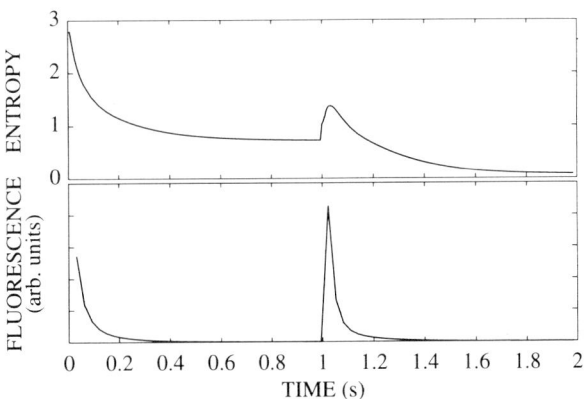

FIG. 12. Entropy and fluorescence as functions of time for the same parameters as in Fig. 11.

ping state exists (cf. Fig. 11(d)); however, this may be circumvented using different pulse sequences.

This procedure also works well in two and three dimensions. As an example, we have chosen to simulate the preparation of a Fock state in two dimensions which would correspond, for example, to the motion of an ion in the ring plane of a Paul trap. Starting with an initial thermal distribution ($\langle n_x \rangle = 4$, $\langle n_y \rangle = 4$) as shown in Fig. 13(a), an intermediate (nonclassical) state is reached after the fivefold application of the following pulse sequence: (i) five pulses with a laser along the x direction with $\delta_{1,x} = -\nu_x$; (ii) five pulses with $\delta_{1,x} = +\nu_x$; (iii) five pulses with a laser along the y direction with $\delta_{1,y} = -\nu_y$; and (iv) five pulses with $\delta_{1,y} = +\nu_y$. Of course, each pulse on the metastable transition is individually followed by a pulse on the appropriate strong transition to provide for the optical pumping and the population transfer. The entire sequence to reach the state shown in Fig. 13(b) requires a total interaction time of 0.74 sec. Eventually, application of 50 more pulse sequences as described before yields, after approximately 7.4 sec, the two-dimensional Fock state indicated in Fig. 13(c).

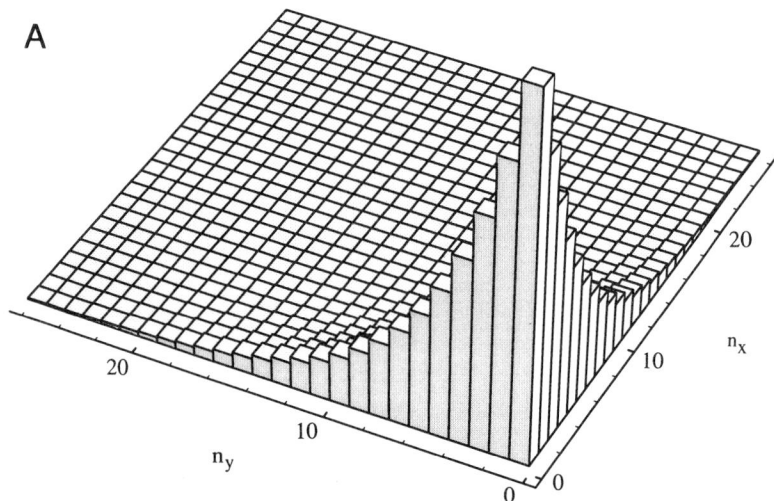

FIG. 13. Two-dimensional state distributions: (a) initial thermal distribution; (b) intermediate distribution; (c) final Fock state distribution. For parameters and pulse sequence, see text.

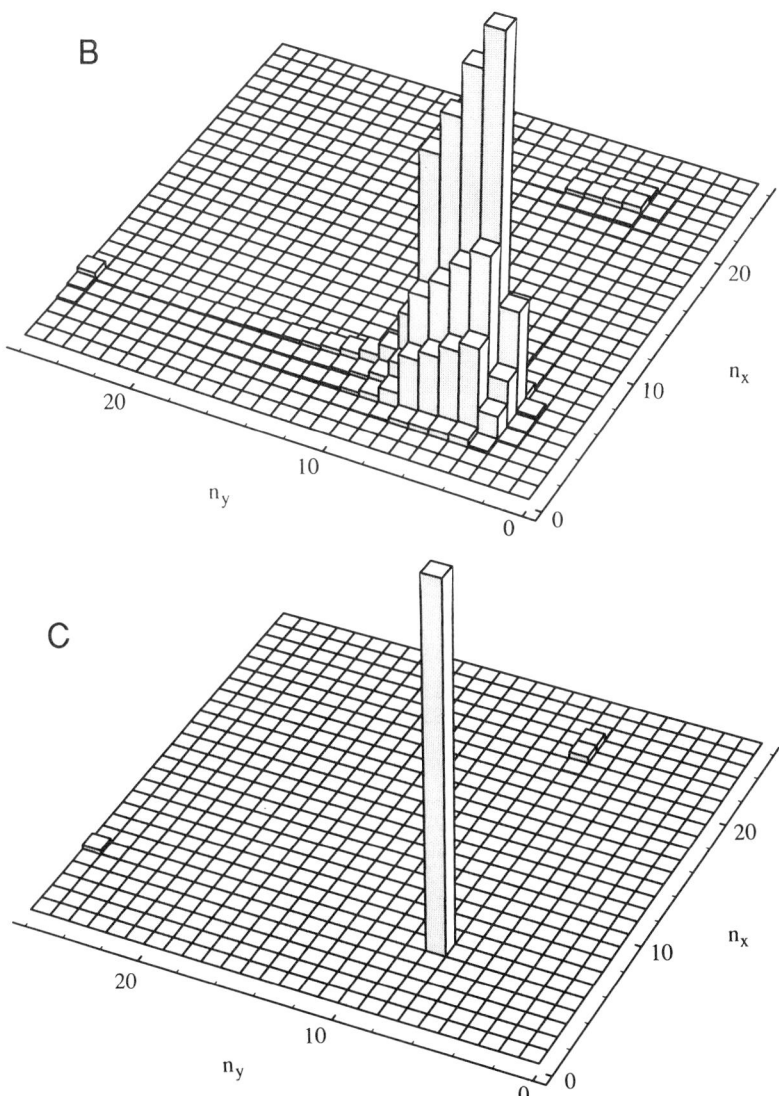

FIG. 13. *(continued)*

V. Preparation of Squeezed States

Squeezed states possess the property that one quadrature phase has reduced fluctuations compared with the ordinary vacuum. The ideal squeezed state of a harmonic oscillator is defined as (see, for example, Loudon and Knight, 1987)

$$|\alpha, \epsilon\rangle = D(\alpha)S(\epsilon)|0\rangle \tag{47}$$

Here $\mathscr{D}(\alpha) = \exp(\alpha a^\dagger - \alpha^* a)$ is the displacement operator, and $\mathscr{S}(\alpha) = \exp(\frac{1}{2}\epsilon^* a^2 - \frac{1}{2}\epsilon a^{\dagger 2})$ is the "squeeze" operator, with $\epsilon = re^{i\theta}$ (r is the squeezing parameter). Defining quadrature phase operators $a_{1,2}$ by $a = (a_1 + ia_2)e^{i\theta/2}$ [proportional to the position and momentum operators given in (2)], their vacancies in the squeezed state (47) can be evaluated as $(\Delta a_{1,2})^2 = e^{\mp 2r}/4$, from which it is clear that the noise in one quadrature can be reduced below the standard quantum limit $(\Delta a_{1,2})^2 = 1/4$, and that the state (47) is a minimum uncertainty state, i.e., $\Delta a_1 \Delta a_2 = 1/4$.

In this section we review two methods to prepare a squeezed state of the form (47) in the motion of a trapped ion. The first is based on a sudden change of the trap frequency ν, whereas the second is based on multichromatic laser excitation of the ion.

A. Sudden Change of the Trap Frequency

It has been known for some time that sudden changes in the frequency of a harmonic oscillator can lead to squeezed states of the oscillator (see, for example, Janszky and Yushin, 1986), and the possible application of this phenomenon in ion traps has been considered (Graham, 1987; Heinzen and Wineland, 1990; Agarwal and Kumar, 1991). A brief and simple description of such an effect is as follows.

Let us denote the (vibrational) state of the ion by $|\Phi\rangle$, and let us assume that the frequency of the trap is suddenly changed from ν to ν' (Heinzen and Wineland, 1990). According to the well-known "sudden approximation" of quantum mechanics, the state of the ion will remain practically unchanged. However, since the harmonic trap has now changed its shape, the eigenstates of the harmonic trap will change, too. Thus, it is convenient to reexpress the state $|\Phi\rangle$ in the new basis that diagonalizes the new Hamiltonian.

The Hamiltonian after the sudden change of frequency is

$$H'_{tp} = \frac{P^2}{2M} + \frac{1}{2}M\nu'^2 R \equiv \nu' a'^\dagger a' \tag{48}$$

where the new creation and annihilation operators are defined by

$$R = \sqrt{\frac{1}{2M\nu'}}\,(a' + a'^{\dagger}), \qquad P = i\sqrt{\frac{M\nu'}{2}}\,(a'^{\dagger} - a') \qquad (49)$$

Equating Eqs. (2) and (49), one can derive the following relations

$$a' = a\cosh(r) + a^{\dagger}\sinh(r) \qquad (50a)$$

$$a'^{\dagger} = a^{\dagger}\cosh(r) + a\sinh(r) \qquad (50b)$$

where $\cosh(r) = \frac{1}{2}[(\nu'/\nu)^{1/2} + (\nu/\nu')^{1/2}]$. Using the definition of the squeeze operator S, we can reexpress these relations as

$$a' = S(-r)^{\dagger} a S(-r), \qquad a'^{\dagger} = S(-r)^{\dagger} a^{\dagger} S(-r) \qquad (51)$$

According to (51), it is precisely the squeeze operator, $S(-r)$, that transforms the old basis set (Fock states of the harmonic potential of frequency ν) into the new basis set (Fock states of the harmonic potential of frequency ν'). That is, denoting by $|n'\rangle$ the eigenstates of (48), we have $|n'\rangle = S(-r)^{\dagger}|n\rangle$. Thus, for example, if the initial state of the ion is the ground state $|\Phi\rangle = |0\rangle$, after the change in trap frequency the ion will be in the state $|\Phi\rangle = S(-r)|0'\rangle$, i.e., a squeezed state of the form $|\Phi\rangle = |0, -r\rangle$ [compare (47)]. A state of the form (47) with $\alpha \neq 0$ can also be prepared if, after the sudden change in trap frequency, one slightly, and suddenly, shifts the trap center a distance $\alpha(2/M\nu')^{1/2}$. Obviously, if this shift is applied to the initial state $|0\rangle$, a coherent state of the atomic motion will be produced. We note that a more involved method for preparing a coherent state was discussed by Zeng and Lin (1995a).

B. BICHROMATIC EXCITATION

An alternative scheme for preparing squeezed states of motion that does not rely on sudden changes in the trapping potential, but rather employs bichromatic laser excitation of the ion and was put forward by Cirac *et al.* (1993a). A closely related scheme making use of two-photon Raman transitions to realize suitable couplings (in the manner described at the end of Section III.B) has also been proposed by Zeng and Lin (1995b). Here, we describe the proposal of Cirac *et al.* (1993a).

Consider the situation in which a trapped ion oscillates around a *common node* of two standing wave fields: one of frequency $\omega_{L1} = \omega_0 - \nu$, and the other of frequency $\omega_{L2} = \omega_0 + \nu$. These standing wave fields drive the electric-dipole allowed transition $|g\rangle \to |e\rangle$. Following the analysis of

Section II, the master equation that describes this situation in the LDL is

$$\dot{\hat{\rho}} = -i\left[\hat{H}_a + \hat{H}_b, \hat{\rho}\right] + \mathcal{L}\hat{\rho} \tag{52}$$

where

$$\hat{H}_a = g_a |e\rangle\langle g| a + \text{h.c.} \tag{53a}$$

$$\hat{H}_b = g_b |e\rangle\langle g| a^\dagger + \text{h.c.} \tag{53b}$$

and $\mathcal{L}$ is given by (19). We assume the low intensity and strong confinement limits and define $g_{a,b} = \eta_2 \Omega_{a,b} e^{-i\psi_{a,b}}/2$, with Ω_a (Ω_b) and ψ_a (ψ_b) the Rabi frequency and the phase corresponding to the interaction of the ion with the first (second) laser, respectively. The coupling Hamiltonians $\hat{H}_a$ and $\hat{H}_b$ can be associated with cooling and heating of the ion [compare with (10) and (11)], in that they describe processes in which the energy of the ion decreases (terms of H_a proportional to $|e\rangle\langle g| a$) or increases (terms of H_b proportional to $|e\rangle\langle g| a^\dagger$).

To find the steady state solution of the master equation (52), we perform the transformation defined by the squeeze operator $S(\epsilon)$, where now

$$\epsilon = re^{i(\psi_a - \psi_b)}, \quad \tanh(r) = \left|\frac{g_b}{g_a}\right| < 1 \tag{54}$$

(Note that for this transformation to be properly defined, one requires that $|g_a| > |g_b|$). Using the well-known properties of the squeeze operator,

$$S(\epsilon)^\dagger a S(\epsilon) = a \cosh(r) - a^\dagger \sinh(r) e^{i(\psi_a - \psi_b)} \tag{55a}$$

$$S(\epsilon)^\dagger a^\dagger S(\epsilon) = a^\dagger \cosh(r) - a \sinh(r) e^{-i(\psi_a - \psi_b)} \tag{55b}$$

it can be shown that the master equation for the density operator $\rho = S(\epsilon)^\dagger \hat{\rho} S(\epsilon)$ is exactly (40), with $\Omega = |g_a|\cosh(r) - |g_b|\sinh(r)$. From the discussion following (40), it follows that the steady state solution for the state of the ion in the transformed picture is simply $|0, g\rangle$. Inverting the unitary transformation (55), the steady state in the original picture is then

$$|\epsilon, g\rangle = S(\epsilon)|0\rangle \otimes |g\rangle \tag{56}$$

Note that $S(\epsilon)|0\rangle$ is a squeezed vacuum state; i.e., in steady state this configuration yields a *pure* state given by the product of a squeezed state of the quantized motion with the ground internal state $|g\rangle$ of the ion. As we had in the case of an ion at the node of a *single* standing wave in Section III.B, the ion ends up in a state from which it ceases to emit photons. Further, in this state, it is easy to see that $(H_a + H_b)|\epsilon, g\rangle = 0$; i.e., the ion does not see the laser once it has reached this state. Consequently, the generation of a squeezed state coincides with a dark state (Alzetta *et al.*, 1976) for the emitted fluorescence. An analogous dark state, coinciding with the production of a coherent field state, occurs in cavity QED when an externally driven two-level atom is coupled to a lossless cavity mode (Alsing *et al.*, 1992).

Note that for the Lamb–Dicke limit to hold, it is necessary that the mean quantum number, $\langle a'a \rangle$, not be too large. Note also that if, for example, the frequency of the second laser differs from $\omega_0 - \nu$, the resulting Hamiltonian H_b is time dependent, and different results are obtained. In general, for the steady state to be $|\epsilon, g\rangle$, it is necessary for the difference frequency between the two laser fields to be 2ν.

In Fig. 14 we have plotted the fluorescence intensity (which is proportional to the excited state population ρ_{ee}) and the quadrature phase

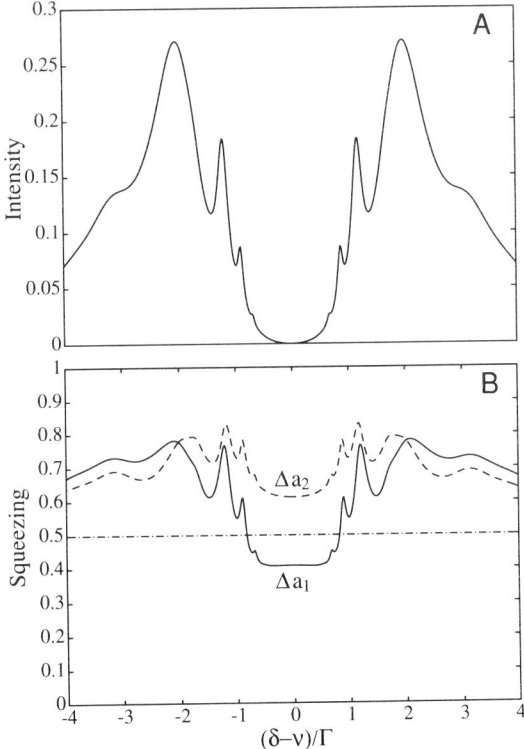

FIG. 14. Fluorescence intensity (a) and quadrature phase variances (b) as functions of the frequency of the second laser standing wave (relative to $\omega_0 - \nu$). The dash–dotted line indicates the standard quantum limit.

variances as functions of the frequency of the second laser with respect to $\omega_0 - \nu$. The results were obtained from a numerical solution of the master equation (52), but allowing for the frequency of the second laser to be varied. At $\omega_{L2} = \omega_0 - \nu$, a dark resonance occurs, accompanied by a reduction in Δa_1 below the standard quantum limit, indicating that a squeezed state has been generated.

Finally, we note that the scheme presented here can be generalized to two or three dimensions, in which case interesting entangled states of two or three harmonic oscillators (each of them corresponding to the oscillations of the ion in two or three orthogonal directions) can be produced in the steady state.

VI. Preparation of Schrödinger Cat States

Up to now, the schemes we have described for the preparation of nonclassical states of motion have been based on the LDL, as well as on the low excitation limit. In this section, we analyze the possibility of preparing nonclassical states of motion of a single ion in a completely different regime (Poyatos et al., 1995). We concentrate on the *strong excitation regime*, whereby the Rabi frequency describing the interaction of the ion with laser field is much larger than the trap frequency ($\Omega \gg \nu$). In this regime, the Hamiltonian describing the interaction is quite different from the simple Jaynes–Cummings model, and it leads to a variety of other possibilities. In particular, Poyatos et al. (1995) have recently shown how one might prepare coherent quantum superpositions of two "macroscopically" distinct states, e.g., superposition states of the form

$$|\Psi\rangle = K(|\alpha\rangle_c + |-\alpha\rangle_c) \tag{57}$$

where K is a normalization constant and $|\alpha\rangle_c$ is a coherent state of the motion. States (57) are referred to as Schrödinger Cat states (Schrödinger, 1935b; Yurke and Stoler, 1986). The method we review here consists of two parts: First, using a laser, one splits into two parts the ion wavefunction; then, one makes a measurement of the *internal* state of the ion in order to project its *external* state onto the state (57). Thus, as in the preparation of Fock states by observation of quantum jumps, the projection postulate of quantum mechanics is one of the ingredients of our method. We will consider a scheme to prepare the state (57) based on the application of laser pulses to the electric-dipole forbidden transition $|g\rangle \to |r\rangle$.

Thus, we consider an ion interacting with a laser beam on resonance with the transition $|g\rangle \to |r\rangle$, $\delta_1 = 0$. In contrast to the analysis of the previous sections, we will consider here that the laser field is in a traveling wave configuration, so that the interaction is described by Hamiltonian (5b). The evolution is readily described if we perform a unitary operation defined by

$$U_\pm = e^{\mp i\eta_1(a+a^\dagger)|r\rangle\langle r|} \tag{58}$$

Using this operator, the states are defined as

$$|\bar\Psi_\pm\rangle = U_\pm|\Psi\rangle \tag{59}$$

and the Hamiltonian becomes

$$\tilde H_\pm \equiv U_\pm H_\pm U_\pm^\dagger = \nu a^\dagger a \pm i\nu\eta_1(a - a^\dagger)|r\rangle\langle r|$$
$$+ \nu\eta_1^2|r\rangle\langle r| + \frac{\Omega_1}{2}\sigma_x \tag{60}$$

where $\sigma_x = \sigma_+ + \sigma_-$. The new terms appearing in the Hamiltonian (60) correspond to the Doppler recoil energy of the internal excited state (i.e., when one photon is absorbed).

The Hamiltonian (60) can be simplified in the strong excitation regime $\Omega_1 \gg \nu$. Let us consider a laser pulse, the duration τ of which fulfills the inequality

$$g\eta\tau \max(\bar n, \eta_1^2) \ll 1 \tag{61}$$

where $\bar n = \langle a^\dagger a\rangle$. Note that in the strong excitation limit, this interaction time can correspond to pulses of area $\Omega_1\tau \sim \pi$. Under condition (61), the first terms in the Hamiltonian (60) will have little effect on the evolution. The Hamiltonian can thus be reduced to $\tilde H = \frac{1}{2}\Omega_1\sigma_x$ (for both cases, $\tilde H_\pm$). The evolution of any initial state is then given by

$$|\Psi_\pm(\tau)\rangle = U_\pm^\dagger e^{-i\Omega_1\sigma_x\tau/2}U_\pm|\Psi(0)\rangle \tag{62}$$

Note that the evolution given by H_- can be obtained from the given by H_+ by simply exchanging $\eta_1 \to -\eta_1$.

In the following, we will need to know the evolution for the case where the initial state of motion is a coherent state $|\alpha\rangle_c$. In this situation, Eq. (62) gives

$$|g\rangle|\alpha\rangle_c \to \cos\left(\frac{\Omega_1\tau}{2}\right)|g\rangle|\alpha\rangle_c - i\sin\left(\frac{\Omega_1\tau}{2}\right)|r\rangle|\alpha \pm i\eta_1\rangle_c, \tag{63a}$$

$$|r\rangle|\alpha\rangle_c \to \cos\left(\frac{\Omega_1\tau}{2}\right)|r\rangle|\alpha\rangle_c - i\sin\left(\frac{\Omega_1\tau}{2}\right)|g\rangle|\alpha \mp i\eta_1\rangle_c \tag{63b}$$

where the upper (lower) sign corresponds to a laser pulse propagating in the positive (negative) R direction. Note that we have assumed square laser pulses. For any other kind of pulse, formulas (63) are still valid, but now one simply replaces $\Omega_1 \tau$ by $\int_0^\tau \Omega_1(t)\,dt$.

A simple way to prepare the state (57) using laser pulses consists of the following three steps:

1. Excite the ion with a $\pi/2$ pulse ($\Omega_1 \tau = \pi/2$) using a laser propagating in the negative R direction. According to (63), this pulse will perform the following transformation

$$|g\rangle|0\rangle_c \rightarrow \frac{1}{\sqrt{2}}(|g\rangle|0\rangle_c - i|r\rangle|-i\eta_1\rangle_c) \tag{64}$$

where $|\alpha\rangle_c$ denotes the coherent state $D(\alpha)|0\rangle$.

2. Excite the ion with another $\pi/2$ pulse ($\Omega_1 \tau = \pi/2$), but now using a laser beam propagating in the opposite direction. The state of the ion will become

$$\frac{1}{2}[(|0\rangle_c - |-2i\eta_1\rangle_c)|g\rangle - i(|i\eta_1\rangle_c + |-i\eta_1\rangle_c)|r\rangle] \tag{65}$$

3. Measure the internal state of the ion using the quantum jump technique. To do this, one can drive the ion with a different laser beam on resonance will the electric-dipole allowed transition $|g\rangle \leftrightarrow |e\rangle$ (see Fig. 1). If fluorescence *is* observed, the state of the ion will be projected onto the part of the wavefunction (65) that contains the state $|g\rangle$. If *no* fluorescence is observed, it will be projected onto the state

$$|\Psi_{sc}\rangle = K(|i\eta_1\rangle_c + |-i\eta_1\rangle_c)|r\rangle \tag{66}$$

where K is a normalization constant.

The state (66) is already of the form (57), i.e., similar to that studied in the cavity QED context (Gea-Banacloche, 1990; Brune et al., 1992). Note that if fluorescence is observed, the state of the motion will be completely modified due to the photon recoil acquired by the ion in each absorption–spontaneous emission cycle, and therefore a steady state superposition of coherent states will *not* be produced. That is, an experiment based on these steps will be successful only in those runs in which no fluorescence is observed (i.e., 50% of the time).

On the other hand, for the state (66) to be considered as a Schrödinger Cat, it should be macroscopic (in the sense that one can distinguish between the states $|\alpha\rangle$ and $|-\alpha\rangle$). However, there are two reasons that the state (66) cannot be directly observable as it stands: (a) Its probability distribution in the position representation has only one peak centered at $\langle \hat{R} \rangle = 0$, and therefore the two parts of such a state ($|i\eta_1\rangle$ and $|-i\eta_1\rangle$) cannot be distinguished by direct observation. (b) Since the ion is in its excited internal state, it cannot be observed by shining light on the $|g\rangle \leftrightarrow |e\rangle$ transition. Problem (b) is easily solved if, just after the state (57) has been produced, one applies a laser π pulse along a perpendicular axis (cf. z) that transforms the excited state into the ground state. Note that to avoid modifying the state of the ion this with this last pulse, it is necessary for the ion to be confined in the Lamb–Dicke limit in the z direction, i.e., $\eta_z = k_L(2M\nu_z)^{1/2} \ll 1$. This may be the case, for example, in a linear ion trap (Raizen et al., 1992; Walther, 1994), for which the transverse directions have trap frequencies much larger than those along the axial direction. The first problem, (a), can be solved by noting that the free evolution of a coherent state is given by $|\alpha(t)\rangle = |e^{-i\nu t}\alpha(0)\rangle$ and, therefore, if after producing the state (57) one waits for a time $t = \pi/2\nu$, the state will become

$$|\Psi_{sc}\rangle = K(|\eta_1\rangle_c + |-\eta_1\rangle_c)|r\rangle \qquad (67)$$

This state has two maxima (in the position representation) centered at $\langle \hat{R} \rangle = \pm 2\eta_1/\sqrt{2M\nu}$, respectively (these maxima correspond to the states $|\pm\eta_1\rangle$, respectively). To be able to observe these two states, the corresponding peaks must be spatially separated by more than a wavelength, which requires that $\eta_1^2 > \pi/2$. For tight traps, that is, for values of η_1 not fulfilling this condition, one can proceed as follows. After step (1), one applies a sequence of π pulses from the left and from the right (n pulses in each direction), in an alternating sequence. It is easy to check that the state after step (3) and free evolution will be

$$|\Psi_{sc}\rangle = K(|(2n+1)\eta_1\rangle_c + |-(2n+1)\eta_1\rangle_c)|r\rangle \qquad (68)$$

For these states, the number of pulses required to observe distinguishable (macroscopic) states is $(2n+1)^2 > \pi/2\eta_1^2$. Note also that now, in condition (61), it is the total time corresponding to all the pulses that enters.

To illustrate the performance of the method presented here, we have plotted in Fig. 15 the state after step (3) for several values of Ω_1/ν and η_1. To produce these plots, we have numerically solved the evolution equations for the ion using the exact Hamiltonian (5b). The figures display the real part of the density operator in the momentum representation $\langle p|\rho|p'\rangle$ (note that the axes are rescaled in terms of $p_0 = \sqrt{M\nu/2}$). Figures 15(a)–(c) correspond to $\eta_1 = 0.5$ and $n = 2$ (i.e., two intermediate

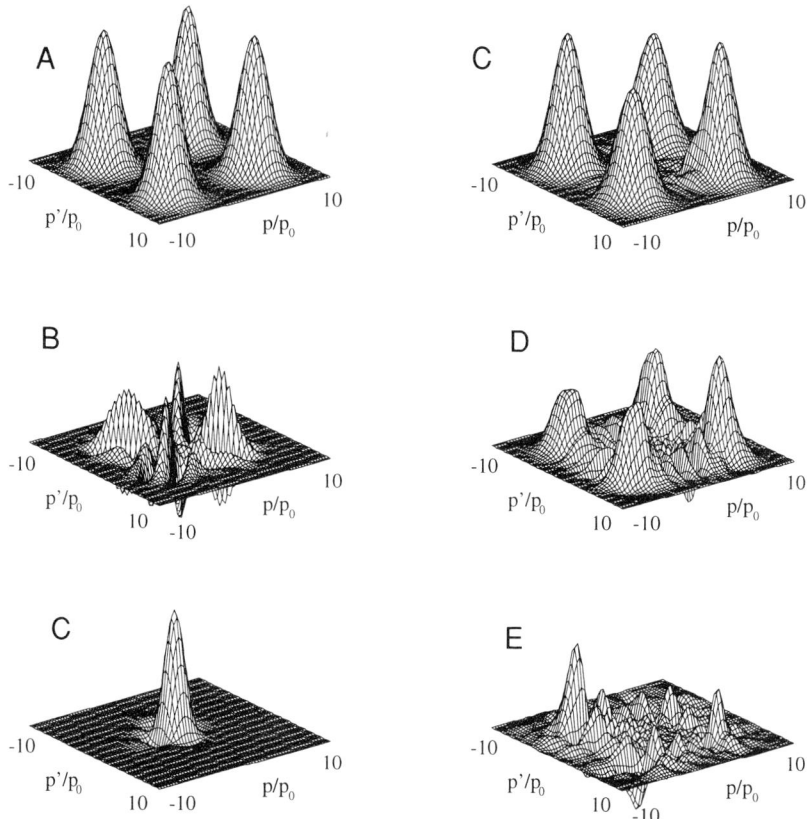

FIG. 15. Real part of the density operator in momentum representation $\langle p| \rho |p'\rangle$ in arbitrary units [note that the axes are rescaled in terms of $p_0 = M\nu/2)^{1/2}$], after the preparation of the superposition state using laser pulses. $\eta = 0.5$, and $n = 2$ (a–c), or $\eta = 2.5$ and $n = 0$ (d–f). Here, $\Omega/\nu = 100$ (a, d), 10 (b, e), or 1 (c, f).

π pulses in each direction), whereas Figs. 15(d)–(f) correspond to $n = 0$ with $\eta_1 = 2.5$. For $\Omega_1/\nu = 100$ (Figs. 15a, d) the method works almost ideally, since condition (61) is satisfied. There are four peaks in the plots, two of them corresponding to the diagonal parts of the density operator ($|i\eta_1\rangle\langle i\eta_1|$ and $|-i\eta_1\rangle\langle -i\eta_2|$), and the other two corresponding to the coherences ($|-i\eta_1\rangle\langle i\eta_1|$ and $|i\eta_1\rangle\langle -i\eta_1|$). These latter peaks confirm that the state is a truly pure state, as opposed to a statistical mixture (in which these two peaks do not show up at all). As soon as the ratio Ω_1/ν is decreased, so that condition (61) is not satisfied, this four-peak structure disappears. Note that the plots corresponding to $n = 2$ are much more sensitive to this condition than those with $n = 0$.

Finally, we mention that one can also prepare Schrödinger Cat states using the adiabatic passage technique, rather than using laser pulses (Poyatos *et al.*, 1995). In such a case, it is the laser frequency that one varies continuously in time in order to produce the required transitions. As with the adiabatic passage scheme, however, it is straightforward to generalize the scheme reviewed here to prepare such nonclassical states in more than one dimension.

VII. Analysis of the Nonclassical States of Motion

In this section we consider how to analyze experimentally the motion of the ion once a nonclassical state has been prepared. We will consider two schemes. The first scheme involves measuring the atomic population inversion as a function of time; the behavior of the inversion reflects the state of the ion motion, exhibiting, for example, collapses and revivals (Blockley *et al.*, 1992; Cirac *et al.*, 1994b). In the second subsection, we describe a method to distinguish a (pure) Schrödinger Cat state from a state corresponding to a statistical mixture with the same phonon distribution (Poyatos *et al.*, 1995).

A. COLLAPSES AND REVIVALS

The Phenomenon of collapses and revivals is well known in quantum optics in the context of the Jaynes–Cummings model (Eberly *et al.*, 1980; Narozhny *et al.*, 1981; Yoo *et al.*, 1981). Cavity quantum electrodynamics provides an experimental realization of the JCM, both in the optical and the microwave domain. With n photons present in the cavity, a two-level atom initially in its ground state undergoes Rabi oscillations with a Rabi frequency proportional to $\sqrt{n}$. Generally, a cavity does not contain a fixed number of photons; rather it is characterized by a density operator with photon number distribution P_n. In this case, Rabi oscillations corresponding to different photon numbers are superimposed and give rise to collapses and revivals in the time evolution population inversion. The structure of these revivals thus depends on the probability distribution P_n and hence provides a measure of the state of the oscillator. Of course, collapses and revivals are also a direct manifestation of the discrete nature of the eigenstates of an electromagnetic field mode. As we have seen, under certain operating conditions the dynamics of a trapped ion interacting with a laser field can be described by the Jaynes–Cummings model, and so it is not surprising that collapses and revivals also occur in the ion trap context. Correspondingly, they can be used to study and characterize the vibrational state distribution of the ion motion.

For an observation of the collapse and revival phenomenon, one can employ a single ion trapped in a harmonic trap and an internal level scheme as indicated in Fig. 1. The two-level transition $|g\rangle \leftrightarrow |r\rangle$ is used to observe the collapse and revival. It is assumed that this transition is driven by a strong standing wave field in the LDL in such a way that the effective coupling constant $\eta_1 \Omega_1 / 2$ leads to several Rabi oscillations during the interaction time (while still satisfying the low excitation conditions). The third level $|e\rangle$ is used for a measurement of the population inversion (on the $|g\rangle \leftrightarrow |r\rangle$ transition) as will be explained.

In a first step, one prepares a particular nonclassical state of the ion motion using one of the schemes previously described. In what follows, we will concentrate on situations in which the initial state of the harmonic oscillator is one of the following: thermal, Fock, coherent, or squeezed state. Obviously, the present approach is valid for any initial state.

Collapse and revival in the three-level atom is then observed by exciting the narrow transition $|g\rangle \rightarrow |r\rangle$ in the following way: (i) After the ion is laser cooled and an initial state distribution $\rho(t_i)$ is prepared, the cooling radiation is turned off. (ii) Then, laser radiation on the narrow (two-level) transition is applied for a certain time τ, with the trap center located at the node of the standing wave laser field. (iii) At the end of the time interval τ, the laser exciting the strong transition is turned on again; the absence or presence of fluorescence from the strong transition will indicate whether or not the ion has made a transition (quantum jump) to the excited metastable level $|r\rangle$ or not. We repeat the same sequence in order to determine the probability $P_r(\tau)$ that after the interaction time τ the ion is in the state $|r\rangle$ [i.e., to determine the population inversion $\langle \sigma_z(\tau) \rangle = 1 - 2P_r(\tau)$]. Repeating this measurement for different interaction times τ, we build up the evolution population inversion as a function of time τ. Depending on the initial state distribution, this population inversion exhibits different collapses and revivals, which provide a signature of the statistical distribution of the initially prepared state $\rho(t_i)$.

Following the considerations of Section II, it is possible to determine the evolution of the mean value of the population inversion. Defining the quantities

$$w^n = \langle n-1, r| \rho |n-1, r\rangle - \langle n, g| \rho |n, g\rangle \tag{69a}$$

$$\rho_{gr}^n = \langle n, g| \rho |n-1, r\rangle \tag{69b}$$

$$\rho_{rg}^n = \langle n-1, r| \rho |n, g\rangle \tag{69c}$$

we can derive equations of motion

$$\frac{d}{dt}\begin{pmatrix} \langle \rho_{rg}^n \rangle \\ \langle \rho_{gr}^n \rangle \\ \langle w^n \rangle \end{pmatrix} = \begin{bmatrix} z^* & 0 & \frac{i}{2}\Omega_n \\ 0 & z & -\frac{i}{2}\Omega_n \\ i\Omega_n & i\Omega_n & 0 \end{bmatrix} \begin{pmatrix} \langle \rho_{rg}^n \rangle \\ \langle \rho_{gr}^n \rangle \\ \langle w^n \rangle \end{pmatrix} \equiv A_n \begin{pmatrix} \langle \rho_{rg}^n \rangle \\ \langle \rho_{gr}^n \rangle \\ \langle w^n \rangle \end{pmatrix} \quad (70)$$

with $z = i[\delta_1 - \nu] + \Gamma'$ and $\Omega_n = \eta_1 \Omega_1 \sqrt{n}$. Here, we have assumed that the laser has phase fluctuations of bandwidth Γ', and the notation $\langle \rho_{ij}^n \rangle$ denotes the average over these fluctuations (Haslwanter et al., 1988).

Using the definition (69a), we see that the population inversion averaged over laser phase fluctuations, $\langle \sigma_z \rangle = \langle \text{Tr}[(|r\rangle\langle r| - |g\rangle\langle g|)\rho] \rangle$, can be written as

$$\langle \sigma_z(\tau) \rangle = \sum_{n=1}^{\infty} \langle w_n(\tau) \rangle - P_g^0 \quad (71)$$

where P_g^0 denotes the initial probability for being in the ground state $|0, g\rangle$. Starting with the two-level ion initially in the ground state $|g\rangle$, we find, for the inversion

$$\langle \sigma_z(\tau) \rangle = -\left(P_0 + \sum_{n=1}^{\infty} P_n \left[\exp(A_n \tau) \begin{pmatrix} 0 \\ 0 \\ 1 \end{pmatrix} \right]_3 \right) \quad (72)$$

where P_n with $n = 0, 1, 2, \ldots$ denotes the initial occupation number of the harmonic oscillator state $|n\rangle$, and the subscript 3 indicates the third element of the resulting vector.

The population inversion (72) is a superposition of (weakly damped) terms exhibiting Rabi oscillations with frequencies Ω_n, weighted with the initial state distribution P_n. This gives rise to collapses and revivals in $\sigma_z(\tau)$ as a function of time τ. We note that only the diagonal elements (and no coherences) of the initial density matrix enter into Eq. (71). The inversion of (71) and the reconstruction of the distribution P_n from the time evolution of $\sigma_z(\tau)$ have been discussed by, e.g., Fleischhauer and Schleich (1993).

Consider now numerical results for the population inversion derived using Eq. (71). We consider the following initial population distributions of the harmonic oscillator:

1. Coherent state distribution of mean excitation number $\langle n \rangle$,

$$P_n^{coh} = \frac{\langle n \rangle^n}{n!} e^{-\langle n \rangle} \quad (73)$$

2. Thermal distribution of mean excitation number $\langle n \rangle$,

$$P_n^{th} = \frac{\langle n \rangle^n}{(\langle n \rangle + 1)^n} \tag{74}$$

3. Squeezed vacuum distribution, with squeezing parameter r,

$$P_{2n}^{sq} = \frac{(2n)!}{(2^n n!)^2} \frac{(\tanh r)^{2n}}{\cosh r} \tag{75}$$

4. Fock state ($|n_0|$) distribution,

$$P_n^{Fock} = \delta_{n,n_0} \tag{76}$$

Figure 16(a) shows the population inversion according to Eq. (72) for an initial coherent state of the trapped ion motion with a mean excitation number $\langle n \rangle = 4$. The Rabi frequency was chosen as $\Omega_1 = 2\pi \times 10^5$ sec^{-1}. Collapse appears after approximately 0.75 msec, and the first revival occurs for $\tau \approx 2.5$ msec. The solid, dash-dotted, and dashed lines indicate the inversion expected for a laser bandwidth of $\Gamma' = 2\pi \times 0$, 100, 1000 sec^{-1} respectively; it is clearly seen that for a bandwidth comparable with the inverse of the interaction time, the revival phenomenon is no longer observed.

Standard laser cooling theories, for example, for Doppler cooling of the ion in the Lamb–Dicke limit, predict a thermal (Bose–Einstein) distribution for the final state (Stenholm, 1986). Figure 16(b) shows the expected population inversion for a thermal state with mean excitation $\langle n \rangle = 5$, as expected for a typical Doppler cooling experiment with trap frequency $\nu = \Gamma/5$. The other parameters are the same as in Fig. 16(a). Again, collapse and revival may be observed, however, with a rather irregular time evolution (Knight and Radmore, 1982). Again, the revival is no longer observed when the laser bandwidth is too large.

Figure 17 shows the population inversion as a function of time for an initial squeezed state. The upper (solid) curve shows a result for a large squeezing parameter ($r = 1.5$), which yields behavior similar to that found for a thermal distribution (Gea-Banacloche et al., 1988). The lower set of curves (dashed, dotted, dash-dotted and solid) show results for decreasing squeezing parameters ($r = 0.8, 0.55, 0.35, 0.2$). A strong dependence on the degree of squeezing is exhibited, which should, in principle, enable one to derive the magnitude of squeezing parameter r from an experiment. The results shown in Fig. 17 were obtained for a laser bandwidth $\Gamma' = 2\pi$

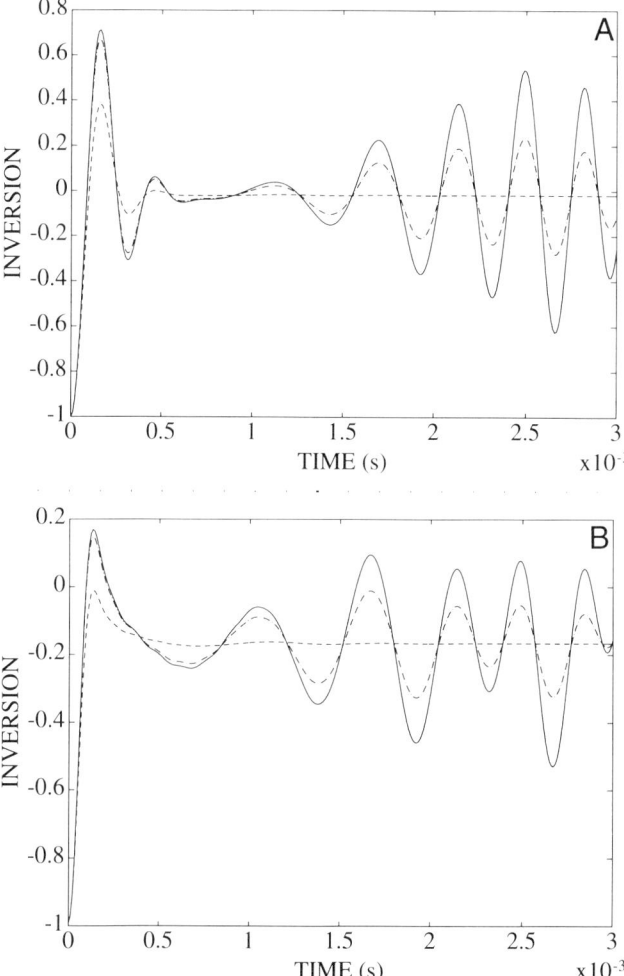

FIG. 16. (a) Population inversion as a function of time for an initial coherent state distribution with $\langle n \rangle = 4$ and for varying laser bandwidth $\Gamma' = 2\pi \times 0$, 100, 1000 sec^{-1} (solid, dash–dotted, dashed curves, respectively). (b) Population of state $|r\rangle$ as a function of time for an initial thermal state distribution with $\langle n \rangle = 5$ and for varying laser bandwidths, parameters are as in (a). In (a) and (b), the Rabi frequency is $\Omega = 2\pi \times 10^5$.

100 sec^{-1}; a similar behavior as in Fig. 16 is observed when varying the laser bandwidth.

In Fig. 18, we plot the results for the Fock state $|n_0 = 4\rangle$. The different curves indicate the behavior for the laser bandwidths considered in Fig. 16. The solid line shows the sinusoidal behavior expected for a Fock state, and

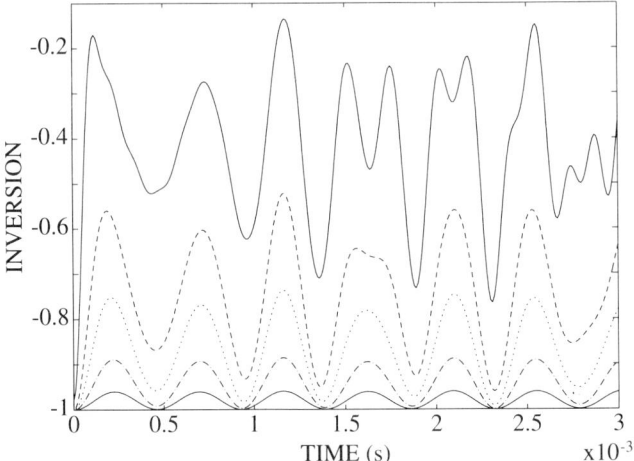

FIG. 17. Population inversion as a function of time for an initial squeezed state distribution with a laser bandwidth $\Gamma' = 2\pi\ 100\ \text{sec}^{-1}$. The (top) solid, dashed, dotted, dash–dotted, and (bottom) solid curves show results for decreasing squeezing parameter $r = 1.5, 0.8, 0.55, 0.35,$ and 0.2, respectively. The Rabi frequency is $\Omega = 2\pi \times 10^5$.

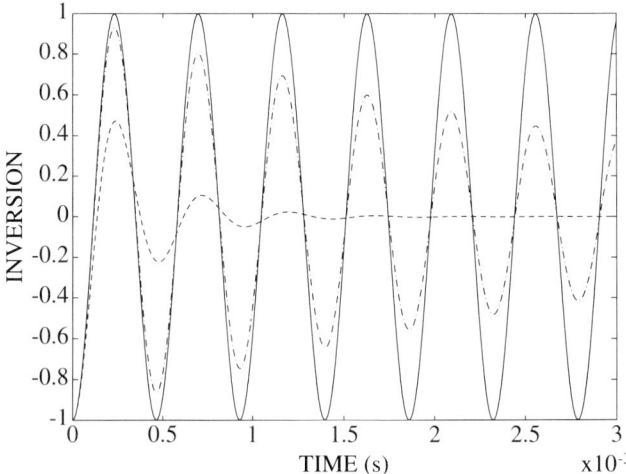

FIG. 18. Population inversion as a function of time for an initial Fock state $n = 4$ and laser bandwidth $\Gamma' = 2\pi \times 0, 100, 1000\ \text{sec}^{-1}$ (solid, dashed, dash–dotted curves, respectively). The Rabi frequency is $\Omega = 2\pi \times 10^5$.

the period of the oscillation is determined by the quantum number n_0. As expected, for larger laser bandwidths, the inversion damps out as a function of time.

As can be seen from Figs. 16–18, the nature of the collapses and revivals depends sensitively on the initial state distribution. These examples demonstrate that the proposed measurement of collapse and revival phenomena provides an interesting and useful tool for the analysis of motional state distributions in an ion trap. However, this also requires knowledge of the laser bandwidth. We emphasize, however, that even in the case of a large laser bandwidth (which would not allow the observation of collapse and revival), the temperature of a trapped ion (in a thermal state) can be determined. This can be easily seen from Eq. (72) in the limit $t \to \infty$ (i.e., $\tau \ll 1/\Gamma'$). In this case, the inversion in Eq. (72) is given simply by the population of the harmonic oscillator state $|0\rangle$, i.e., $\langle \sigma_z(\tau \gg \Gamma'^{-1}) \rangle \to P_0$, and for a thermal distribution, we obtain

$$\langle n \rangle \equiv \bar{n} = \frac{1}{P_0} - 1 \qquad (77)$$

Thus a measurement of P_0 immediately yields the temperature of the single trapped ion. This is indicated in Fig. 19, which shows the expected result for the inversion as a function of time for different mean values of

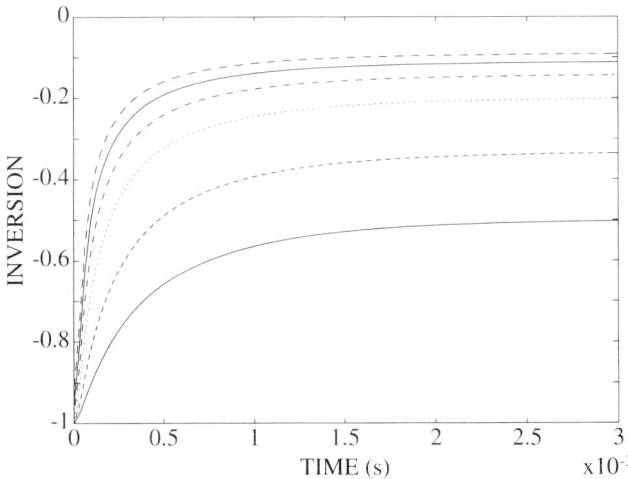

FIG. 19. Population inversion as a function of time for an initial thermal state with $\langle n \rangle = 1, 2, 4, 6, 8, 10$ (from bottom to top) and laser bandwidth $\Gamma' = 2\pi \times 10000$ sec^{-1}. The Rabi frequency is $\Omega = 2\pi \times 10^5$.

the initial thermal distribution. Of course, in an experiment it suffices to take measurements for a single interaction time τ only, and therefore such a measurement would be very simple to perform. Hence, this is a useful alternative to analyzing atomic excitation spectra, which usually require a longer measurement time and can perturb the state distribution of the ion.

So far, we have only taken into account frequency fluctuations. Of course, amplitude fluctuations would also impede the appearance of the collapse and revival effects. However, this problem is not to severe, because the interaction [cf. Eq. (40)] on the sideband involves the small Lamb–Dicke parameter, and hence amplitude fluctuations are accordingly reduced. From our numerical calculations, we estimate that amplitude fluctuations of about 1% do not significantly alter the results obtained.

Finally, it should be noted that selecting other detunings and using other laser configurations (for example, at the antinode with $\delta_1 = -2\nu$, and starting from an internal state of the form $\alpha |g\rangle + \beta |r\rangle$, one can also investigate the behavior of the coherences of the motional distribution. In principle, one could reconstruct the whole density matrix of the state of the ion.

B. Distinguishing between Pure and Mixed States

One of the fascinating predictions of quantum mechanics is the existence of coherent superposition states, such as the Schrödinger Cat state. Given a scheme to prepare such a state, it is also highly desirable to have a way to confirm experimentally that the final state is indeed a pure state of the form (66) rather than an incoherent superposition of the form

$$\rho \propto (|\eta\rangle_c\langle\eta| + |-\eta\rangle_c\langle-\eta|)|r\rangle\langle r| \qquad (78)$$

As for the pure state (66), in this statistical mixture the ion is either to the right (state $|\eta\rangle$) or to the left (scale $-|\eta\rangle$) of the x axis.

Let us state the problem in a different (but equivalent) way. Suppose the ion is in the state (78). Consider the following experiment consisting of four steps:

1. We measure if the ion is to the right. This can be done by using the quantum jump technique with a laser focused on the right side only (after changing the state of the ion in the right hand side from $|r\rangle$ to $|g\rangle$, as indicated in Section III, i.e., using an auxiliary laser propagating along the z direction). If we do not detect the ion, then we know that the ion is on the left side, and its state is

$$|\Psi\rangle = |-\eta\rangle_c |r\rangle \qquad (79)$$

2. Now we wait for a time $\tau = \pi/2\nu$, so that the state of the ion will be $|i\eta\rangle |r\rangle$, and its position will be centered around $x = 0$.

3. We send two $\pi/2$ pulses along the x direction, the first propagating in the negative x direction and the second propagating in the opposite direction. The state of the ion after this pulse sequence is easily calculated using (63), resulting in

$$|\Psi\rangle = K[|r\rangle(|i\eta\rangle_c - |3i\eta\rangle_c) - i|g\rangle(|0\rangle_c + |2i\eta\rangle_c)] \qquad (80)$$

4. We measure the state of the ion using the quantum jump technique. Obviously, we will find that the probability of measuring the ion in its ground state is $1/2$.

We could perform the same experiment again, but now in step (1) measuring if the ion is to the left. If we do not detect the ion, we conclude that the state of the ion is

$$|\Psi\rangle = |\eta\rangle_c |r\rangle \qquad (81)$$

Following the same steps as before, (2) and (3), the state of the ion will become

$$|\Psi\rangle = K[|r\rangle(|-i\eta\rangle_c - |i\eta\rangle_c) - i|g\rangle(|0\rangle_c + |-2i\eta\rangle_c)] \qquad (82)$$

Again, if we measure the state of the ion, we will find it in the ground state with probability $1/2$.

Summarizing, we could state that regardless of the position of the ion (i.e., whether it is to the left or to the right), after performing steps (2) and (3), we will detect the ion in the ground state half of the time, and in the excited state the other half of the time. However, if we take the state (66) and calculate what happens after applying steps (2) and (3), we will obtain the state

$$|\Psi\rangle = K|r\rangle(|-i\eta\rangle_c - |3i\eta\rangle_c)$$
$$- i|g\rangle(2|0\rangle + |-2i\eta\rangle_c + |2i\eta\rangle_c) \qquad (83)$$

With this state, the probability of detecting the atom in the ground state is $3/4$!

Obviously, there is a catch in the preceding argument. To predict that whatever initial state we have for the ion we will find it in the ground state with probability $1/2$, we have used a property that is foreign to quantum mechanics, namely realism (see, for example, Bell, 1987). It is known that quantum mechanics is not a realist theory, and the apparent contradiction that we have explained in the previous paragraph shows this fact. The state (66) is a state that can only be described by quantum mechanics, and there is no classical (realist) analog. Note that the state (78) can indeed be described by a realist theory; therefore, if we perform steps (2) and (3) on this state, we will detect the ion in its ground internal state with probability 1/2. Therefore, the experiment outlined in this section [steps (2) and (3)] allows one to distinguish between a pure state and a statistical mixture.

VIII. Conclusions

In this chapter, we have reviewed a number of schemes that produce nonclassical states of the motion of a single ion trapped in a harmonic potential. In particular, we have described ways of preparing Fock states, squeezed states, and Schrödinger Cat states, as well as methods for the analysis of such states. Most of these schemes are based on the interaction of a laser field with the ion, in such a way that photon absorption and emission processes and their mechanical effects produce, under appropriate conditions, the desired states of motion. These schemes highlight the dynamics of a very fundamental system in quantum mechanics, namely, the system composed of a harmonic oscillator coupled to a two-level system. Furthermore, the internal structure of the ion enables one to manipulate two contrasting effects to achieve the desired goal: When the laser excites a dipole-allowed transition, dissipation, in the form of atomic spontaneous emission, plays a significant role and can be used for laser cooling or to perform measurements using the quantum jump technique; when the laser excites a dipole-forbidden transition, there is no dissipation, and coherent quantum phenomena dominate the dynamics of the system. Hence, single ions in traps offer unique and fascinating situation in which to examine fundamental quantum mechanics from a variety of viewpoints.

Note added in proof. After this review was written, in a series of remarkable experiments (Monroe, 1995; Meekhof, 1996; Monroe, 1996), Wineland and collaborators have produced several nonclassical states of motion (including Fock states, squeezed states, and Schrödinger cats), measured collapse and revival of the population inversion, and demonstrated a fundamental quantum logic gate.

References

Adams, C. S., Sigel, M., and Mlynek, J. (1994). *Phys. Rep.* **240**, 143.
Agarwal, G. S., and Kumar, S. A. (1991). *Phys. Rev. Lett.* **67**, 3665.
Alsing, P. M., Cardimona, D. A., and Carmichael, H. J. (1992). *Phys. Rev. A* **45**, 1793.
Alzetta, G., and Goszini, P. (1976). *Nuovo Cimento Soc. Ital. Fis. B* **36B**, 5.
Appasamy, B., Siemers, I., Stalgies, Y., Eschner, J., Blatt, R., Neuhauser, W., and Toschek, P. E. (1995), unpublished.

Applied Physics (1992). Special Issue on Quantum Noise Reduction in Optical Systems, Vol. B55, p. 189.
Bardroff, P. J., Mayr, E., and Schleich, W. P. (1995). *Phys. Rev. A* **51**, 4963.
Bell, J. S. (1987). "Speakable and Unspeakable in Quantum Mechanics." Cambridge Univ. Press, Cambridge, UK.
Bergquist, J. C., Hulet, R. G., Itano, W. M., and Wineland, D. J. (1986). *Phys. Rev. Lett.* **57**, 1699.
Berman, P., ed. (1994). "Cavity Quantum Electrodynamics," Adv. Mol. Opt. Phys., Suppl. 2. Academic Press, San Diego.
Bernardot, F., Nussenveig, P., Brune, M., Raimond, J. M., and Haroche, S. (1992). *Europhys. Lett.* **17**, 33.
Blatt, R. (1992). *In* "Fundamental Systems in Quantum Optics, Les Houches Summer School, Session LIII" (J. Dalibard, J. M. Raymond, and J. Zinn-Justin, eds.) p. 253. Elsevier, Amsterdam.
Blatt, R., Cirac, J. I., and Zoller, P. (1995). *Phys. Rev. A* **52**, 518.
Blockley, C. A., and Walls, D. F. (1993). *Phys. Rev. A* **47**, 2115.
Blockley, C. A., Walls, D. F., and Risken, H. (1992). *Europhys. Lett.* **17**, 509.
Brune, M., Haroche, S., Lefevre, V., Raimond, J. M., and Zagury, N. (1990). *Phys. Rev. Lett.* **65**, 976.
Brune, M., Haroche, S., Raimond, J. M., Davidovich, L., and Zagury, N. (1992). *Phys. Rev. A* **45**, 5193.
Caldeira, A. O., and Leggett, A. J. (1985). *Phys. Rev. A* **31**, 1059.
Carmichael, H. J. (1995) *In* "Coherence and Quantum Optics VII" (J. H. Eberly, L. Mandel, and E. Wolf, eds.). Plenum, New York.
Cirac, J. I., and Parkins, A. S. (1994). *Phys. Rev. A* **50**. R4441.
Cirac, J. I., and Zoller, P. (1993). *Phys. Rev. A* **47**, 2191.
Cirac, J. I., and Zoller, P. (1994). *Phys. Rev. A* **50**, R2799.
Cirac, J. I., and Zoller, P. (1995). *Phys. Rev. Lett.* **74**, 4091.
Cirac, J. I., Ritsch, H., and Zoller, P. (1991). *Phys. Rev. A* **44**, 4541.
Cirac, J. I., Blatt, R., Zoller, P., and Phillips, W. D. (1992). *Phys. Rev. A*. **46**, 2668.
Cirac, J. I., Parkins, A. S., Blatt, R., and Zoller, P. (1993a). *Phys. Rev. Lett.* **70**, 556.
Cirac, J. I., Blatt, R., Parkins, A. S., and Zoller, P. (1993b). *Phys. Rev. Lett.* **70**, 762.
Cirac, J. I., Parkins, A. S., Blatt, R., and Zoller, P. (1993c). *Opt. Commun.* **97**, 353.
Cirac, J. I., Garay, L. J., Blatt, R., Parkins, A. S., and Zoller, P. (1994a). *Phys. Rev. A* **49**, 421.
Cirac, J. I., Blatt, R., Parkins, A. S., and Zoller, P. (1994b). *Phys. Rev. A* **49**, 1202.
Cirac, J. I., Blatt, R., and Zoller, P. (1994c). *Phys. Rev. A*. **49**, R3174.
Cirac, J. I., Schenzle, A., and Zoller, P. (1994d). *Europhys. Lett.* **27**, 123.
Cirac, J. I., Lewenstein, M., and Zoller, P. (1995). *Phys. Rev. A* **51**, 1650.
Davidovich, L., Maali, A., Brune, M., Raimond, J. M., and Haroche, S. (1993). *Phys. Rev. Lett.* **71**, 2360.
Davidovich, L., Zagury, N., Brune, M., Raimond, J. M., and Haroche, S. (1994). *Phys. Rev. A* **50**, R895.
Dehmelt, H., Janik, G., and Nagourney, W. (1985). *Bull. Am. Phys. Soc.* [2] **30**, 111.
Diedrich, F., Bergquist, J. C., Itano, W. M., and Wineland, D. J. (1989). *Phys. Rev. Lett.* **62**, 403.
Eberly, J. H., Narozhny, N. B., and Sanchez-Mondragon, J. J. (1980). *Phys. Rev. Lett.* **44**, 1323.
Eichmann, U., Bergquist, J. C., Bollinger, J. J., Gilligan, J. M., Itano, W. M., Wineland, D. J., and Raizen, M. G. (1993). *Phys. Rev. Lett.* **70**, 2359.

Ekert, A. (1994). *In* "Proceedings of International Conference of Atomic Physics '94", (S. Smith, C. Wieman, and D. Wineland, eds.) (to be published).
Eschner, J., Appasamy, B., and Toschek, P. E. (1995). *Phys. Rev. Lett.* **74**, 5300.
Filipowicz, P., Javanainen, J., and Meystre, P. (1986). *Phys. Rev. A* **34**, 3077.
Fleischhauer, M., and Schleich, W. P. (1993). *Phys. Rev. A* **47**, 4258.
Garraway, B. R., and Knight, P. L. (1994). *Phys. Rev. A* **50**, 2548.
Gea-Banacloche, J. (1990). *Phys. Rev. Lett.* **65**, 3385.
Gea-Banacloche, J. (1991). *Phys. Rev. A* **44**, 5913.
Gea-Banacloche, J., Schlicher, R. R., and Zubairy, M. S. (1988). *Phys. Rev. A* **38**, 3514.
Gheorghe, V. N., and Vedel, F. (1992). *Phys. Rev. A* **45**, 4828.
Gilbert, S. L., and Wieman, C. E. (1993). *Opt. Photon. News.* **4**, 8.
Glauber, R. J., (1992). *In* "Foundations of Quantum Mechanics," (T. D. Black, M. M. Nieto, H. S. Pilloff, M. O. Scully, and R. M. Sinclair, eds.), p. 23. World Scientific, Singapore.
Goetsch, P., Graham, R., and Haake, F. (1995). *Phys. Rev. A* **51**, 136.
Graham, R. (1987). *J. Mod. Opt.* **34**, 873.
Grynberg, G., Lounis, B., Verkerk, P., Courtois, J.-Y., and Salomon, C. (1993). *Phys. Rev. Lett.* **70**, 2249.
Haroche, S. (1992). *In* "Fundamental Systems in Quantum Optics, Les Houches Summer School, Session LIII" (J. Dalibard, J. M. Raymond, and J. Zinn-Justin, eds.), p. 771. Elsevier, Amsterdam.
Haslwanter, Th., Ritsch, H., Cooper, J., and Zoller, P. (1988). *Phys. Rev. A* **38**, 5652.
Heinzen, D. J., and Wineland, D. J. (1990). *Phys. Rev. A* **42**, 2977.
Herkommer, A. M., Akulin, V. M., and Schleich, W. P. (1992). *Phys. Rev. Lett.* **69**, 3298.
Holland, M. J., Walls, D. F., and Zoller, P. (1991). *Phys. Rev. Lett.* **67**, 1716.
Itano, W. M., and Wineland, D. J. (1982). *Phys. Rev. A* **25**, 35.
Itano, W. M., Bergquist, J. C., and Wineland, D. J. (1987). *Science* **237**, 612.
Janszky, J., and Yushin, Y. Y. (1986). *Opt. Commun.* **59**, 151.
Javanainen, J., Lindberg, M., and Stenholm, S. (1984). *J. Opt. Soc. Am. B* **1**, 111.
Jaynes, E. T., and Cummings, F. W. (1965). *Proc. IEEE* **51**, 89.
Jefferts, S. R., Monroe, C., Bell, E. W., and Wineland, D. J. (1995). *Phys. Rev. Lett.* **51**, 3112.
Joos, E., and Lindner, A. (1989a). *Z. Phys. D: At. Mol. Clusters* **11**, 295.
Joos, E., and Lindner, A. (1989b). *Z. Phys. D: At. Mol. Clusters* **11**, 301.
Journal of Modern Optics (1987). Special Issue on Squeezed States of Light, Vol. 34, p. 709.
Journal of Modern Optics (1992). Special Issue on the Physics of Trapped Ions, Vol. 39, p. 192.
Journal of the Optical Society of America (1987). Special Issue on Squeezed States of Light, Vol. B4, p. 1450.
Journal of the Optical Society of America (1989). Special Issue on Laser Cooling and Trapping, Vol. B6, p. 2020.
Kasevich, M., and Chu, S. (1992). *Phys. Rev. Lett.* **69**, 1741.
Kimble, H. J. (1992). *In* "Fundamental Systems in Quantum Optics, Les Houches Summer School, Session LIII" (J. Dalibard, J. M. Raymond, and J. Zinn-Justin, eds.). Elsevier, Amsterdam.
Knight, P. L., and Radmore, P. M. (1982). *Phys. Lett. A* **90A**, 342.
Kochan, P. F. (1995). Doctoral Dissertion, University of Oregon, Evgene.
Krause, J., Scully, M. O., and Walther, H. (1987). *Phys. Rev. A* **36**, 4547.
Landauer, R. (1995). *Proc. R. Soc. Lonson, Ser. A* (in press).
Lindberg, M., and Javanainen, J. (1986). *J. Opt. Soc. Am. B* **3**, 1008.
Lindberg, M., and Stenholm, S. (1984). *J. Phys. B: At. Mol. Phys.* **17**, 3375.
Loudon, R., and Knight, P. L. (1987). *J. Mod. Opt.* **34**, 709.
Marzoli, I., Cirac, J. I., Blatt, R., and Zoller, P. (1994). *Phys. Rev. A* **49**, 2771.

Matos Filho, R. L., and Vogel, W. (1994). *Phys. Rev. A* **50**, R1988.
Meekhof, D. M., Monroe, C., King, B. E., Itano, W. M., and Wineland, D. J. (1996). *Phys. Rev. Lett.* **76**, 1796.
Meystre, P. (1987). *Opt. Lett.* **12**, 669.
Meystre, P. (1992). *Prog. Opt.* **30**, 261.
Meystre, P., Slosser, J. J., and Wilkens, M. (1990). *Opt. Commun.* **79**, 300.
Monroe, C., Meekhof, D. M., King, B. E., Jefferts, S. R., Itano, W. M., and Wineland, D. J. (1995). *Phys. Rev. Lett.* **75**, 4717.
Monroe, C., Meekhof, D. M., King, B. E., and Wineland, D. J., to be published.
Nagourney, W., Sandberg, J., and Dehmelt, H. (1986). *Phys. Rev. Lett.* **56**, 2797.
Narozhny, N. B., Sanchez-Mondragon, J. J., and Eberly, J. H. (1981). *Phys. Rev. A* **23**, 236.
Neuhauser, W., Hohenstatt, M., Toschek, P. E., and Dehmelt, H. G. (1978). *Phys. Rev. Lett.* **41**, 233.
Neuhauser, W., Hohenstatt, M., and Toschek, P. E. (1980). *Phys. Rev. A.* **22**, 1137.
Parkins, A. S., Marte, P., Zoller, P., and Kimble, H. J. (1993). *Phys. Rev. Lett.* **71**, 3095.
Parkins, A. S., Marte, P., Zoller, P., Carnal, P., and Kimble, H. J. (1995). *Phys. Rev. A* **51**, 1578.
Phoenix, S. J. D., and Barnett, S. M. (1993). *J. Mod. Opt.* **40**, 979.
Poyatos, J. F., Cirac, J. I., Blatt, R., and Zoller, P. (1995). *Phys. Rev. A* (in press).
Raimond, J. M., Brune, M., Lepape, J., and Haroche, S. (1989). *In* "Laser Spectroscopy IX" (M. S. Feld, J. E. Thomas, and A. Mooradian, eds.), p. 140. Academic Press, San Diego, CA.
Raizen, M. G., Thompson, R. J., Brecha, R. J., Kimble, H. J., and Carmichael, H. J. (1989). *Phys. Rev. Lett.* **63**, 240.
Raizen, M. G., Gilligan, J. M., Bergquist, J. C., Itano, W. M., and Wineland, D. J. (1992). *Phys. Rev. A* **45**, 6493.
Rempe, G., Wlather, H., and Klein, N. (1987). *Phys. Rev. Lett.* **58**, 353.
Rempe, G., Schmidtkaler, F., and Walther, H. (1990). *Phys. Rev. Lett.* **64**, 2783.
Rempe, G., Thompson, R. J., Brecha, R. J., Lee, W. D., and Kimble, H. J. (1991). *Phys. Rev. Lett.* **67**, 1727.
Sanchez-Mondragon, J. J., Narozhny, N. B., and Eberly, J. H. (1983). *Phys. Rev. Lett.* **51**, 550.
Sauter, Th., Neuhauser, W., Blatt, R., and Toschek, P. E. (1986). *Phys. Rev. Lett.* **57**, 1696.
Schrödinger, E. (1935a). *Naturwissenschaften* **23**, 807.
Schrödinger, E. (1935b). *Naturwissenschaften* **23**, 823, 844.
Shore, B. W., and Knight, P. L. (1993). *J. Mod. Opt.* **40**, 1195.
Sleator, T., and Weinfurter, H. (1994). *In* "IQEC Technical Digest 1994," OSA Tech. Dig. Ser., Vol. 9, p. 140. OSA, Washington, DC.
Sleator, T., and Weinfurter, H. (1995). *Phys. Rev. Lett.* **74**, 4087.
Slosser, J. J., Meystre, P., and Braunstein, S. L. (1989). *Phys. Rev. Lett.* **63**, 934.
Smithey, D. T., Beck, M., Faridani, A., and Raymer, M. G. (1993). *Phys. Rev. Lett.* **70**, 1244.
Stenholm, S. (1986). *Rev. Mod. Phys.* **58**, 699.
Thompson, R. J., Rempe, G., and Kimble, H. J. (1992). *Phys. Rev. Lett.* **68**, 1132.
Toschek, P. E. (1984). *In* "New Trends in Atomic Physics, Les Houches Summer School, Session XXXVIII" (G. Grynberg and R. Stora, eds.), Vol. 1, p. 381. North-Holland Publ., Amsterdam.
Toschek, P. E., and Neuhauser, W. (1989). *J. Opt. Soc. Am. B* **6**, 2220.
Vogel, K., Akulin, V. M., and Schleich, W. P. (1993). *Phys. Rev. Lett.* **71**, 1816.
Walther, H. (1992). *Phys. Rep.* **219**, 263.
Walther, H. (1994). *Adv. At. Mol. Opt. Phys.* **32**, 379.
Weitz, M., Young, B. C., and Chu, S. (1994). *Phys. Rev. Lett.* **73**, 2563.

Wells, A. L., and Cook, R. J. (1990). *Phys. Rev. A* **41**, 3916.
Wineland, D. J., and Dehmelt, H. G. (1975). *Bull. Am. Phys. Soc.* [2] **20**, 637.
Wineland, D. J., and Itano, W. M. (1979). *Phys. Rev. A* **20**, 1521.
Wineland, D. J., and Itano, W. M. (1987). *Phys. Today*, June, p. 34.
Wineland, D. J., Itano, W. M., and VanDyck, R. S., Jr. (1983). *Adv. At. Mol. Phys.* **19**, 135.
Wineland, D. J., Itano, W. M., Bergquist, J. C., and Hulet, R. G. (1987). *Phys. Rev. A* **36**, 2220.
Wineland, D. J. Bollinger, J. J. Itano, W. M., Moore, F. L., and Heinzen, D. J. (1992). *Phys. Rev. A* **46**, R6797.
Wineland, D. J. Bollinger, J. J. Itano, W. M., and Heinzen, D. J. (1994). *Phys. Rev. A* **50**, 67.
Yoo, H. I., Sanchez-Mondragon, J. J., and Eberly, J. H. (1981). *J. Phys. A* **14**, 1383.
Yurke, B., and Stoler, D. (1986). *Phys. Rev. Lett.* **57**, 13.
Zeng, H., and Lin, F. (1994). *Phys. Rev. A* **50**, R3589.
Zeng, H., and Lin, F. (1995a). *Phys. Lett* **201A**, 139.
Zeng, H., and Lin, F. (1995b). *Phys. Rev. A* **52**, 809.
Zhu, Y., Gauthier, D. J., Morin, S. E., Wu, Q., Carmichael, H. J., and Mossberg, T. W. (1990). *Phys. Rev. Lett.* **64**, 2499.
Zurek, W. H. (1991). *Phys. Today* **10**, 36.

THE PHYSICS OF HIGHLY CHARGED HEAVY IONS REVEALED BY STORAGE / COOLER RINGS

P. H. MOKLER[*,†] AND TH. STÖHLKER[*,‡]

*GSI–Darmstadt, Darmstadt, Germany; †University of Geissen, Germany; and ‡Institut für Kernphysik, University of Frankfurt, Germany

I. Introduction	297
II. The Physics of Highly Charged Heavy Ions	300
A. The Structure of Heavy Few-Electron Ions	300
B. Interaction Processes for Highly Charged Heavy Ions	308
III. Storage and Cooler Rings for Heavy Ions	316
A. The Production of Highly Charged Heavy Ions	316
B. Survey of Heavy Ion Storage Rings	319
C. Cooling and Experimental Facilities at the ESR	323
IV. Charge Changing Processes	332
A. Recombination Processes in the Cooler	332
B. Electron Capture Processes in the Gas Jet Target	339
C. Ionization and Excitation Processes	346
V. Atomic Structure Studies	348
A. The Ground State Lamb Shift	348
B. Doubly Excited States	353
C. Rydberg States	356
D. Hyperfine Interactions	358
E. β Decay into Bound States	360
VI. Future Developments	362
References	365

I. Introduction

With the successful commissioning of storage and cooler rings for bright beams of very heavy ions near the threshold of the last decade of this century, not only did a prosperous development in heavy ion accelerator technology (Möller, 1994; Pollock, 1991; Franzke, 1988) come to its present summit, but also fundamental fields in heavy ion physics were opened widely for exciting explorations. Now, essential aspects in this area are accessible, aspects one only dared to dream of another decade ago. In the meantime, great progress already has been made in the fundamental physics in this field (Schuch, 1993; Bosch, 1993; Müller, 1994; Lindgren et al., 1995a). This is particularly true for achievements in the atomic physics of highly charged heavy ions. In this chapter, we present a review of the current advances in this rapidly developing field.

There are two general domains to be considered in the atomic physics of highly charged heavy ions: the fields of collisions and of atomic structure. Both aspects have to be explored equally, as they are strongly interconnected. One has to investigate the interaction processes to know, for instance, the population of excited states to help answer questions on the atomic structure; and conversely, one has to know the structure to understand the interactions. In both the fields, fundamental principles can be studied uniquely. This is in particular true for the heaviest ion species with only a few—or even zero—electrons left.

In these very heavy few-electron ions, extremely strong central fields are present, which are probed by the residual inner electrons. For the highest nuclear charges or atomic numbers Z, relativistic and even quantum electrodynamical (QED) effects may dominate the atomic structure of these ions, because both effects increase roughly with Z^4, whereas the binding increases only with Z^2. For the heaviest species, higher order QED contributions can be probed that are not accessible to experiments for low-Z ions (cf. Lindgren et al., 1995b; Mokler et al., 1994; Indelicato, 1990; Johnson and Soff, 1985). Moreover, due to the shrinking of the electron wavefunctions with Z, there is an appreciable overlap with the nucleus, and strong hyperfine effects can be tested uniquely (Klaft et al., 1994). These effects increase roughly with Z^3. The interaction of the atomic and nuclear parts may even be so large that new channels in radioactive decay can be opened or closed. A typical example is the recently measured β decay into bound atomic states (Jung et al., 1992). And, Rydberg states, which are practically free of nuclear influences, also provide deep insight into the atomic structure of high-Z, few-electron ions (Borneis et al., 1994; Schüssler et al., 1995a, b).

In collisions, especially at high velocities, $\beta = v/c$, subtle effects of relativity may be probed (Grieser et al., 1995). Here, in particular, magnetic and spin-flip effects can also be studied; and the interaction processes at high central fields may deviate considerably from those involving low-Z ions (Stöhlker et al., 1995; Rymuza et al., 1993). This is not only true for the typical excitation and ionization and the different capture and recombination processes, but also for quasi-molecular processes at lower collision velocities (Schuch et al., 1988). By the study of superheavy collision systems with initial vacancies in inner shells, quasi-atoms in the superheavy nuclear charge region ($100 < Z_{\text{eff}} < 185$) can be tested (for a review, see, e.g., Mokler and Liesen, 1978). There, the strongest possible fields determine the atomic structure.

All these interesting and fundamental effects can be tested sensitively with the new tools: the heavy ion storage/cooler rings. Ions up to naked uranium ($Z = 92$) can, after acceleration and stripping, be accumulated to high intensities in storage rings. By cooling—mainly by cold electrons—the phase-space density can be increased, so that high luminosity beams of highly charged heavy ions can be provided for experiments (for a review of cooling techniques, see, e.g., Möhl, 1988). The extremely well-defined energy of the ions is precisely fixed by the cooling technique. On the other hand, the velocity of these projectiles can be changed actively to the requirements of the experiments; in particular, the highly charged ions can be decelerated. With these tools, the interaction of the heavy ions with photons (laser), with electrons, and with atoms or even with ions can be studied selectively. Because of the long storage times for the heavy ions, lifetime measurements for both atomic and nuclear states also become accessible (Kluge, 1995; Andersen et al., 1993; Irnich et al., 1995; Schmidt et al., 1994). And, last but not least, the storage rings can be tuned as high resolution mass spectrometers, giving important new information on the nuclear and atomic structure of these heavy ions. Here, however, only the first steps had been taken on the long way toward the extreme precision known in measurements in low-Z ion traps (cf. Kluge, 1995; Thompson, 1990).

In the following, we first explain the challenging physics of highly charged heavy ions (Section II) to prepare for the introduction to and a survey of storage rings and their cooling properties (Section III). Then, new results on important charge changing processes—recombination and capture reactions—will be reviewed (Section IV), before the basic results on the atomic structure are discussed (Section V). Here, the ground state transitions give direct insight into Lamb shift contributions. We also present results from doubly excited states, from Rydberg states, and from the ground state hyperfine splitting, elucidating the QED structure contributions. The radioactive decay gives another access to the interplay between atomic shells and the nucleus. Finally, we look toward future developments in the physics of highly charged ions and storage/cooler rings (Section VI).

II. The Physics of Highly Charged Heavy Ions

A. THE STRUCTURE OF HEAVY FEW-ELECTRON IONS

1. *Isoelectronic Sequences*

The binding of electrons to nuclei gives the most direct access to the structure of ions or atoms and reveals most clearly the fundamental principles of physics. Hence, the precise spectroscopy of transitions, mainly between two bound atomic levels, was from the earliest days in atomic physics the tool to study these principles. First, Bohr's atomic model, discussed already in 1913, was roughly in agreement with experimental findings. There, at least for H-like species, the binding energy of the one electron increases with the square of the atomic number Z according to the Rydberg formula

$$E_B = -\text{Ry} \frac{Z^2}{n^2} \qquad (1)$$

where n is the main or shell quantum number. Later, quantum mechanics (1926), both in the nonrelativistic (Schrödinger) and the relativistic (Dirac) regimes, was excellently confirmed by spectroscopic investigations. (For historical details, the reader is referred to textbooks, see, e.g., Bethe and Salpeter (1957) or more specifically, Labzowky et al. (1993).) Finally, Lamb and Retherford (1947) discovered tiny deviations in their spectra from hydrogen atoms, the Lamb shift, which is caused by quantum electrodynamical effects. For light atomic species, such as hydrogen, the complete theory is now excellently confirmed by ultrahigh precision spectroscopy experiments; see, for instance, the Doppler-free two-photon laser experiment of Weitz et al. (1992). For heavy ions, such precision is not feasible. However, because of the increase in the central nuclear potential, the effects on the atomic structure are dramatically increased, and higher order effects can be studied uniquely in heavy ions.

Within an isoelectronic sequence, the electron–nucleus interaction grows in strength with increasing atomic number, considerably changing the atomic structure. Electron–electron interaction is less and less important, whereas relativistic (REL) and quantum electrodynamic (QED) influences are increasingly important (see, e.g., Mokler et al., 1994). These influences are overlayed onto the general Z^2 dependence for the binding energies. Most clearly, these effects can be studied for few-electron ions, in particular for H- and He-like ions. This is seen in Fig. 1; there, the reduced binding energies E_{L_i}/E_{GS} are plotted as functions of the atomic number for the various L shell levels.

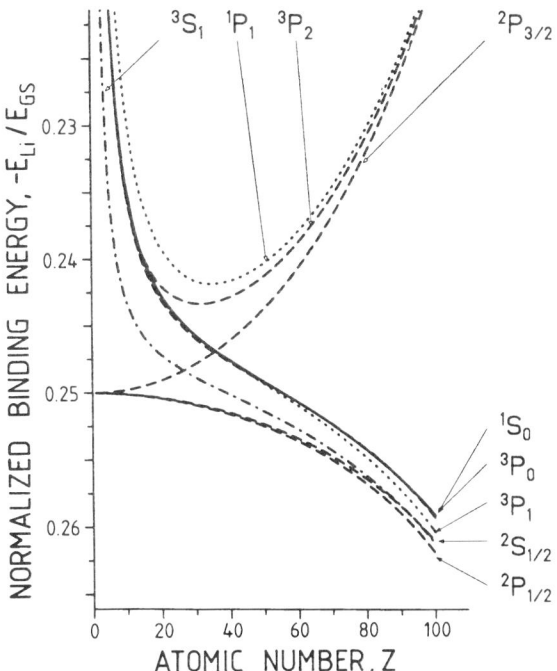

FIG. 1. Reduced L shell binding energies, E_{L_i}/E_{GS}, normalized to the ground state binding energy in hydrogenic ions. The reduced energies are plotted as functions of the atomic number Z for H- and He-like ions. The corresponding reduced K–L transition energies are given by $1 - E_{L_i}/E_{GS}$.

2. Hydrogen-like Ions

For low-Z hydrogenic ions, the reduced L shell binding energy is $-\frac{1}{4}$, in accordance with the Rydberg formula. In this reduced representation, the L–K transition energy corresponds to $\frac{3}{4}$. With increasing central atomic field (Z), the L shell splits up into the different j states with $j = \frac{1}{2}$ and $\frac{3}{2}$. This fine structure splitting is caused by relativistic effects and increases with the fourth power of the atomic number. In the reduced representation in Fig. 1, the fine structure splitting is proportional to Z^2. However, already evident there, we find at high Z a tiny splitting of the $^2S_{1/2}$ and $^2P_{1/2}$ levels, which cannot be described by the Dirac equation. This is the classical Lamb shift, lowering the absolute value for the binding of s states slightly. The total Lamb shift is caused mainly by quantum electrodynamic effects, and it increases with Z^4. The ground state, $1s^2S_{1/2}$, is naturally even more influenced by QED effects, as the $1s$ electron probes more closely the strong part of the central potential than does a $2s$ electron (see also Fig. 6 later).

The main QED contribution to the Lamb shift (Johnson and Soff, 1985) at low Z is the self-energy (SE) of the bound electron, i.e., the emission and reabsorption of a virtual photon by the electron in the strong field. At higher Z, the vacuum polarization (VP) contributes increasingly to the Lamb shift. The vacuum polarization can be characterized in the lowest order as the coupling of the electron to a virtual electron–positron pair created in the strong central potential. For heavy Z, naturally, the finite size of the nucleus influences the binding energy of the electron, via the central potential (to be discussed). This effect is usually included in the Lamb shift. Johnson and Soff (1985) did extended calculations on the Lamb shift of H-like ions from all over the periodic table. They explicitly calculated the aforementioned first order QED graphs, estimated the higher order terms, and also added the nuclear size effects. In total, the Lamb shift can be characterized by the formula

$$L_{ns} = \frac{\alpha}{\pi} \frac{(Z\alpha)^4}{n^3} F(Z\alpha) mc^2 \qquad (2)$$

where n is the shell quantum number, α the fine structure constant, and the function F can be written as a series expansion in $Z\alpha$. In Fig. 2, the ground state Lamb shift for H-like systems is plotted in units reduced by Z^4; i.e., the dependence of the $F(Z\alpha)$ function is given. The separate contributions from self-energy, vacuum polarization, and from the finite nuclear size are also shown in Fig. 2.

To get a feeling for the absolute energy range covered by Fig. 2, we give rough energies for a hydrogen atom and for a H-like U^{92+} ion in comparison: The Lyman-α transition energies are about 10 eV and 100 keV, respectively; the L shell fine structure splitting corresponds to 45 μeV and 4.56 keV, the $2s$–$2p_{1/2}$ Lamb shift to 4.4 μeV and 75 eV, and the ground state Lamb shift to 35 μeV and 458 eV, respectively (Johnson and Soff, 1985).

Beyond about $Z = 40$, higher order terms in the $Z\alpha$ expansion start to dominate the self-energy (see, e.g., Indelicato, 1990). The higher order aspects can only be tested by highly charged, very heavy atoms. Actually, second order graphs are now accessible to calculations (cf. Mohr, 1994) and will give deep insight into QED physics in strong fields. This will particularly be true when corresponding high precision experimental data becomes available.

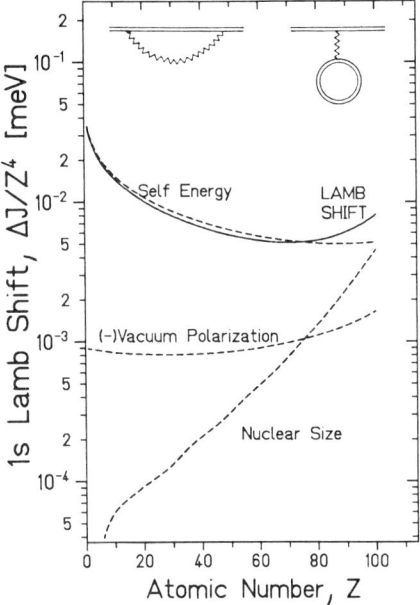

FIG. 2. Relative contributions $\Delta J/Z^4$ to the ground state Lamb shift in H-like ions according to Johnson and Soff (1985). The graphs for the first order QED contributions, self-energy (SE, left), and vacuum polarization (VP, right) are inserted at the top of the figure. SE, VP, and nuclear size (NS) contributions are separately shown.

3. *Helium-like Ions*

He-like ions are the simplest multielectron systems. They represent an ideal testing ground for our understanding of relativistic and quantum electrodynamic effects in many-body systems. Unlike hydrogenic species, there is no exact solution for the structure of He-like ions, and the theory of two-electron ions has to consider, aside from the dominant Coulomb term, effects from electron–electron correlations such as the Breit interaction, screening of the Lamb shift, and higher order radiative corrections. Because of the electron–electron interaction, the binding energies are lower than for H-like systems (i.e., the normalized ground state transition energies in Fig. 1 are increased), and the triplet levels are more tightly bound than the corresponding singlet levels. As the electron–electron interaction decreases with $1/Z$ relative to the electron–nucleus interaction with increasing nuclear charge, the He-like level structure for high-Z ions approaches an H-like character. Consequently, the *LS* coupling scheme, valid for low-Z ions, loses meaning, and one has to consider a pure *jj* coupling.

There is substantial progress in the theory of two-electron systems, in particular for the ground state of these ions (Lindgren et al., 1995a). On the basis of many-body Lamb shift calculations, the two-photon QED contributions to the electron–electron interaction (nonradiative QED, screened vacuum polarization, and self-energy) can now be calculated without any approximation, and the theoretical precision for the ground state energies in He-like ions is now as high as the one for one-electron systems (Persson et al., 1995).

In Fig. 3, the various contributions to the ground state binding energies in He-like ions are plotted using the tabulated values from Drake (1988). Although these data were evaluated by means of the $Z\alpha$ expansion, which is known to be incomplete at the level $(Z\alpha)^4$, the figure depicts the general scaling of the most important effects that contribute to the binding energies: relativistic effects including the full Breit interaction, the mass polarization, the nuclear size effect, and the various QED effects. Note that the last is essentially dominated by one-electron QED terms.

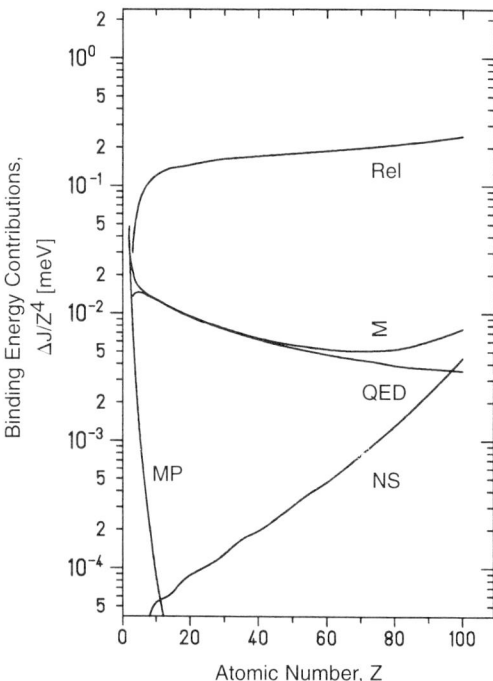

FIG. 3. Relative contributions $\Delta J/Z^4$ to the ground state ionization potential in He-like ions plotted as a function of atomic number Z according to Drake (1988).

4. Forbidden Transitions

Beside the ground state and the transition energies, the lifetimes of the excited levels are of particular interest for atomic structure investigations. In low-Z hydrogenic systems, the $2\,^2S_{1/2}$ level is metastable and decays by the nonrelativistically allowed $2E1$ decay. For very-high-Z ions, however, the $M1$ spin-flip transition is about an order of magnitude faster than the $2E1$ decay mode, leading to a prompt decay of the $2\,^2S_{1/2}$ level. The general scaling of the transition modes for hydrogenic systems and the level scheme are given in Fig. 4.

For H-like heavy ions, such as U^{91+}, all ground state transitions can be considered in the experiment as prompt. The Lyman α_2 transition

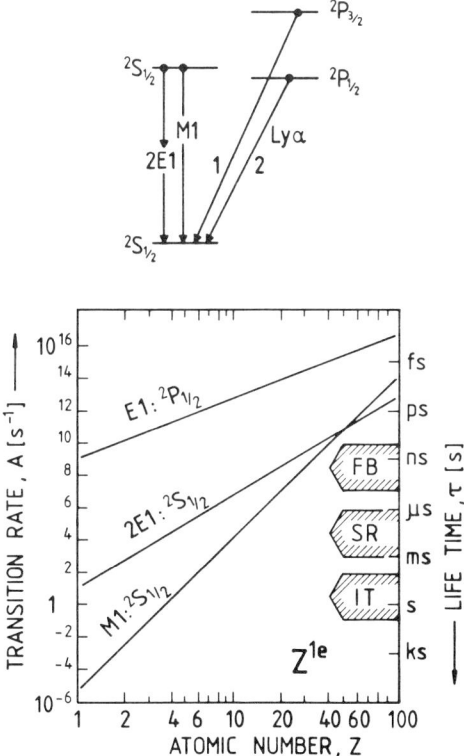

FIG. 4. The level scheme of the excited L shell levels in H-like ions (top) along with their transition rates, which are plotted as a function of Z (bottom). There, the lifetime ranges typically accessible for fast beams (FB), storage rings (SR), and ion traps (IT) are also given.

($2\,^2P_{1/2} \to 1\,^1S_{1/2}$) is here blended by the $M1$ decay ($2\,^2S_{1/2} \to 1\,^1S_{1/2}$). For convenience, we marked in the figure the lifetime regions typically accessible to fast beam spectroscopy (FB ~ nsec), to long lifetime measurements in storage rings (SR ~ μsec to msec) and those in ion traps (IT ~ sec). Depending on the experimental goals, these regions can be extended considerably, e.g., in ion traps down the nanosecond, and in storage rings up to the time scale of years.

For He-like ions the level scheme is more complex; see Fig. 5. Going along the isoelectronic sequence, the atomic structure changes drastically, as is illustrated for the cases of Ar^{16+} (LS coupling) and U^{90+} (jj coupling). Also given in the figure are the decay rates of the nonrelativistically forbidden transitions modes as well as of the $2E1$ decay. As seen from the figure, the increase of the nuclear charge and, therefore, of relativistic effects engenders a drastic increase in the decay rates of all forbidden transitions ($^3P_1 : Z^{10}(E1)$, $^3S_1 : Z^{10}(M1)$, $^3P_2 : Z^8(M2)$, $^1S_0 : Z^6(2E1)$, $^3P_0 : Z^6(E1M1)$) (cf. Marrus and Mohr, 1978). Consequently, for U^{90+}, all excited L shell states decay predominately to the ground state by direct transitions. Only in the case of the $^3P_2$ and $^3P_0$ levels must $\Delta n = 0$ transitions also be considered (decay to the $^3S_1$ state). Moreover, for a fast beam spectroscopy we emphasize that all ground state transitions from the L shell can be considered as prompt except for the $^3P_0$ decay. This is true for all heavy He-like species, such as U^{90+}.

5. Interaction with the Nucleus

As already mentioned, the structure of the nucleus may cause important corrections for the binding energies, mainly of the s electrons. Due to the radial shrinking of the wavefunctions with Z and the increase of the nuclear radius, an appreciable overlap of the innermost electrons with the nucleus is present for heavy ions. In Fig. 6, the radial parts $(r\Psi)^2$ of the $1s$ (solid line), $2s$ (dashed-dotted line), $2p_{1/2}$ (dashed line), and $2p_{3/2}$ (dotted line) wavefunctions of hydrogenic Bi^{82+} (Soff, private communication, 1995) are depicted, and the radius of ^{209}Bi ($r = 7.125$ fm) is also marked.

Because of the finite overlap of the s electron wavefunctions with the nucleus, the nuclear charge distribution not only influences the Dirac eigenvalue, it leads also to important corrections for the QED contributions itself. This is in particular true for the $1s$ level of high-Z ions, where the total finite nuclear size correction is larger than the vacuum polarization and approaches already the size of the self-energy contribution (see Fig. 2). Therefore, an ultimate test of QED corrections in high-Z systems requires a detailed understanding of the nuclear structure, i.e., of the nuclear charge distribution.

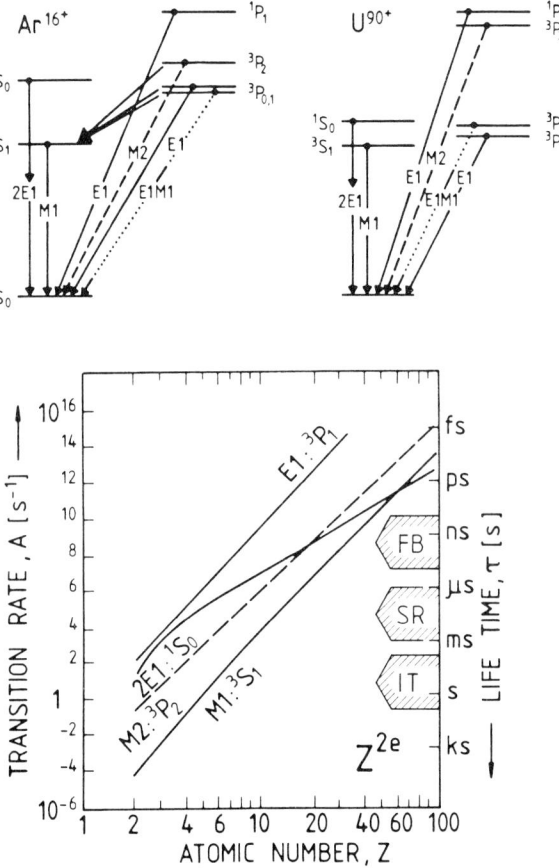

FIG. 5. Level diagrams of the excited L shell levels in He-like ions for Ar^{16+} and U^{90+} (top). The corresponding transition rates are given, in addition, as a function of atomic number Z (bottom).

Moreover, the magnetic moment distribution of the nucleus (Bohr–Weisskopf effect) (Bohr and Weisskopf, 1950) may have a considerable impact on the atomic structure. For the case of nuclei with an unpaired number of neutrons or protons (e.g., ^{209}Bi), the nuclear spin I couples with the angular momentum of the atomic core J. The resulting total angular momentum F causes the (in general, small) hyperfine structure splitting, which is approximately given by

$$\Delta E_{HFS} \sim \text{const} \cdot \frac{Z^3}{n^3} g_I \qquad (3)$$

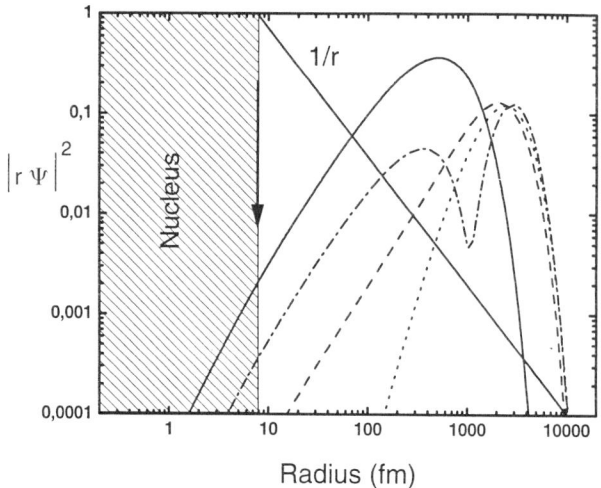

FIG. 6. The radial parts $(r\Psi)^2$ of the $1s$ (solid line), $2s$ (dashed-dotted line), $2p_{1/2}$ (dashed line), and $2p_{3/2}$ (dotted line) wavefunctions of hydrogenic Bi^{82+} given as functions of the distance from the center of the nucleus. The mean square radius of ^{209}Bi of $r = 7.125$ fm is indicated (G. Soff, 1995).

where g_I denotes the g factor of the nucleus. As depicted in Fig. 7, the ground state splitting approaches the eV region for H-like ions beyond $Z = 50$, which is accessible to laser spectroscopy. A precise measurement of this splitting not only gives access to the nuclear magnetic moment, it also probes higher order QED effects, in particular, the magnetic part of the radiation field.

B. Interaction Processes for Highly Charged Heavy Ions

1. Capture and Recombination

Interaction with the Photon Field. QED experiments dealing with few-electron systems require a detailed understanding of the interaction processes of highly charged ions with matter. More generally, the precise knowledge of all possible atomic projectile charge-exchange processes constitutes the basis for accelerator–storage ring–based research as well as for the operation of electron beam ion sources/traps (EBISs/EBITs). In the following, we discuss briefly the most important electron capture mechanisms for bare, H- and He-like heavy ions colliding with atoms or electrons, i.e., radiative electron capture (REC) or radiative recombination

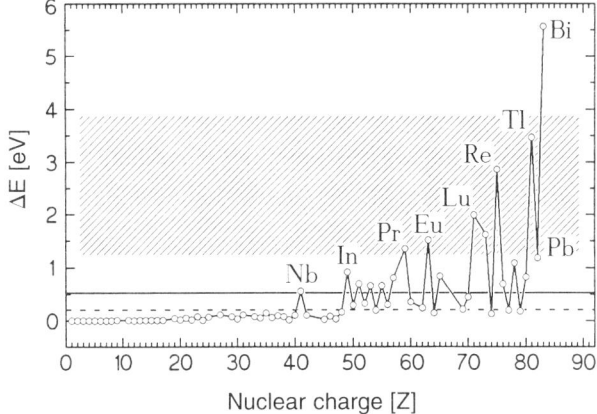

FIG. 7. The hyperfine structure transition energy in H-like ions plotted for some nuclei as a function of atomic number Z. The shaded area depicts the regime that is accessible by laser spectroscopy (T. Kühl, private communication, 1995).

(RR), resonant transfer and excitation (RTE) or dielectronic recombination (DR), and nonradiative capture (NRC) or three-body recombination (TR). Here, we treat first the radiative processes mediated by the coupling of the electron with the photon field of the nucleus.

In collisions of highly charged ions with free electrons, an electron may undergo a direct transition into a bound state of the ion via simultaneous emission of a photon with the energy $\hbar\omega = E_{kin} + E_B$, where E_{kin} denotes the kinetic electron energy and E_B the electron binding energy in its final bound state (see Fig. 8). This process is the inverse of the fundamental photoelectric effect (Stobbe, 1930). Because of the requirements of momentum and energy conservation, this radiative recombination (RR) process constitutes the only possible recombination process between bare ions and free electrons in an isolated environment, e.g., in low density plasmas. Similar to the RR process, target electrons also can be captured radiatively into a bound state of the projectile. If the kinetic energy of the electron greatly exceeds the initial binding energy of the target electron (impulse approximation), one may disregard the latter, so that this radiative electron capture process (REC) is equivalent to the RR mechanism. In addition, one has to take into account the initial momentum distribution by folding the Compton profile of the target electron into the cross section for radiative recombination.

As early as 1930, Stobbe derived within the framework of the nonrelativistic dipole approximation a general result for RR into an arbitrary (n, l)

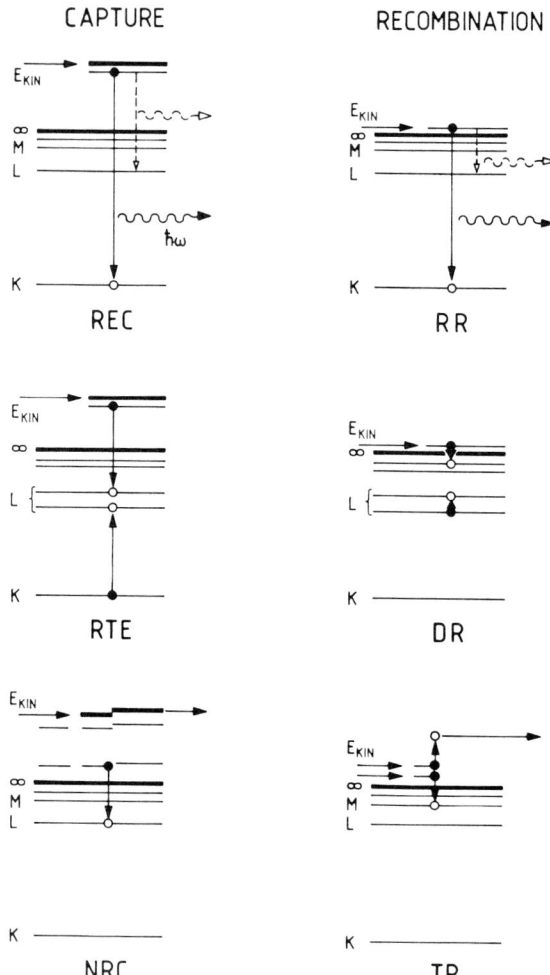

FIG. 8. Right side: Recombination processes that take place in collisions of free electrons with highly charged ions (RR: radiative recombination; DR: dielectronic recombination; TR: three-body recombination). On the left side of the figure, the corresponding processes for collisions between highly charged ions and bound electrons are given (REC: radiative electron capture; RTE: resonant transfer and excitation; NRC: nonradiative electron capture).

state of a bare ion. The result of Stobbe is of the form

$$\sigma_{nl}^{\text{Stobbe}} = \frac{\pi^2}{3} \alpha^3 a_0^2 \left(\frac{\nu^3}{1+\nu^2}\right)^2 \left\{(l+1)\left[C_{nl}^{l+1}(\nu)\right]^2 + l\left[C_{nl}^{l-1}\right]^2\right\} \quad (4)$$

where $\nu = \alpha Z/n\beta$ is the Sommerfeld parameter, $\beta = v/c$, and a_0 is the Bohr radius. The quantities $C_{nl}^{l\pm 1}$ are the dipole matrix elements for transitions between bound states (n,l) and continuum states with angular momenta $l \pm 1$. An estimate for REC cross sections for light target atoms can be obtained from (4) by multiplying σ^{Stobbe} by the number Z_T of quasi-free electrons in the target (Kleber and Jakubassa, 1974). If only capture into the projectile K shell is considered, Eq. (4) simplifies to the well-known expression (Bethe and Salpeter, 1957)

$$\sigma_{1s}^{\text{Stobbe}} = 9.165 \times \left(\frac{\nu^3}{1+\nu^2}\right)^2 \frac{e^{-4\nu \arctan(1/\nu)}}{1 - e^{-2\pi\nu}} \times 10^{-21} \text{ cm}^2 \quad (5)$$

We emphasize that the projectile charge number Z and the velocity v enter the Stobbe cross sections not separately, but only in the combination Z/v occurring in the Sommerfeld parameter ν. The Sommerfeld parameter is simply related (nonrelativistically) to the "adiabaticity parameter" η regularly used to describe collision processes by (Madison and Merzbacher, 1975)

$$\eta = \frac{1}{\nu^2} = \frac{E_{\text{kin}}}{E_B} \quad (6)$$

This means that the Stobbe formula provides a universal cross section scaling law, valid for all nonrelativistic ion–atom collision systems. Note that exact relativistic RR (REC) calculations (Ichihara *et al.*, 1994) have demonstrated that this simple dipole approximation is even applicable for high projectile charges (e.g., uranium) and relativistic beam energies up to a few hundred MeV/u.

RR is the dominant loss process in storage rings equipped with electron cooler devices and is hence of great practical importance. In a cooler, the relative kinetic energy of the free electrons with respect to the ions is extremely small, characterized by the electron temperature; see Fig. 8. REC into heavy projectiles opens via the time reversal the relativistic regime of the photoeffect, which is not accessible even at the third generation synchrotron facilities. Here, the relative energy of the quasi-free electrons is normally close to the size of the K shell binding energy ($\eta \geq 1$, cf. Fig. 8).

Electron–Electron Interaction Processes. Beside the interaction of a free or quasi-free electron with the radiation field of a fast moving projectile, the electron–electron interaction may also give rise to capture of a free electron into a bound projectile state. In this dielectronic recombination (DR) process (Burgess, 1964), the energy gain by electron capture into the projectile is used to excite radiationlessly, an initial projectile electron, forming in general a doubly excited projectile state (which in the end stabilizes radiatively). Consequently, DR requires the presence of at least one projectile electron in the initial state, and it represents the time reversal of the important Auger effect. Because of energy conservation, this process can occur only at certain resonance energies E_{res}, which correspond precisely to the energy transfer needed for projectile excitation (ΔE); i.e., $E_{res} + E_d = \Delta E$, where E_d denotes the intermediate state populated by the recombination process (cf. Fig. 8). Applying the adiabaticity parameter definition used in (6), we find, e.g., for the case of a *KLL* resonance (Auger notation: Excitation from the K to the L shell via resonant capture into the L shell) the general nonrelativistically correct resonance condition (Mokler and Reusch, 1988)

$$E_{res}^{KLL} = \tfrac{1}{2}\eta \qquad (7)$$

In a given resonance, the DR cross section may drastically enhance the total recombination cross section, and a measurement of the cross section strength, in particular of subshell resonances, provides a unique tool for the investigation of electron–electron interaction effects such as the Breit interaction (Zimmerer *et al.*, 1990). Also, the spectroscopy of the resonance energies can be applied for detailed atomic structure studies, for instance, QED investigations in few-electron systems (Spies *et al.*, 1995).

Resonant transfer and excitation (RTE), which occurs in ion–atom collisions, can be treated within the impulse approximation as a DR process, and it refers to initially quasi-free, instead of free, electrons. In this process, which has been studied systematically for low- as well as for high-Z ions up to uranium (Graham *et al.*, 1990; Kandler *et al.*, 1995a), the resonance profiles are broadened by the Compton profiles of the quasi-free target electrons. However, unlike DR, this broadening does in general not allow one to experimentally resolve the fine structure splitting of the ions, except for high-Z ions. We add that with RTE high lying resonance states such as *KLL* resonances ($\Delta n = 1$) can be probed, whereas with cooler devices normally only low lying resonances ($\Delta n = 0$) are accessible (cf. Fig. 8).

Three-Body Interaction. Three-body recombination, (TR), which may take place in electron–ion collisions, is caused by transfer of the "recombination energy" onto a second electron present during the collision (see Fig. 8). Consequently, this process requires extremely high electron densities, and it can only contribute significantly to the total recombination cross section at low collision velocities (Stevefelt et al., 1975). With respect to the present experimental conditions for electron cooler devices, electron targets, and electron beam–ion sources, this process can generally be neglected. This is in complete contrast to the electron capture process mediated by three-body interaction in ion–atom collisions, the Coulomb or nonradiative electron capture (NRC) (Eichler, 1990). Here, the third particle involved in the collision is the target atom, and the electron is transferred radiationlessly from a bound target state into a bound state of the projectile, where the target recoil receives the access momentum. At medium and high collision energies ($\eta \geq 1$), only the most tightly bound target electrons determine the total NRC reaction cross section. In general, a precise theoretical description is difficult to perform, as the Coulomb field of the projectile leads to distortions of the atomic wavefunctions in the target even at infinite distances. However, already the simple first order OBK approximation (Oppenheimer, 1928; Brinkman and Kramers, 1930) gives the rough cross-section scaling dependence:

$$\sigma_{\text{NRC}} \sim \frac{Z^5 Z_T^5}{E_{\text{kin}}^5} \tag{8}$$

Note that for relativistic collision conditions the energy dependence in (8) has to be modified; there, it approaches asymptotically a E_{kin}^{-1} form. The strong scaling of the NRC cross section with the nuclear charge of the target implies that at not too high energies this capture process is by far the dominant one in collisions of high-Z projectiles with medium- or high-Z target atoms (Eichler, 1990).

2. *Excitation and Ionization*

Electron Loss. Beside the nonresonant electron capture processes REC and NRC, projectile ionization is the most important charge-exchange channel in collisions between few-electron projectiles and neutral target atoms. In the relativistic energy domain and for high-Z projectiles, the K shell ionization is of particular importance, as the interplay between electron capture and K shell ionization processes crucially determines the production yield for bare ions.

Within the nonrelativistic first order perturbation theory [plane wave born approximation, PWBA (Rice et al., 1977), or the semiclassical approximation, SCA (Rösel et al., 1982)] the K shell ionization cross sections follow the simple law (cf. Madison and Merzbacher, 1975)

$$\sigma_{\text{ion}} \sim \text{const} \cdot \frac{Z_T^2}{Z_P^4} f\left(\frac{v}{v_k}\right) \tag{9}$$

Here, $f(v/v_k)$ is a slowly varying function of η that has its maximum close to collision velocities v equal to the velocity of the active K shell electron, v_k, i.e., at an adiabaticity parameter $\eta \approx 1$. For K shell target ionization by fast moving ions, the same equation is valid, and one has simply to consider the reversed collision system. Taking into account correct relativistic bounds and continuum wavefunctions, this simple scaling law is abolished. For collision velocities around $\eta = 1$, however, these corrections are of minor importance.

Note that, within the framework of the first order perturbation theory, the ionization cross section is independent of the sign of the charge of the ionizing particle, and hence, the cross sections for electron and proton impact are the same. This is true for high energies where the collision velocity greatly exceeds the velocity v_k. However, at velocities $v \sim v_k$ one has a threshold behavior for electron impact; i.e., only at energies $v \geq v_k$ is the momentum transfer by electron impact sufficient to promote the bound K shell electron into the continuum.

In Fig. 9, the calculated K-shell ionization cross sections for H-like uranium projectiles colliding with protons (solid lines) and electrons (dash–dotted lines) are given versus the adiabaticity parameter η. For calculation of the electron impact ionization cross sections, the Lotz formula was used (Lotz, 1968); the proton impact cross sections were calculated by applying the PWBA formalism (Anholt, 1979). For both cases, two curves are plotted in the figure. The curves that approach an almost constant cross section value represent the pure nonrelativistic predictions. The other curves, which show an increasing cross section with increasing η value, reflect the transverse ionization term as well (Anholt, 1979). Here, for the case of the PWBA calculation, corrections due to the relativistic bound state wavefunctions are also taken into account. The experimental data given in addition were measured at the Super-EBIT for electron impact (Marrs et al., 1994a) and at the BEVALAC for $U^{91+} \rightarrow M_y$ collisions (Meyerhof et al., 1985a). For the latter case, the simple Z_T^2 scaling law for ionization was applied to compare the measured value with the predictions for proton impact. We emphasize that, for electron impact, precise fully relativistic ab initio calculations are available that are in excellent agreement with the experimental value.

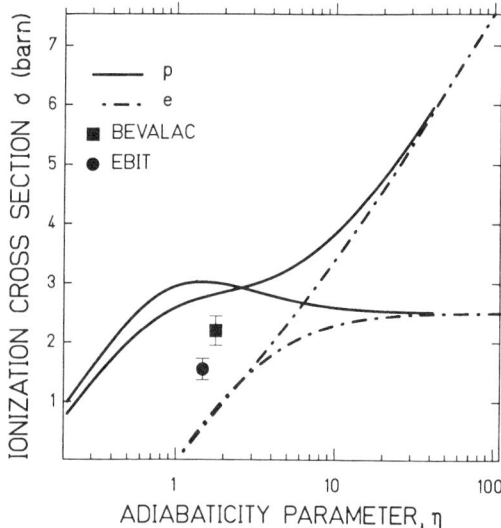

FIG. 9. Ionization cross sections for H-like uranium projectiles colliding with protons (solid lines) and electrons (dash–dotted lines) plotted versus the adiabaticity parameter η (see text).

Excitation Mechanism. Ionization is mediated by the time variation of the distortion from the collision partner seen by the electron of concern. The same interaction can also yield an excitation of a projectile electron of a highly charged ion instead of an ionization. The same difference in the cross section behavior between electron and proton impact is present at small η, whereas at high scaled velocities both cross sections approach each other asymptotically. At small η, we have for electron impact excitation a specific threshold behavior, and for ion impact a smooth increase with η (cf. Fig. 9).

In the adiabatic collision domain ($\eta < 1$), the monopole part of the perturbation generally dominates excitation (and ionization). In contrast, in the high velocity region ($\eta \geq 1$), the dipole contribution is dominant. In the relativistic region, higher multipole contributions, in particular magnetically induced transitions, also have to be considered. More clearly, at low energies $s \rightarrow s$ transitions dominate, as do $s \rightarrow p$ transitions at high energies.

Beyond the various capture and recombination processes discussed previously, which are the main processes for populating excited states in highly charged heavy ions, we have also to consider laser-induced population processes. However, as can be read from Figs. 4 and 5, not only the transition energies increase with Z; the transition rates also increase dramatically. As powerful laser light is only available in a very limited energy region (<10 eV), laser-induced transitions can only bridge gaps between very close levels, i.e., intershell transitions in light ions, transitions in high Rydberg states of medium heavy ions, or transitions in hyperfine ground state levels of very heavy ion species. Moreover, the radiative recombination, which is a free–bound transition, can also be stimulated by a laser field. Additionally, laser-induced transitions provide extremely narrow resonances, giving an ideal probe to test the qualities of ion beams such as their absolute velocities. Also, laser cooling of the lighter heavy ions is a powerful tool where it is applicable (Schröder *et al.*, 1990; Hangst *et al.*, 1991). Here, we also mention that at high velocities even the theory of special relativity can be tested by laser resonance excitation of fast light ions in the ring (Kühl *et al.*, 1994). In Fig. 10, the various laser-stimulated population possibilities for particular excited levels are summarized. The radiation emitted during deexcitation gives additional spectroscopic information about the level of interest.

III. Storage and Cooler Rings for Heavy Ions

A. THE PRODUCTION OF HIGHLY CHARGED HEAVY IONS

Atoms can be ionized by photon interaction, by electron impact, or by bombardment of heavy particles such as ions or atoms. To ionize inner shell electrons in heavy atoms, correspondingly high energy or momentum transfers are involved. For photons, the necessary high fluxes at high energies are presently not available for efficient ionization of large atom quantities. Only heavy particle or electron impact techniques can provide

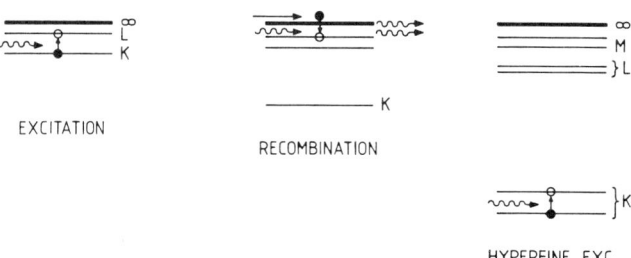

FIG. 10. Laser stimulated processes: excitation for light ions (left); stimulated recombination in medium heavy ions (center); induced ground state hyperfine transitions in very heavy ions (right).

sufficient numbers of highly charged heavy ions for experiments. In the previous section, it was shown that at equivalent collision velocities with $\eta \geq 1$, the cross sections for ionization of the last inner shell electron are comparable for electron or heavy particle impact. In particular, for $\eta \approx 1$ and due to the Z^2 dependence of the inner shell ionization, heavy ion impact is superior to electron impact. For both cases, experimental cross sections reduced to the same charge for distortion (electrons and protons) are given in Fig. 9 for one typical example in uranium. In the case of heavy particle bombardment, the classical roles of projectile and target atom is interchanged: *Fast* heavy projectiles are ionized by bombardment through target atoms in a stripper (inverted collision system).

Both ionization methods, direct electron impact and heavy particle impact by stripping, have been applied successfully to produce ultimately charged, very heavy ions. The electron impact ionization is exploited by the EBIT (electron beam ion trap) technique, where trapped ions are progressively ionized to higher charges by continuous electron bombardment (Levine *et al.*, 1989; Knapp *et al.*, 1993; Marrs *et al.*, 1994b). In contrast, ionization by heavy particle impact is utilized in the cooler ring technique; here, in an injector, heavy ions are accelerated to high energies and then stripped before they are accumulated in a storage ring (see, e.g., Pollock, 1991). Both methods are schematically shown in Fig. 11.

In the EBIT device, the ions are confined by magnetic and electric fields in the trap, and the ions are continuously bombarded by an intense beam of electrons. If the electron energy is sufficiently high, the trapped ions will successively be ionized until they have lost even their last electron—provided recombination and loss processes are not too fast. In the SUPER-EBIT device in Livermore, the electron beam energy can be turned up to about 200 keV, which even in uranium is well above the ionization energy for the last K shell electron (Marrs *et al.*, 1994a). A small number of bare U^{92+} ions have already been produced in the trap. The advantage of having the ions almost at rest in the trap is counterbalanced by the drawback of the environment of ions in different ionization stages.

In storage rings, which can also be considered as traps for ions at high energies, this drawback of charge-state mixing is clearly avoided; however, the high ion velocities may cause via the Doppler effect restrictions for precision experiments (to be discussed). On the other hand, really high intensities of the heaviest and ultimately charged ions can be provided by the acceleration–stripping–storing technique. In the storage ring at the heavy ion accelerator facility at GSI in Darmstadt, up to 10^8 bare U^{92+} ions have been stored already for experiments in the ESR (Experimental Storage Ring) (B. Franzke, private communication, 1995). To strip the

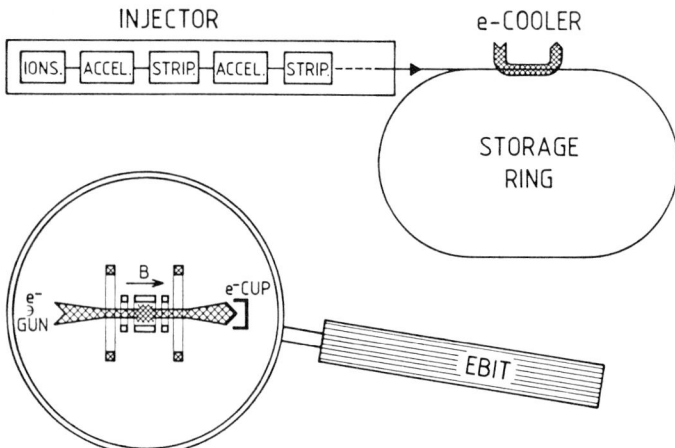

FIG. 11. The two methods to produce highly charged heavy ions. Top: The storage ring technique, where ions are accelerated to high energies and then stripped to high charge states before they are stored and cooled in a huge ring. Bottom (within the magnifying glass): The EBIT technique, where ions stored in a small trap are continuously ionized by a high energy electron beam to high ionization stages.

heavy ions to the necessary high charge states before injection into the ring, one must accelerate them to velocities corresponding to about the orbital velocities of the most strongly bound electrons of concern. As can be read from Fig. 9, this is about equal to the electron beam velocity for electron impact ionization at the threshold of concern. For fast U ions, energies between 250 and 500 MeV/u are needed to produce sufficiently high fractions of completely stripped ions. In this case, the acceleration–stripping technique has to be applied in several stages, as is indicated in Fig. 11.

The stripping technique will certainly increase the emittance of the ion beam; it will become more divergent, i.e., slightly hotter. Such a stored beam will further heat up by intrabeam scattering and distant collisions with the remnant gas molecules in the ring. Hence, the advent of efficient beam cooling techniques was crucial for the prosperous development of heavy ion storage rings (Möhl, 1988). By various beam cooling techniques —mainly electron cooling—stored heavy ion beams can be cooled down to low emittances. Thus, even for high intensity beams, all highly charged ions have almost exactly the same velocity vector, and the ions are confined to a small beam diameter. The cooling techniques fix the ion velocity with extreme stability over long time periods, also counterbalancing the energy loss of the ions colliding with remnant gas particles. Cooling is the prerequisite for heavy ion storage rings providing high luminosity beams of ultimately charged heavy ions even for precision spectroscopy experiments. Moreover, in a ring the charge states of the ions are well defined and accessible to experiments before and after any reaction.

Both methods for producing highly charged heavy ions have advantages and drawbacks. It is evident that for acceleration of heavy ions to high velocities and confining them by magnets in a ring, a large scale facility is compulsory, whereas for the electron impact technique, a relatively small arrangement is sufficient. However, as stripping is more efficient than electron impact ionization, mainly because the higher particle densities in the stripper foils, quite large quantities of highly charged heavy ions can be provided in a well-defined charge state in a storage ring. Because of continuous cooling, the Doppler effect can accurately be taken into account. The drawback of the mixed charge-state distribution in EBIT will be overcome by extracting the ions and retrapping pure charge-state populations of ions; see the RETRAP project (Schneider *et al.*, 1994). The disadvantage of the high velocities in storage rings may be overcome by actively decelerating the ions in the ring; ultimately, these decelerated ions can be extracted and, after some further steps stored, in a smaller ring or maybe even in a trap; see the HITRAP proposal initiated by Poth (Beyer *et al.*, 1990). In the following, we restrict the discussion to the results concerning storage rings for very heavy ions.

B. Survey of Heavy Ion Storage Rings

After the tremendous success with the initial cooling experiments at CERN (Bell *et al.*, 1981), numerous particle storage rings have been designed and built for various applications. For a complete survey, we have to refer to the literature (see, e.g., Pollock, 1991; Franzke, 1988; Bosch, 1993; Schuch, 1993; Müller, 1994; Larsson, 1996). Here, we concentrate only on storage and cooler rings for really heavy ions. These are the facilities at GSI in Darmstadt, at MPI (Max-Planck-Institut für Kernphysik) in Heidelberg, at the MSI (Manne Siegbahn Institute) in Stockholm, and at ISA (Institute for Synchrotron Radiation, Aarhus). These storage rings are called ESR = Experimental Storage Ring (Franzke, 1987), TSR = Test Storage Ring (Jaeschke *et al.*, 1990), CRYRING = CRYogenic ion source/injector/RING (Herlander, 1991), and ASTRID = Aarhus Storage RIng Denmark (Möller, 1990), respectively. The corresponding beam guidance magnets can bend beams to closed orbits up to magnetic rigidities for the swift ions of 10, 1.5, 1.4, and 2 Tm, which ultimately determine the size of the rings. In Fig. 12(a), sizes and arrangements for these four heavy ion storage rings are compared. They have circumferences of 108.4, 55.4, 51.6, and 40.0 m, respectively.

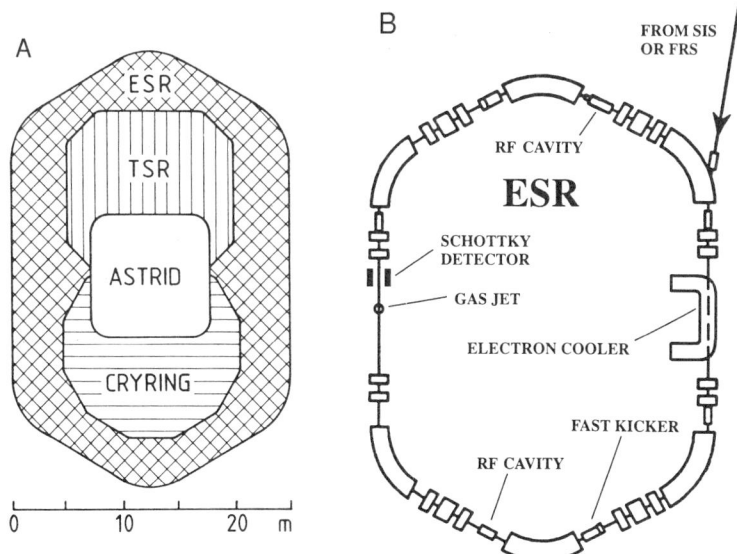

FIG. 12. A survey of storage rings for very heavy ions. (a) Sizes and general configurations for the presently running heavy ion storage rings shown in comparison. (b) The layout of the ESR storage and cooler ring. The beam guidance system, i.e., bending magnets, quadrupoles, and hexapoles, is indicated; the systems for beam handling, kicker, rf cavities, and the electron cooler are assigned; the Schottky noise pick-up detector and the supersonic gas jet target are also shown.

The ion injectors used are for ESR the heavy ion synchrotron facility SIS, for TSR a tandem postaccelerator facility, for CRYRING a cryogenic ion source in connection with a linear accelerator (radio frequency quadrupole structure), and for ASTRID an isotope separator. For ASTRID, an injection from the available tandem accelerator is also planned; ASTRID is additionally used as an electron storage ring at ISA (Institute of Synchrotron Radiation, Aarhus University). Among these facilities, only the SIS–ESR arrangement can store the heaviest ions, e.g., bare U^{92+}. In Fig 12(b), the configuration of the ESR is displayed in more detail. The main components are identified in the figure legend.

To strip U ions down to the K shell, one needs extremely high energies, between 250 and 500 MeV/u, cf. Fig. 9; this demands adiabaticity parameters η between 1 and 2. After using a stripper, we find a charge-state distribution for the ions that depends critically on η and, via the interplay between ionization and capture, also on the Z number of the stripper material. For very heavy ions at these energies, a Cu stripper may be best. Nevertheless, for comparison we display in Fig. 13 the equilibrium charge

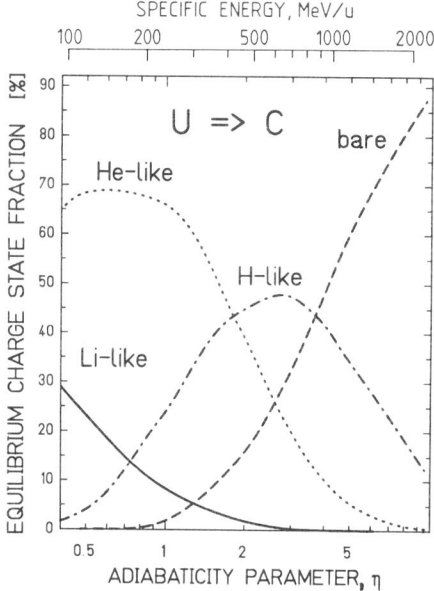

FIG. 13. Equilibrium charge state fractions for U ions behind a C stripper as a function of the scaled kinetic energy, i.e., adiabaticity parameter η. For calculations, see the code given by Stöhlker et al. (1991).

state fractions for bare, H-, He-, and Li-like U ions for a C stripper as a function of the general adiabaticity parameter. The calculations were performed using the code of Stöhlker (Stöhlker et al., 1991). In general, around $\eta = 1$ the K shell is opened, and the bare ion fraction starts to evolve. At about $\eta = 2$—for U, at about 500 MeV/u—the H-like ion fraction is already greater than the He-like ion fraction.

The η representation chosen in Fig. 13 is in general approximately valid for all ions over the whole periodic table, giving only minor variations with the atomic numbers for ion species and stripper material. Therefore, we can use such a representation to show the ranges of operation for the various rings more generally. In Fig. 14, these ranges are plotted in the coordinate system η, Z, i.e., adiabaticity parameter and atomic number of the stored ions. As in a nonrelativistic consideration the orbital velocities for the K shell electrons of concern are proportional to Z and η is proportional to $1/Z^2$, the specific ion energies follow straight lines in a double logarithmic plot.

Figure 14 demonstrates once more that ESR is the only storage ring that can store all heavy ions up to bare U^{92+} ($\eta > 1$). The typical range of

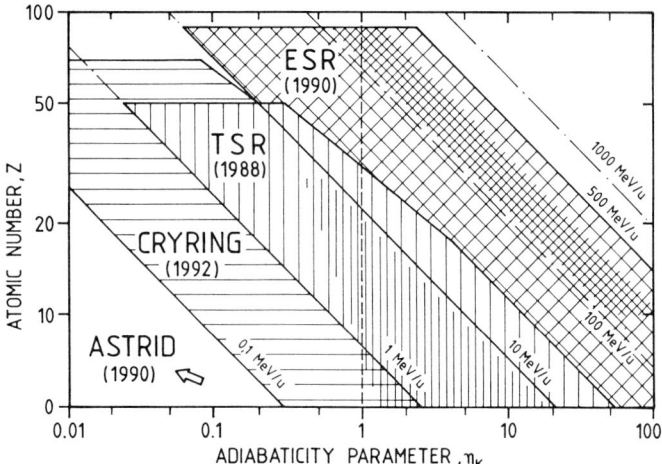

FIG. 14. Ranges of operation for the presently running heavy ion storage rings in the parameter space of atomic number Z of the ion and scaled ion energy η. Specific ion energies, in MeV/u, are indicated for several "diagonals." The years when the storage rings came into operation are given in parentheses.

operation (densely cross-hatched area) is between 100 and 350 MeV/u. From the η range, it is clear that predominately bare and few-electron ions starting from Ne upward to U can be stored. Moreover, after injection, the stored ions can be actively decelerated in the ESR, so that its final range will cover specific energies down to about 10 MeV/u. Presently, bare U^{92+} ions have already been decelerated in a first experiment from about 360 down to 50 MeV/u with an efficiency of 90%, where at the lowest energy the cooling was switched on again (Stöhlker, 1995).

The TSR operates typically below about 10 MeV/u; see the densely hatched area in Fig. 14. For high-Z ions ($Z \geq 25$), η becomes smaller than 1, and it is difficult to produce few-electron ions for injection. The TSR is the first heavy ion ring that came into operation, in the year 1988. The CRYRING started operation in 1992; for a while, mostly light heavy ions had been used there. As for ASTRID the injection from the tandem accelerator is not yet finished; only moderately charged or light ions from the mass separator have been injected at very low η. On the other hand, ASTRID is designed also for high rigidity ions; and indeed, extremely heavy species, such as charged C_{60} clusters, have successfully been stored there (Andersen et al., 1995a). Also, the structure and collision dynamics of negative ions and molecules were studied there intensively (T. Andersen et al., 1993; L. H. Andersen et al., 1995b). The results for light heavy ions, for heavy clusters, and for negative ions and molecules are very exciting; however, we concentrate in this chapter only on the challenging physics of the very heavy, highly charged ions. Therefore, we report mainly on the results from the ESR, with some discussion of results from the other heavy ion storage rings, especially from the TSR.

C. Cooling and Experimental Facilities at the ESR

1. *Ion Beam Cooling Techniques*

For all ion storage rings, beam cooling is *the* crucial issue. In general, there are three different beam cooling techniques (for a review, see, e.g., Möhl, 1988):

1. Stochastic cooling
2. Electron cooling
3. Laser cooling

The principles of these techniques will briefly be described here; for details, we refer to the literature.

In stochastic cooling, the mean position of a small bunch of particles in the phase space is measured, and then this bunch is transferred by active kickers to the correct spot, i.e., to the center of the emittance ellipse. Stochastic cooling is particularly appropriate for hot beams, i.e., beams with large emittances. This is, for instance, important if reaction products from nuclear reactions or exotic particles are accumulated in a ring (Van der Meer, 1972). At he ESR, stochastic cooling will be installed in 1996 in order to cool efficiently hot fragments injected from the fragment separator FRS placed between SIS and ESR.

In electron cooling, the ions dive in the cooler into a bath of cold co-moving electrons (cf. Poth, 1990a). The electron beam is merged by magnetic guidance fields with the ion beam, whence both beams have the same mean longitudinal velocity. Due to the variance in the velocities (temperature), the ions transfer their excess temperature via elastic scattering onto the cold electrons. The cooling works best if the initial ion temperature is not too hot compared with the electron beam temperature. Electron cooling is ideal for accelerator beams injected after stripping into a storage ring. The electron coolers are the workhorses for all the ion storage rings giving low emittance beams with high luminosities (Wolf, 1988; Poth, 1990b).

Ions can also interact efficiently with photons from laser beams as long as a proper excited state in the ion is available. By photon absorption, a certain unidirectional momentum is transferred to the ion, whereas the reemission is isotropic. Therefore, a laser can transfer, on average, momentum to ions of a certain velocity class. This effect can be used also in combination with other techniques to squeeze the ions almost into one velocity class. Ultracold ion beams have been produced by laser cooling (Schröder et al., 1990; Hangst et al., 1991). However, as the energy range for lasers is very limited, only low-Z ions have been cooled at present. Laser cooling seems not to be a general cooling technique for highly charged very heavy ions, even though it is extremely efficient where it can be applied.

For stored beams of heavy, highly charged ions, electron cooling is *the* cooling technique. In the ESR, the electron gun can accelerate electrons up to 270 keV, corresponding to a specific ion energy of about 500 MeV/u. The electron beam, with a total current up to 5 A, is merged at the bottom of the U-shaped magnetic guidance field with the circulating beam over a length of 2 m, corresponding to about 2% of the ring circumference (Spädtke et al., 1991). The ions traverse the cooling section about 10^6 times per second.

2. *The ESR Facility*

The ESR facility with its devices is shown in Fig. 15. After injection (1), the ions are cooled in the electron cooler (2) (Spädtke et al., 1991). The intensity of the stored beam is measured by a current transformer, and the beam position can be probed by different pick-up devices (neither device is shown in the figure). From the frequency analysis of the Schottky noise, picked up by the electrode of the Schottky detector (see also Fig. 12(b)), one can deduce the longitudinal momentum distribution of the circulating particles. A corresponding frequency spectrum is shown in the inset in Fig. 15 for the case of bare U^{92+} injected at 295 MeV/u into the ring. The noncooled beam has a longitudinal momentum spread $\Delta P/P$ of about 1×10^{-3} (FWHM), whereas a continuous cooling with 250-mA electrons reduces this spread to 1×10^{-5} (Eickhoff et al., 1991). This inserted plot drastically shows the necessity and efficiency of electron cooling.

In the cooler, electrons may recombine with ions, changing their charge state. These ions will be separated from the original ion beam in the next magnet downstream and can then be detected in a position-sensitive detector mounted behind a thin window in a movable pocket (3) (Klepper et al., 1992). The size of the beam spot for the recombined ions in this detector reflects the transverse momentum spread of the circulating beam. In Fig. 16, an example is given for an He-like U^{90+} beam circulating in the ring at 250 MeV/u. Unfolding the detector resolution of 1.2 mm, we can

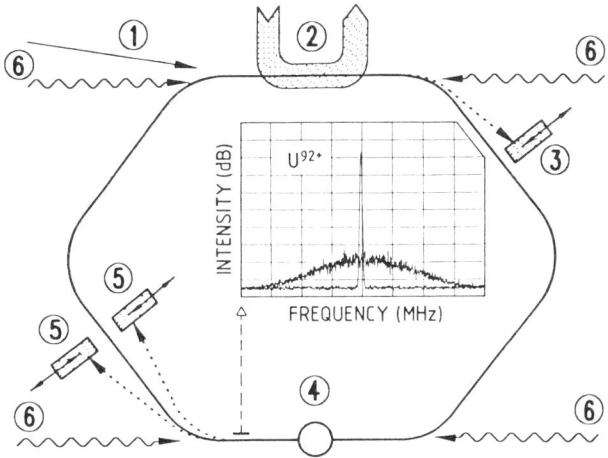

FIG. 15. The main devices of the ESR relevant for experiments are shown: (1) ion beam injection; (2) electron cooler; (3) particle detector for ions recombined in the cooler section; (4) gas jet target; (5) particle detectors for charge-changed (ionization/capture) ions; (6) laser beams to merge the ion beam at the cooler or gas target. The inset shows the Schottky frequency spectrum for the circulating ions (relative units "dB"). For the case of 295-MeV/u U^{92+}, the longitudinal momentum distribution is displayed for a noncooled and a continuously cooled beam.

deduce for this case an emittance of 0.05π mm · mrad (Bosch, 1989, 1993). Within the logarithmic representation of Fig. 16, we see also some electron capture events from collisions with the remnant gas particles in the dipole and, additionally, some rare events where finally two electrons are captured in the straight section around the cooler of the ESR.

In the long straight section opposite the cooler, see Fig. 15, we have the experimental area of the ESR with a supersonic gas jet target (4) (Gruber et al., 1989). Gas jets of, e.g., H_2, CH_4, N_2, Ar, with particle densities of 10^{11} to 10^{13} cm^{-2} can be provided for experiments. Behind the next bending magnet, the ions that have changed their charge by either ionization or capture can be monitored by position-sensitive detectors mounted in movable pockets (5) (Klepper et al., 1992). The gas target itself is easily accessible for detectors, in particular for photon and X-ray detectors.

In both straight sections of the ESR, the cooler section and the gas target section, collinear and/or counterpropagating laser beams (6) can be merged with the ion beam, and in the cooler section additionally with the electron beam; see Fig. 15. Thus, excitation or recombination can be induced by laser beams.

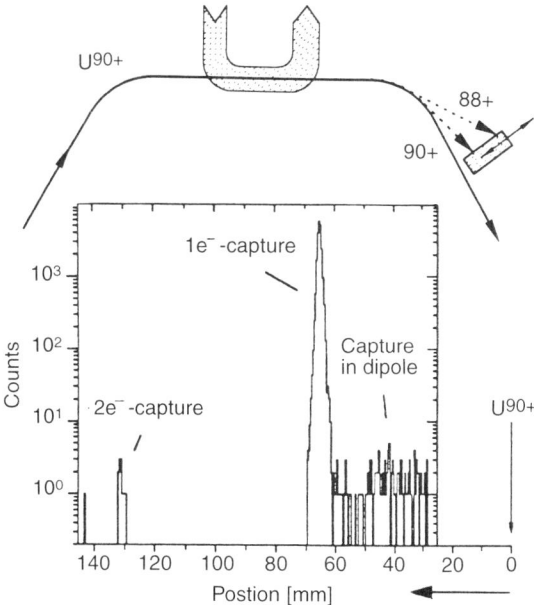

FIG. 16. Position and charge spectrum of ions having changed their charge in the cooler by recombination, for the case of He-like U^{90+} ions circulating at 250 MeV/u in the ESR (Bosch, 1993).

3. Storage of Heavy Ions

For stable, long term storage of swift heavy ions in a ring, a continuous cooling of the ion beam is indispensible. By intrabeam scattering, i.e., ion–ion interaction, the beam will spread out in phase space incessantly. By interaction with remnant gas atoms in the vacuum, the ions will lose energy by small-angle scattering, which additionally increases the beam divergence. Both effects will finally destroy the stored beam if it is not cooled continuously. Schottky spectra shown in Fig. 17 demonstrate clearly what happens to a beam if the cooling is switched off, for the case of a stored bare Ar^{18+} beam: The beam becomes hotter and hotter and permanently loses energy (Spädtke et al., 1991). At the high ion energy of 250 MeV/u that is used, energy loss and scattering cross sections for collisions with the remnant gas atoms are tiny for the Ar ions (the mean vacuum in the ring was at that time 10^{-10} mbar, a factor of 10 worse than present).

As nuclear cross sections are small compared with atomic scattering or charge-exchange cross sections, nuclear reactions do normally not influ-

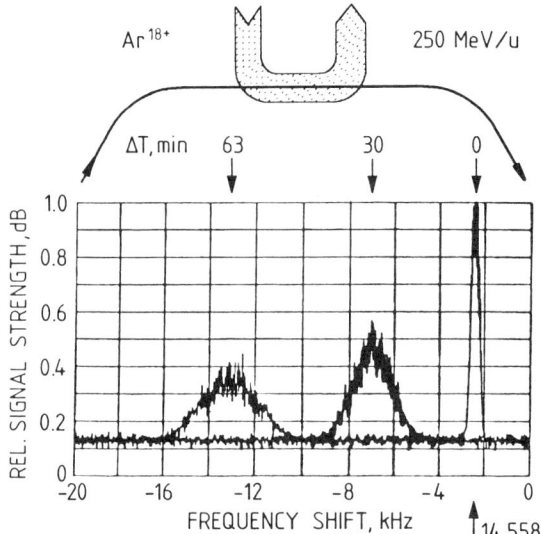

FIG. 17. Propagation of the Schottky signal for a stored 250-MeV/u Ar^{18+} beam after switching off the electron cooling. Three elapsed times are displayed: $t = 0$, 30, and 63 min after switching off (Spädtke et al., 1991).

ence the lifetime of a stored ion beam. Also, at high ion velocities, characterized by $\eta > 1$, the beam losses by charge exchange with the remnant gas atoms (10^{-11} mbar) are not dominant, although they are already large. The most important loss channel is electron–ion recombination in the cooler. These recombined ions will be separated in the next downstream magnet behind the cooler and can be monitored by the corresponding particle detector (cf. Fig. 15).

For recombination, we have normally to consider only the RR. These cross sections, σ_{RR}, are proportional to Z^2 and to the phase-space overlap of electron and ion beams, $\langle \sigma \cdot v_e \rangle$ (Bosch, 1993). For the rate of recombined ions, we have

$$\lambda_{RR} = \langle \sigma \cdot v_e \rangle n_e \sim Z^2 n_e \tilde{n}^{-1}(kT_e)^{-1/2} \qquad (10)$$

where v_e, n_e, and T_e are the electron velocity, density, and transverse temperature; $\tilde{n}$ is the mean ionic shell into which the recombination occurs. The measured recombination rates for bare and H-like heavy ions agree roughly with the predictions. From these rates, one can deduce the lifetimes for the corresponding ion beams. The measured lifetimes of stored heavy ion beams are indeed mainly determined by radiative recombination.

In Fig. 18, the measured half-lives of heavy bare ion beams stored in the ESR are plotted as a function of Z. The Z^{-2} dependence predicted for radiative recombination is evident. At a current of 100 mA for the cooling electrons and specific ion energies between 250 and 300 MeV/u, we find lifetimes on the order of a day for light ions and around an hour for the heaviest species, such as U^{92+}.

We now summarize the characteristics for stored ion beams that at present can routinely be expected at the ESR. Assuming the typical specific ion energies mentioned (200 to 300 MeV/u) and a low electron current of 100 mA for cooling, we have approximately the following relations:

Number of stored ions: $N(Z) \approx 5 \times 10^{11} \cdot Z^{-2}$
Momentum spread: $\Delta P/P \approx (N \cdot 10^{-20})^{1/3}$
Storage times: $\tau_{1/2} \approx Z^{-2} \cdot 10^4$, in hours

These rough equations give only typical numbers. In particular, the number of stored ions will be improved in future. Also, the given storage times only are guidelines. If one decelerates the ions in the ring, the capture in the remnant gas of the vacuum also has to be taken into account. For instance, for a bare U^{92+} beam decelerated to 50 MeV/u, the lifetime was reduced to the order of 10 min at 50 mA cooling current (Stöhlker, 1995a). Still, radiative recombination was the dominant loss channel at the experimental conditions used. The momentum spread given is mainly determined by the counterbalance between cooling forces and intrabeam scattering.

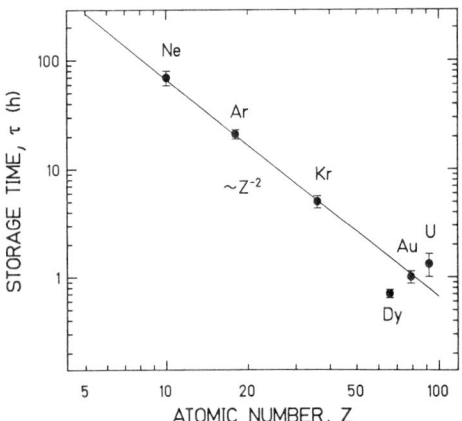

FIG. 18. Half-lives of bare heavy ion beams stored in the ESR as a function of the atomic number Z of the ions. The data are normalized to an electron current of 100 mA (Bosch, 1993).

4. Special Applications

The recombination in the cooler can also be utilized to transfer the stored ion beam into a beam with ions down-charged by one unit, provided the recombination particle detector (device (3) in Fig. 15) is pulled out of the way. The momentum acceptance of the ESR is about 3.6%, so that for the heaviest ions up to three beams of adjacent charge states can circulate in the ring at the same time. All the beams may match in the electron cooler and will be cooled simultaneously to the same velocity. The slow transfer by recombination into a beam with ions down-charged by one unit can be used to extract continuously a cooled beam to an external experimental area (Mokler *et al.*, 1989). However, it can also be used to breed internally an adjacent charge state for a particular experiment. This was done, for instance, for the measurements of dielectronic recombination in Li-like ions such as Au^{76+} or U^{89+}, where at high energies He-like ions have originally been injected and the Li-like fraction was bred in the ring (Müller, 1992). This technique, where the ions had been injected at a high energy, provided a more intense beam of Li-like ions than the direct method.

What is true for the electron target should also be valid for the gas jet target. There, we may have both ionization and capture, and we can transfer the ions to both neighboring charge states. Measuring, for instance, capture cross sections in the gas target with the downstream capture detector, it may be reasonable to transfer the fraction of further ionized ions in the cooler back to the desired charge state by recombination and then use these ions again. For the method, see the example given in Fig. 19 using initially injected bare Pb^{82+} ions at 277 MeV/u (Kandler *et al.*, 1996). The bare ion beam is confined to a narrower orbit, which does not see the gas target; by recombination, it is gradually transferred into H-like ions hitting the gas target; the down-charged ions are monitored in the corresponding particle detector, whereas the ionized fraction feeds the original bare fraction again. Naturally, the recombination of H-like into He-like ions in the cooler also can be detected by the recombination monitor. The whole evolution of the various beam intensities is governed by differential rate equations, where the appropriate cross sections have to be plugged in. This is represented in Fig. 19 by the solid lines. These calculations are compared with the measured total circulating current ($Pb^{82+} + Pb^{81+}$) and the He-like rate measured by the capture detector, which is proportional to the intensity of the circulating H-like beam. In the figure, the data have been normalized to the calculations in absolute height. An almost perfect agreement results for the evolution of charge fractions with time.

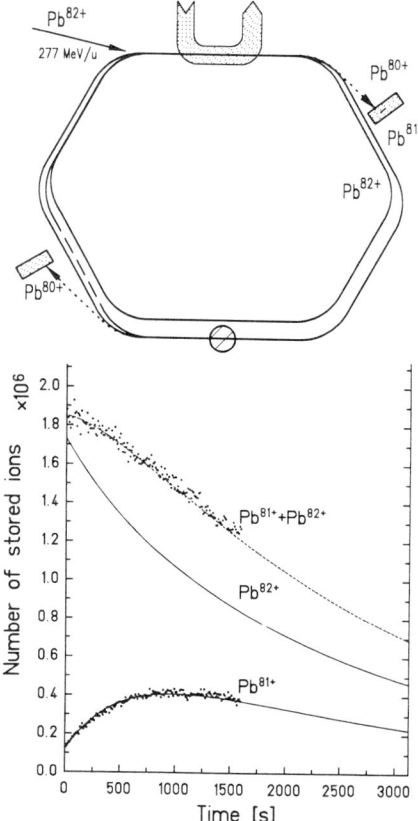

FIG. 19. Breeding of charge-state neighboring ion beams. Bare 277-MeV/u Pb^{82+} ions are injected, and H-like ions are bred; behind the gas target and behind the cooler, the He-like ion fractions are monitored. The intensity of the circulating ion current and the capture rate behind the gas target are compared with theoretical predictions for the evolution of the various ion fractions (Kandler et al., 1996).

As the different ion beams are confined according to their charges to different tracks in the storage ring, i.e., of slightly differing circumferences, and as all beams have exactly the same velocity because of the cooling, the revolution frequency is different for the differently charged ions. Therefore, the evolution of the various fractions in the ring can also be read from the Schottky frequency spectrum, which shows isolated lines for the different charge states.

From this, it is clear that storage rings are not only traps for high velocity ions at a very specific charge state; they can also be used as extremely sensitive filters for their momentum. As the velocity of the ions is accurately fixed by cooling, storage rings can also be utilized as precise mass spectrometers distinguishing, for instance, between isotopes or even isomeric states (Bosch, 1993). In Fig. 20, an example is given: The Schottky noise spectrum shows a stored beam of bare $^{163}\text{Dy}^{66+}$ ions together with neighboring nuclei produced by nuclear reactions in the N_2 gas target. H-like $^{163}\text{Ho}^{66+}$ ions are created by bound-state β decay from bare Dy ions; these ions are then stripped off in the gas target to the bare ions seen (Jung et al., 1992). From the evolution of the intensities for the various peaks with time, reaction cross sections or lifetimes can be deduced. The Schottky noise diagnosis is extremely sensitive: Even a single circulating highly charged ion may be probed. As the reaction products have tiny differences in their masses, and hence in their momenta, the reaction products do not feel the same intrabeam scattering as the original ions do in the high intensity beam. Therefore, the Schottky signals for the reaction products show extremely narrow lines. At the ESR, the mass resolution of this method is tested to $\Delta M/M \approx 10^{-6}$ (Kluge, 1995).

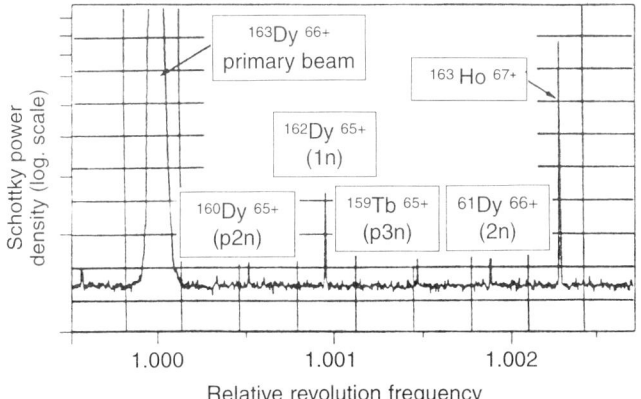

FIG. 20. Frequency spectrum of the Schottky noise picked up for a stored and cooled beam of bare $^{163}\text{Dy}^{66+}$ ions (broad line). After switching on the gas target, narrow lines show up; these are signals from neighboring nuclei made by nuclear reactions. The $^{163}\text{Ho}^{67+}$ line results from bound-state β decay of the bare Dy ions into H-like Ho ions, which are then stripped in the gas target into the circulating bare ions observed (Bosch, 1993).

IV. Charge Changing Processes

A. RECOMBINATION PROCESSES IN THE COOLER

As was emphasized in the previous section, cooling of ions by electrons is the prerequisite for successful operation of heavy ion storage rings. Recombination losses in an electron cooler represented one of the most fundamental design issues for heavy ion storage rings. We show in Fig. 21 the estimated cross section ratio for radiative recombination and for cooling (cf. Kienle, 1985). For bare ions, this ratio is plotted as a function of the atomic number assuming different equivalent electron energies $T(\text{rel})$ relative to the ions, corresponding to different temperatures. For ion beams injected from the accelerator, we have to assume for $T(\text{rel})$ typically a value of a few electron volts, for reaction fragments a value of more than 10 eV, and for a cooled ion beam we have to assume $T(\text{rel})$ to

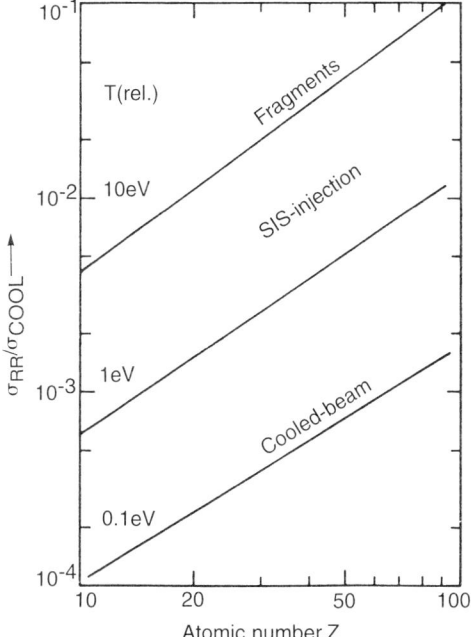

FIG. 21. Calculated cross section ratios for radiative recombination and electron cooling as a function of the atomic number Z for bare ions at different equivalent electron temperature $T(\text{rel})$ relative to the ions (cf. Kienle, 1985).

be around 0.1 eV. From the figure it is evident that, even for the heaviest ions, the recombination losses are small compared with the cooling effect; this is in particular true for an already cooled ion beam.

In general, the half-lives of stored beams can reliably be estimated by using theoretical cross sections for radiative recombination; compare Fig. 18. For an exact comparison of measured recombination rates with theory, however, one has to convolute the cross sections with the actual velocity distributions for electron and ion beams. Here, not only do the velocity distributions have to be known exactly, but all possible field inhomogenities in the magnetic solenoid field of the cooler, small misalignments between the beams, reionization of electrons in high n states in the next downstream dipole magnet by the motional electric field seen by the ions, and other subtle effects also have to be included into the calculations. Under cooling conditions, often slightly too large recombination rates are found, in particular for heavy ions. Beyond $Z = 5$, enhancement factors between 1.5 and 2 have been reported from the TSR (Wolf et al., 1993). Large enhancement factors, up to at least 40, have been found for very heavy ions such as U^{28+}, also in single-pass experiments. (to be discussed) (Uwira et al., 1995). More detailed studies are needed of this important issue.

As the cooling forces are strong, it is difficult to measure the recombination rate as a function of the relative velocities between the beams in a storage ring. For slightly detuned velocities, the beam velocity is almost instantaneously pulled toward the electron velocity. Hence, single-pass experiments will be more accurate near matching velocities. Using the ASTRID cooler target off-line at the Aarhus tandem van de Graaff accelerator, Andersen and Bolko (1990) found excellent agreement in the dependence of the radiative recombination rate on the detuning energy with predictions according to Bethe and Salpeter (1957) or to Stobbe (1930) for bare ions up to F^{9+}. In Fig. 22, this energy dependence is shown. The radiative recombination strongly decreases with increasing relative energy between electrons and ions.

For heavier ions, the radiative recombination rates also agree fairly well with predictions, especially at nonvanishing relative velocities. For matching velocities, a considerable enhancement over the predictions is found for heavy species. This is in particular true for dressed heavy ions, such as U^{28+} (Schennach et al., 1991; Uwira et al., 1995). For illustration, we show in Fig. 23 the recombination rate for Li-like Ar^{15+} ions measured with a very dense electron target in a single-pass experiment at the UNILAC, GSI–Darmstadt (Schennach et al., 1994). In the enlarged spectrum at the top, the expected radiative recombination is shown as a dashed line. At

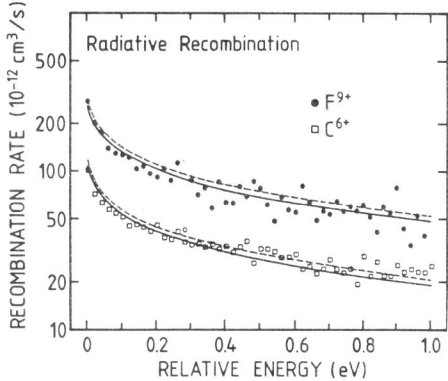

FIG. 22. Radiative recombination rates for bare C^{6+} and F^{9+} as functions of the relative energy between ion and electron beam according to Andersen and Bolko (1990). (For the flattened velocity distributions in the electron target, we have temperatures of $kT_\perp = 150$ meV and $kT_\parallel = 2$ meV.) The rate predictions according to Bethe and Salpeter (1957) and Stobbe (1930) are shown as dashed and solid lines, respectively.

cooling, i.e., at zero relative energy, the recombination rate is found to be a factor of four larger than predicted.

On top of the radiative recombination we find strong resonance contributions from dielectronic recombination (DR); see Fig. 23, bottom. There are two DR resonance series, an excitation of the $2s_{1/2}$ electron into the $2p_{1/2}$ and the $2p_{3/2}$ levels, respectively, and a resonant capture to the nl level series (or vice versa). Using an isolated resonance approximation, an overall good agreement is found for the rates and the position of the dielectronic resonances on top of the radiative recombination continuum.

As the natural widths of the DR resonances are small compared with the temperatures involved with the electron beam, the shapes of the resonances are mainly determined by the actual velocity distributions. The resonances are asymmetric; the low energy side is mainly determined by the transverse temperature, and the high energy side reflects more the longitudinal temperature of the electron beam (Andersen et al., 1989, 1990b). Because of the acceleration of the electrons in the gun, a so-called flattened velocity distribution is seen for the target electrons (Poth, 1990a). The resonance at about 1.2 eV can be fitted best with electron temperatures of $kT_\parallel = 2.4$ meV and $kT_\perp = 260$ meV (full curve). For calculating the radiative recombination contribution, these temperatures have been used; see dashed curve in the top spectrum of Fig. 23. At truly matching velocities, an appreciable excess in the recombination rate is found.

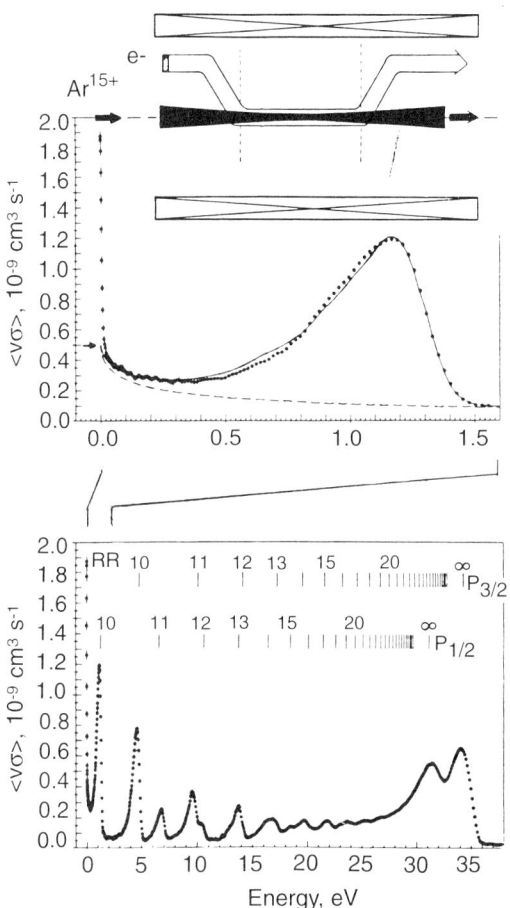

FIG. 23. Recombination rate for Li-like Ar^{15+} ions ($1s^2 2s_{1/2}$) measured in a single-pass experiment through a dense electron target, with temperatures ($kT_\parallel$ = 2.4m eV and $kT_\perp$ = 260 meV). The dielectronic recombination resonances $1s^2 2p_{1/2} nl$ and $1s^2 2p_{3/2} nl$ are seen on top of the radiative recombination continuum. In the spectrum in the top part of the figure, the energy region up to a relative energy of about 1.5 eV is enlarged. There, the DR resonance is fitted by convoluting the resonance with the actual temperatures of the electron target. Correspondingly, the radiative radiation continuum is calculated (dashed curve). The inset at the very top shows the arrangement of the electron target used; for details, see Schennach et al. (1994).

Three-body recombination seems not to be responsible for this enhancement. On the other hand, the strong central potential of the highly charged ions may locally increase the electron density in the target. Moreover, a stimulated radiative recombination caused by the electromagnetic radiation of the neighboring electrons from their cyclotron motion and by the black body radiation may also contribute to this enhancement.

To measure the recombination at small relative energies in a cooler at an in-ring experiment, one has to switch the relative energy continuously between both the beams from zero (cooling) to the measuring value (detuning) and back again. The switching has to be fast enough so that on the one side the ion beam stays cooled, and on the other side the ion energy is not changed by the dragging force of the electron beam toward the measuring energy. The cycle is typically in the 20-msec region. This method gives a perfect normalization of the recombination to the rate at cooling. The first in-ring measurements have been performed in this way at the TSR for highly charged heavy ions (Kilgus et al., 1990). In Fig. 24, a measurement from the TSR is shown for Li-like Cu^{26+} ions (Kilgus et al., 1992). Once more, on top of the radiative recombination continuum, decreasing with the relative energy between the beams, we see the resonances for the $\Delta n = 0$ series with $1s^2 2s_{1/2} + e \rightarrow 1s^2 2p_{1/2} nl$ and

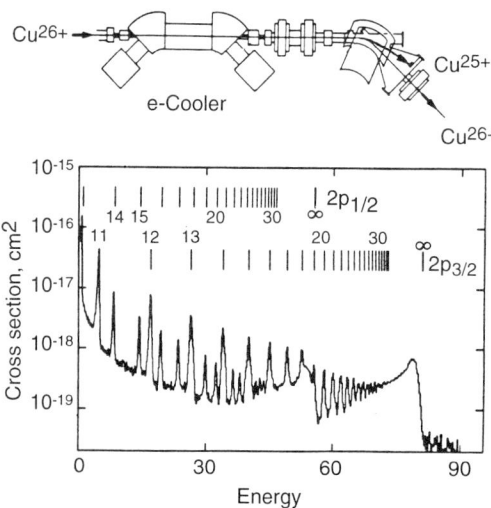

FIG. 24. Recombination rate measured for Li-like Cu^{26+} ions at the TSR. On top of the radiative recombination continuum, resonances from dielectronic recombination appear for $\Delta n = 0$ transitions, $1s^2 2s_{1/2} + e \rightarrow 1s^2 2p_{1/2} nl$ and $1s^2 2p_{3/2} nl$ (Kilgus et al., 1992).

$1s^2 2p_{3/2} nl$. Up to the series limit, good agreement is found for the resonance energies and for the cross sections. The same is true for $\Delta n = 1$ transitions, i.e., for the $1s^2 3lnl'$ resonances (Kilgus *et al.*, 1992, 1993).

For the $\Delta n = 0$ resonances in Li-like ions, it is evident that for heavier ions more resonances are resolved, and that the two series limits split rapidly with increasing atomic number Z. This splitting gives in the limit ($n \to \infty$) the fine structure splitting between the $p_{1/2}$ and the $p_{3/2}$ levels in Li-like ions, whereas the absolute energy positions of the series limits give additionally the splitting to the $2s_{1/2}$ level. An appreciable part of the last splitting is caused for heavy species by QED contributions. As will be shown in Section V, the positions of the resonances within these series can be utilized for a precise spectroscopy, in particular of very heavy ions. At the ESR, dielectronic resonances have been studied for Li-like Au^{76+} ions (to be discussed) (Spies *et al.*, 1992). For these heavy ions, good agreement also was found between experiment and theories applying fully relativistic codes including Breit term, QED corrections, and nuclear size effects.

As pointed out, good agreement between resonance strengths and experimental rates have been found for $\Delta n = 0$ transitions in Li-like ions even for the heaviest species. This indicates that the electron–electron interaction is reasonably well understood—within the available accuracies—if one includes all relativistic and all other effects. On the other hand, the electrons involved for Li-like ions do not probe sensitively the innermost part of the strong central potential of the atomic nucleus (cf. Fig. 6). For such a test, one has to measure $\Delta n = 1$ resonances involving a K shell electron, as was already done at the TSR for lighter ions (Kilgus *et al.*, 1990). For heavy ions, where the relativistic influence can best be probed, the resonance energies are so high that the cooler cannot technically be detuned fast enough. For those measurements, an independent second electron target probably has to be installed in a heavy ion storage ring. Such a second electron target is currently projected for the ESR. Then, the current–current interaction can be measured under truly relativistic conditions (Zimmer *et al.*, 1990; Chen, 1990).

To demonstrate the drastic influence of the relativistic and the Breit interaction on the dielectronic recombination, we show in Fig. 25 the dependence of the involved *KLL* Auger rates as a function of the atomic number Z for the case of final He-like ions (Zimmerer, 1992; Zimmerer *et al.*, 1991). In the nonrelativistic approach (NRW) or in a $1/Z$ expansion, the Auger rate is independent of Z. Using relativistic wavefunctions, the Auger rate increases beyond about Xe considerably, yielding for U on the average an enhancement factor of about 1.5. On top of this, the Breit interaction gives an additional increase of the same order of magnitude. For selected resonances, the enhancement can increase up to an order of

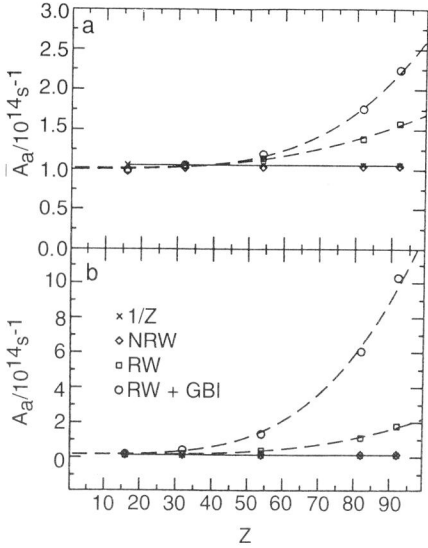

FIG. 25. KLL Auger rates for final He-like ions as a function of the atomic number Z according to Zimmerer (1992): (a) mean Auger rates; (b) Auger rate for the $2s_{1/2}2p_{1/2}\,^3P_0$ doubly excited state. NRW, 1/Z, RW, and GBI mean nonrelativistic approach, 1/Z expansion, relativistic calculation, and Breit interaction, respectively.

magnitude compared with the nonrelativistic approach. This is shown for the $2s_{1/2}2p_{1/2}\,^3P_0$ resonance for final two-electron systems.

At present, these interesting questions can only be studied by applying quasi-free electron targets, i.e., electrons confined to bound states by target atoms (Graham *et al.*, 1990; Kandler *et al.*, 1995b, 1996). Unfortunately, the bound electrons have a certain intrinsic velocity distribution (Compton profile), so that no isolated resonances can be resolved by those measurements. Nevertheless, reasonable agreement is found with predicted cross sections in an experiment for He-like U^{90+} projectiles. At the SIS facility, coincidences between the "recombined" Li-like U^{89+} projectiles and the stabilizing X rays have been measured for this resonant transfer and excitation process (Kandler *et al.*, 1995a, b). In Fig. 26, the $K\alpha_1$ and $K\alpha_2$ excitation function for KLL RTE are shown for this case. Three resonance groups for the various jj' doubly excited L states ($\frac{1}{2}\frac{1}{2}$, $\frac{1}{2}\frac{3}{2}$, $\frac{3}{2}\frac{3}{2}$) are resolved. Due to the Breit interaction, the reaction strength for $jj' = \frac{1}{2}\frac{1}{2}$ is enhanced by a factor of two over the nonrelativistic approach. Moreover, the experimental value for this resonance group overshoots even the relativistic calculations, whereas the other two groups are in accordance with theory.

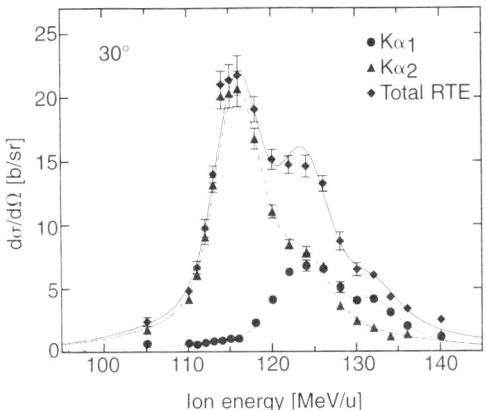

FIG. 26. Measured differential $K\alpha_1$, $K\alpha_2$ excitation function for KLL-RTE observed in $U^{90+} \rightarrow C$ collisions. The total $K\alpha = K\alpha_1 + K\alpha_2$ yield is also given (Kandler et al., 1995a).

B. Electron Capture Processes in the Gas Jet Target

The investigation of projectile charge-exchange processes in encounters of high-Z ions with neutral target atoms not only provides a test of the theory of atomic collision dynamics, but it is also of general importance for the operation of heavy ion accelerators and storage rings. The ESR is the only storage ring equipped with a gas jet target. Because of the high luminosity of the ESR, the gas jet target provides the unique possibility of studying projectile charge exchange in encounters of high-Z ions with low density matter. This is in particular true for the relativistic collision velocity regime and for low-Z targets. At such collision conditions, where the charge-exchange cross sections are comparably low, one has to rely in standard single-pass experiments on the application of solid targets with typical densities of 10^{23} particles/cm³. The latter has to be compared with the densities of the gaseous targets at the ESR of about 10^{13} particles/cm³, which exclude the possible influence of multistep processes (solid target effects) on the charge-exchange processes to be studied. In the following, we will concentrate on the discussion of the electron capture processes and particularly of the radiative capture process.

In fast collisions of bare high-Z projectiles with low-Z target atoms, REC is the only relevant charge changing process. This is illustrated in Fig. 27, where the total one-electron projectile pick-up cross sections are plotted versus target atomic number Z_T for 295-MeV/u bare uranium ions (Stöhlker et al., 1994). The measured data are compared with theoretical predictions for the two competing electron capture processes, i.e., for

REC and NRC. For NRC the relativistic eikonal approximation (Eichler, 1985) was applied, whereas for REC the Stobbe theory (Stobbe, 1930) was used. As seen in the figure, very good agreement between the predicted Z_T cross section dependence and the experimental data is observed ($\sigma_{NRC} \sim Z_T^5$, $\sigma_{REC} \sim Z_T$; compare also Section II). In particular, the experimental data confirm the prediction that for $Z_T < 13$ the NRC contribution can be neglected. It is emphasized that the presentation in Fig. 27 includes data from solid as well as from gaseous targets. Obviously, for such bare, high-Z projectiles, target density effects do not influence the electron capture probability. This observation can be explained using the electron capture cross section dependence on the final projectile main quantum number, n_f. At high beam energies, the cross sections for both capture processes, REC and NRC, scale with $1/n_f^3$. Moreover, the fast decay rates of the excited levels in such high-Z systems contribute significantly to the vanishing of solid state effects.

As has been discussed in Section II, the Stobbe cross section for REC depends only on the adiabaticity parameter η of the collision. This provides a convenient method for a systematic comparison of cross sections gained for different projectile–target combinations and collision

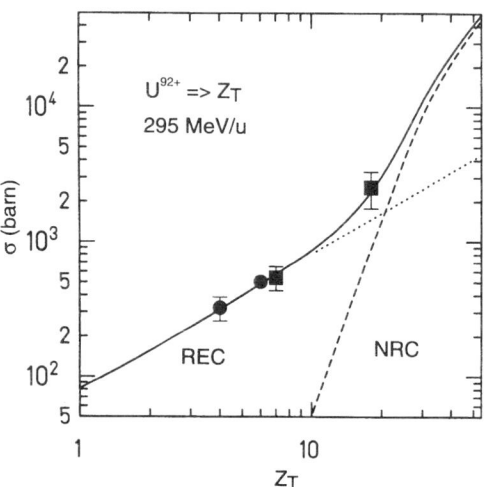

FIG. 27. Total electron capture cross sections for bare U^{92+} ions at 295 MeV/u colliding with gaseous targets (■, $U^{92+} \to N_2$, Ar) and with solid targets (●, $U^{92+} \to$ Be, C) (Stöhlker et al., 1994). For both cases, the cross sections per target atom are given. The results are compared with the theoretical cross section predictions for the NRC and the REC processes (dashed and dotted line, respectively). The sum of both cross sections is given by the solid line.

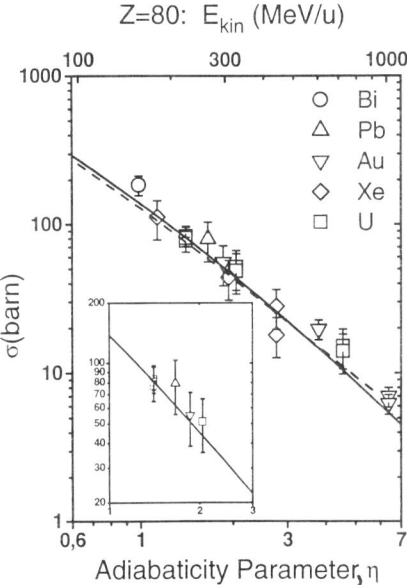

FIG. 28. Total electron capture cross sections per target electron measured for bare high-Z ions ($Z \geq 54$) in collisions with light target atoms (the data from the ESR gas jet are depicted separately in the inset). The results are plotted as a function of the η parameter (lower x axis) and are compared with the result of a relativistic exact calculation for $Z = 80$ (dashed line) as well as with the prediction of the nonrelativistic dipole approximation (solid line). For the particular case of $Z = 80$, the cross section values also are given as a function of beam energy (upper x axis).

energies. In order to show this for a large number of collision systems, it is useful to define an "adiabaticity parameter" η by connecting it to the Sommerfeld parameter ν through the nonrelativistic relation (cf. Eq. (6) in Section II)

$$\eta = \frac{1}{\nu^2} \approx 40.31 \times \frac{E_{\text{kin}} \, (\text{MeV/u})}{Z^2} \qquad (11)$$

In Fig. 28, all available total charge-exchange cross sections related to REC, normalized to the number of quasi-free target electrons, are plotted versus the η parameter (for references, see Stöhlker et al., 1995a). In addition, the upper x axis of the figure shows the energy scale for the particular case of $Z = 80$. The predictions of the nonrelativistic dipole approximation are given in the figure by the solid line, whereas the dashed line depicts the results of the rigorous relativistic calculations performed for $Z = 80$ (Ichihara et al., 1994). The error bars displayed account for both the statistical and the systematic uncertainties of the individual measurements. The data cover a projectile Z regime from $Z = 54$ to 92, and beam energies ranging from 80 to 1000 MeV/u. They were gained in solid target experiments performed at the BEVALAC (Meyerhof et al., 1985b; W. E. Meyerhof, private communication, 1994) and the SIS accelerator (Stöhlker et al., 1995a) as well as in experiments utilizing gaseous targets at the ESR. The latter are given separately, in the inset of the figure.

In general an excellent agreement between experiment and theory is found for the total cross sections; see Fig. 28. The experimental data are, within the error bars, not sensitive to the slight cross section variations predicted by the relativistic theory for different Z systems using a common nonrelativistic η parameter, especially for $\eta \leq 3$. Based on these findings, it is evident that the simple dipole approximation can be applied to get reliable cross section estimations even for high-Z projectiles, but not too high collision energies. In the high energy regime (e.g., 500 MeV/u for $Z = 80$ projectiles), the application of the exact relativistic theory is certainly more appropriate. Here, the results of the latter theory performed for $Z = 80$ already deviate from the predictions of the nonrelativistic approach. In fact, the result of the latter approach seems to underestimate the measured cross sections at higher η values. We emphasize that a nonrelativistic velocity description was used for the foregoing definition of η. Applying a correct relativistic velocity description, the dipole approximation would already fail completely for high-Z ions at intermediate η parameters. Obviously, by using the nonrelativistic scaling law, several relativistic factors cancel arbitrarily.

Analyzing the projectile photon emission associated with electron capture, shell differential cross sections can be obtained. In Fig. 29, we present the efficiency-corrected X-ray spectra, taken in coincidence with the down-charged ions at the internal gas target of the ESR. The two spectra correspond to capture from N_2 molecules into bare U^{92+} and H-like U^{91+} ions, which were taken at a collision energy of 295 MeV/u and a detection angle of 132° (Stöhlker et al., 1993, 1995a). The energy range displayed in the figure covers the X-ray energy regime relevant for REC transitions into the ground state and into excited states, as well as for the characteristic projectile transitions, i.e., the Lyα and $K\alpha$ emission. All the REC contributions appear well resolved in the figure. A comparison of the spectra in middle and bottom of Fig. 29 manifests a reduction by a factor of two for the K-REC intensity for the H-like projectiles due to the partially blocked K shell. In contrast, REC into excited projectile states is

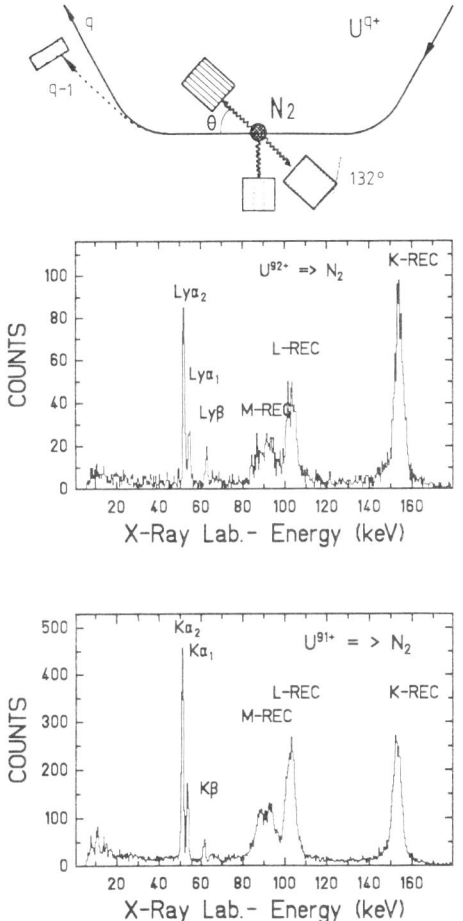

FIG. 29. X-ray spectra measured at 132° and corrected for the energy dependent detector efficiency for incident U^{91+}, U^{92+} ions on N_2 at 295 MeV/u. The spectra were accumulated at the ESR gas target in coincidence with the down-charged ions.

equivalent for both cases. From these X-ray spectra, one can deduce shell differential as well as angular differential REC cross sections.

Also, the total K-REC cross sections, determined by the K X-ray emission, follow closely the nonrelativistic prediction over an extremely wide η range, $0.1 < \eta < 10$, and for projectiles from $Z = 8$ to 92. In Fig. 30, one angular differential K-REC cross section value measured for bare uranium in collisions with N_2 target molecules is compared with the result of exact relativistic calculations (Ichihara et al., 1994; Eichler et al., 1995) as well as with the nonrelativistic $\sin^2 \theta_{lab}$ distribution. According to the relativistic description, the differential cross sections for K-REC show a pronounced deviation from symmetry around $\theta_{lab} = 90°$, and the maximum is shifted slightly toward the forward direction. This behavior is essentially associated with the occurrence of magnetic (spin-flip) transitions, which are not considered by a nonrelativistic approach. This spin-flip contribution to the angular distribution is depicted separately in Fig. 30 by the shaded area. For the particular case of radiative capture into a projectile s state, only spin-flip transitions can produce nonvanishing cross sections at 0° and 180°. The latter follows directly from angular momentum conservation laws for electric multipole transitions (Eichler et al., 1995). Note that there is no other case at present in which spin-flip effects in relativistic atomic collisions have such an unambiguous fingerprint. Within the total experimental uncertainty, a fair agreement between experiment and relativistic theory is found for the absolute values. Here, the prediction of the dipole approximation (dotted line in Fig. 30) overestimates considerably the experimental value at 132°.

Angular differential REC cross sections are extremely sensitive to details of the atomic wavefunctions and therefore provide detailed insight into the atomic structure of one- and few-electron systems. For such studies, high-Z ions are of particular interest. Here, the large fine structure splitting provides direct experimental access even for the investigation

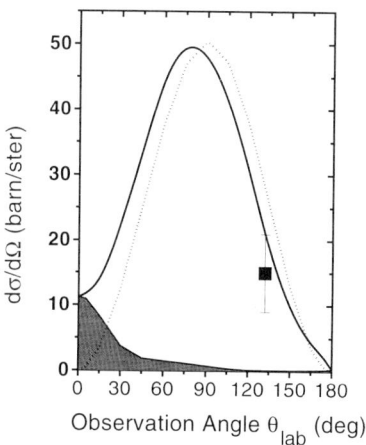

FIG. 30. K-REC angular distribution for 295-MeV/u $U^{92+} \rightarrow N$ collisions. The dotted line gives a $\sin^2\theta_{lab}$ distribution; the solid line represents correct relativistic calculations, and the shaded area depicts the spin-flip contribution (Stöhlker, 1995b).

of the angular distributions associated with REC into the different j sublevels of the L shell. The potential of such challenging studies was demonstrated for the first time experimentally at the GSI Fragment Separator for 89-MeV/u $U^{90+} \to C$ collisions (Stöhlker *et al.*, 1994). At this low projectile energy the L-REC lines split already into the two j-components. The results obtained in this experiment are depicted in Fig. 31. There, the measured angular distributions for REC into the $j = \frac{1}{2}$ and $\frac{3}{2}$ levels of the L shell of He-like uranium, normalized to the sum of both contributions, are plotted in comparison with the predictions of the exact relativistic theory.

The data for capture into the $j = \frac{1}{2}$ levels show a considerable bending of the angular distribution in the forward direction, whereas the distribution for capture into the pure p state ($j = \frac{3}{2}$) exhibits a strong enhancement at backward angles. As can be seen from the figure, the measured very pronounced forward/backward asymmetry is in excellent agreement with the result of the rigorous theoretical treatment (Ichihara *et al.*, 1994), whereas the dipole approximation fails completely in describing the experimental findings (Hino and Watanabe, 1987). The experimental results confirm in particular the predictions of the exact relativistic theory that the REC angular distributions depend crucially on the angular momentum of the final projectile state.

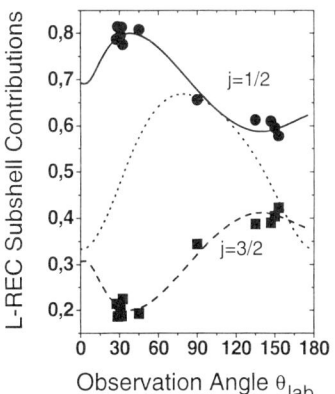

FIG. 31. Experimental angular distribution (89 MeV/u $U^{90+} \to C$ collisions) for REC into the $2s_{1/2}$ and $2p_{1/2}$ states (solid circles) and for REC into the $2p_{3/2}$ level (solid squares) in comparison with exact relativistic calculations; $j = \frac{1}{2}$: full line, $2p_{3/2}$: dashed line (Stöhlker, 1995b). All data given in the figure are normalized to the sum of both contributions. In addition, the result of the nonrelativistic dipole approximation for capture into the $j = \frac{1}{2}$ levels is given (dotted line).

C. IONIZATION AND EXCITATION PROCESSES

At the relativistic energy regime covered by the ESR, the K shell ionization process for initially H- and He-like ions is of particular interest. Up to now, no systematic investigations of this process have been performed at the ESR, although the low density gas jet target provides ideal experimental conditions for such studies. Only during commissioning of the ring and during some atomic physics experiments could a few K shell ionization data be gained for H-like Dy, Au, and Pb projectiles colliding with N_2 or Ar molecules/atoms (Stöhlker, 1993). As has been pointed out in Section II, the projectile K shell ionization cross sections follow, within a strict nonrelativistic treatment, a universal scaling law. By considering the correct relativistic bound and continuum wavefunctions, this simple scaling is abolished. However, for beam energies below 300 MeV/u, the predictions of a relativistic treatment of this process deviate, depending on the projectile charge, from those of the nonrelativistic approach by only about 20%, which in general is within the systematic uncertainty of the experimental data. Within the overall experimental uncertainties of the ESR data, a general agreement with the first order perturbation theories is found. We emphasize that these findings seem to be in disagreement with results of solid target experiments, where systematic deviations of about 50% were observed with respect to relativistic SCA calculations (Rymuza et al., 1993). This is depicted in Fig. 32, where the ratio between solid target data for K shell ionization of high-Z projectiles (per K shell electron) and the predictions of relativistic SCA calculations (Rösel et al., 1982) is given as a function of the reduced velocity v/v_k. Note that only the data of collision systems that fulfill the requirements of first order perturbation theory are considered. From the figure, a general enhance-

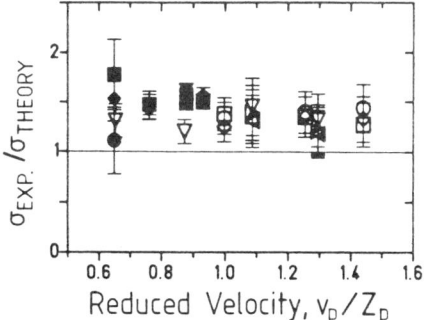

FIG. 32. The ratio between the experimental K shell ionization cross sections for high-Z projectiles and relativistic SCA calculations (solid line) (Rymuza et al., 1993).

ment of the experimental data compared with theory is obvious. Therefore, further studies of this process at the ESR are urgently needed to clarify this point.

The Coulomb excitation of a projectile electron is mediated by exactly the same reaction mechanism in ion–atom collision as in projectile ionization, except that the active electron is excited into a bound and not into a continuum projectile state. The formation of excited projectile states via direct Coulomb excitation can be studied uniquely for one- and two-electron high-Z ions by the observation of the radiative decay of the excited levels to the ground state. Owing to the large L subshell splitting in such ions, the cross section for ground state excitation into the various L shell sublevels can be unambiguously determined. In particular, due to the very large $2s_{1/2} \to 1s_{1/2}$ $M1$ decay width, $E0$ excitation to the $2s_{1/2}$ can be detected directly. The same holds true for the $E1$ excitation mode. Up to now, no K shell excitation experiments have been performed for high-Z ions except a pilot experiment conducted for H- and He-like Bi ions using solid targets at the SIS facility (Stöhlker et al., 1992). For such investigations, the ESR gas jet target provides excellent experimental conditions, and experiments aiming at precise study of the excitation process in the relativistic regime are in preparation. We add that within these planned experiments, the detection of the characteristic Lyα/$K\alpha$ photons can be performed in coincidence with a registration of the recoil momentum of the target atoms. This will provide a first direct measurement of the impact parameter dependence for the various excitation modes for high-Z projectiles.

In collisions between highly charged heavy ions and low-Z targets, the process of resonant transfer and excitation is—in addition to the interest in relativistic effects discussed already—of practical importance for all atomic structure and collision experiments. Considering the Compton profile of the target electrons, RTE might be an important $K\alpha$ production process even outside the RTE resonance energies, and it may also contribute significantly to the total projectile charge-exchange cross section. This can be read from Fig. 26, where the measured KLL RTE excitation functions is plotted for the case of initially He-like U^{90+} projectiles colliding with C atoms (Kandler et al., 1995a). In general, the whole RTE resonance regime should be avoided by K excitation experiments, as the $K\alpha$ transitions are blended by the $K\alpha$ hypersatellites produced by RTE. Also cross section measurements of the nonresonant processes are of course problematic within this resonance regime. Moreover, near the resonances the total capture cross section can increase considerably. Hence, by collisions with remnant gas molecules in the ring (mainly H_2), the lifetime for stored ions can be reduced there.

V. Atomic Structure Studies

A. THE GROUND STATE LAMB SHIFT

By capture and recombination processes, in particular for single-electron atomic reactions, excited levels in highly charged, very heavy ions can selectively be populated. Probing the subsequent radiative transitions with high spectroscopic resolution will give detailed information on the atomic structure for strong central fields. All inner shell transitions ($L-K$) for the heaviest species with only one and, to a great deal, also with two electrons can be considered as prompt (cf., e.g., Mokler et al., 1994). Only in He-like systems may the $1s_{1/2}2p_{1/2}{}^3P_0$ state show a delayed decay pattern compared with the flight velocity of the ions (Marrus et al., 1989, 1995). These inner shell transitions have energies in the 100-keV region (e.g., for U), and high resolution X-ray spectroscopy seems to be the adequate vehicle for precision structure studies. However, crystal spectrometers have in reality such small overall detection efficiencies that until now their use was not possible. Solid state detectors, for example, intrinsic Germanium diodes, Ge(i), have reasonably good resolution and an almost perfect detection efficiency, only determined by the detector sizes. At present, results based on solid state detectors are available for the heaviest ultimately charged ions (to be discussed).

A further severe obstacle for a fast beam precision spectroscopy is the Doppler effect. The radiation from the emitting fast projectiles has to be transformed by the Lorentz transformation into the laboratory system, and further, the finite solid angle of the detector causes a Doppler broadening for the observed line radiation. Moreover, the divergence of the beam and the uncertainty in the knowledge of the absolute beam velocity contribute to the final uncertainty in the desired transition energies (cf. Mokler et al., 1995a). For instance, at a specific energy of 300 MeV/u and for an observation angle of 0°, the projectile emission is blue-shifted by a factor of 2.2; at 180° backward observation, we have a red shift of a factor of 0.45. For this case, the unshifted line is found at about 70° in the laboratory. At a laboratory angle of about 48°, the sensitivity to the beam velocity is minimal, whereas the changes in Doppler shift with observation angle are maximal there. For 0° observation, the opposite is the case. To overcome the variation in Doppler shift with observation angle, segmented or "granular" detectors are used. Thus, large solid angles can be covered without losing resolution by Doppler broadening. Each segment has to be corrected individually with its correct Doppler shift before all the spectra can be added up to one high resolution center of mass (CM) spectrum with good statistical significance (Stöhlker et al., 1992). Because of the active and redundant use of the Doppler effect, this fast beam technique is called Doppler-shift-assisted ion spectroscopy (Mokler et al., 1995a). This is to some extent one modern version of the classical dispersive Doppler-tuned edge spectroscopy of fast ions beams (Schmieder and Marrus, 1973; Lupton et al., 1994).

Excited projectile states can be produced efficiently either by radiative recombination in the electron cooler or by electron capture in the gas jet target of the ESR. Complementary experiments are performed at both areas. The cooler, with its bulky magnetic coils, is only accessible to an almost 0° or 180° observation of the emitted radiation. In contrast, such an observation is not possible at the gas jet in its present design. However, the range from 30° to 150° on both sides of the gas target can be used for X-ray detection, giving ample of angular redundancy. Naturally, at both areas single electron events are assured by the corresponding coincidences of the X-ray emission with the down-charged ions detected behind the corresponding next downstream dipole magnet. In Fig. 33, the methods and the resulting spectra are compared.

From Fig. 33 it is obvious that the two methods give complementary information. First, the population of the projectile L-shell levels is quite different: In the cooler, we have low energy electron–ion collisions preferring more the high angular momentum states; whereas at the gas target, high energy ion–atom collisions, mainly REC, favor the low angular momenta. Correspondingly, the intensity ratio of the Lynman–α doublet is at variance for both cases. For the cooler target, furthermore, the radiative recombination line to the ground state gives directly the ground state binding energy. The analogous REC line at the gas target is due to the high relative kinetic energy of the quasi-free electrons outside the range of the shown spectrum (Fig. 5.1b). However, the dependence of the K-REC peak on beam velocity also can be used for a precise spectroscopy of the ground state binding energy (Mokler et al., 1991). Finally, in the electron cooler, one has to deal with an appreciable Bremsstrahlungs background at lower X-ray energies caused by scattered electrons. The coincidence spectrum at the gas target is absolutely clean down to the lowest X-ray energies, allowing also a spectroscopy of higher shell transitions as is shown by the Balmer lines in the displayed spectrum.

For the case of transitions in H-like U^{92+}, we display at enlarged scale the region for the Lyman transitions for both cases in Fig. 34. At the top of the figure, the corresponding transitions are indicated. The top spectrum (lab energies) was taken under 0° at the cooler for a 321-MeV/u coasting bare U^{92+} beam (Beyer et al., 1995). The radiative recombination line as well as the Lyman–α transitions are clearly visible at strongly Doppler-shifted energies. The Lyman lines show a strong low energy tail from a

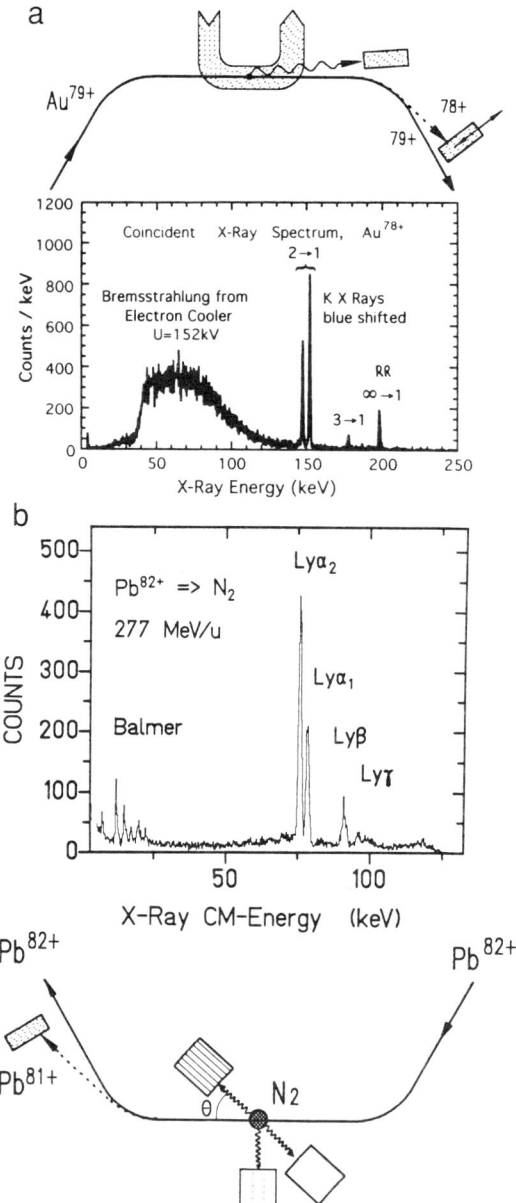

FIG. 33. Comparison of projectile X-ray spectroscopy at the electron cooler and the gas jet section of the ESR. In both cases, single electron events are assured by coincidences between X rays registered by a Ge(i) detector and the down-charged ions. (a) The near 0° arrangement at the electron cooler and a spectrum taken for a coasting bare Au^{79+} beam at 278 MeV/u (Liesen et al., 1994). (b) The Doppler-shift-assisted, multiangle spectroscopy at the gas jet target; bare Pb^{82+} ions from a coasting beam capture an electron from N_2 gas molecules (Mokler et al., 1995a).

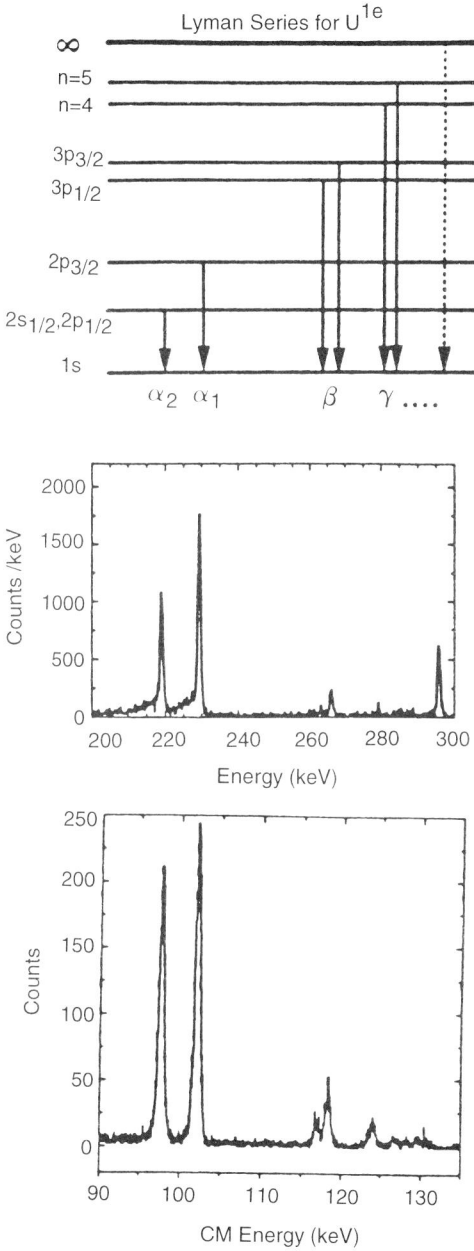

FIG. 34. Lyman spectra for H-like U^{92+} measured at the cooler section (top spectrum) (Beyer *et al.*, 1995), and at the gas jet target (bottom spectrum) (Stöhlker, 1995a; Mokler *et al.*, 1995b). A coasting bare U^{92+} beam of 321 and 68 MeV/u, respectively, was used. In the upper spectrum lab energies and in the lower one CM X-ray energies are given. For convenience, the Lyman decay scheme is displayed at the very top.

cascade feeding via recombination to high lying Rydberg states. This delayed emission involves different geometries and, hence, a different Doppler relevant detection angle. The spectrum at the bottom was taken at the gas target for a cooled U^{92+} beam decelerated in the ring to 68 MeV/u and colliding with N_2 molecules (Stöhlker, 1995a; Mokler et al., 1995b). The Lyman lines are symmetric; even the Lyman-β transitions split clearly into two lines for different j values in the feeding M shell. The Lyman-α_2 line is, in contrast to the single transition in the Lyman-α_1 line, composed of two transitions: the $2p_{1/2}$–$1s_{1/2}$ $E1$ radiation and the $2s_{1/2}$–$1s_{1/2}$ $M1$ transition; cf. Section II.A.

Determining the exact transition energies for the various lines and subtracting the theoretical expectation values for the corresponding transitions from a solution of the Dirac equation, one ends up with an experimental value for the $1s$ ground state Lamb shift. The best value for the ground state Lamb shift in H-like uranium is presently 470 ± 16 eV (Beyer et al., 1995), which has to be compared with theoretical values of 465.5 ± 2 eV (Persson et al., 1996) and 464.6 ± 0.6 eV (Mohr, 1994). Within the errors, there is excellent agreement between experiment and QED expectations. On the other hand, the experimental accuracy is not yet good enough to test the theory sensitively. An evaluation of the more recent and statistically more significant data from the gas target, shown at the bottom of Fig. 34, may improve the experimental situation slightly (Stöhlker, 1995a). However, an experimental accuracy better by an order of magnitude is needed to compete with the theoretical uncertainties. Here, more advanced spectroscopic techniques have to be applied in the future. Another way to tackle this problem is to investigate transitions into or within higher shells. There, not only are the QED contributions smaller, the transition energies are also dramatically reduced, possibly permitting a more accurate determination of the QED terms. However, the higher main quantum number wavefunctions (e.g., in the L shell) do not probe so sensitively the strongest parts of the strong central potentials.

Independently, the recent advances in the ground state Lamb shift measurements at the heaviest H-like ions are a tremendous success in this challenging field. In Fig. 35, these results are summarized and compared with theory. The predictions of Johnson and Soff (1985) are excellently reproduced by the recent experimental findings. In the figure, the data—reduced by the Z^4 scaling—are compared with the $F(Z\alpha)$ function discussed in Section II.A (see also Eq. (2)). The full points in the figure are the results from the synchrotron/storage ring facility SIS/ESR for Dy (Beyer et al., 1993), for Au (Beyer et al., 1994), for Pb (Mokler et al., 1995a), for Bi (Stöhlker et al., 1992), and for U (Stöhlker et al., 1993; Beyer et al., 1995). For the other data, see the references given in Mokler et al.

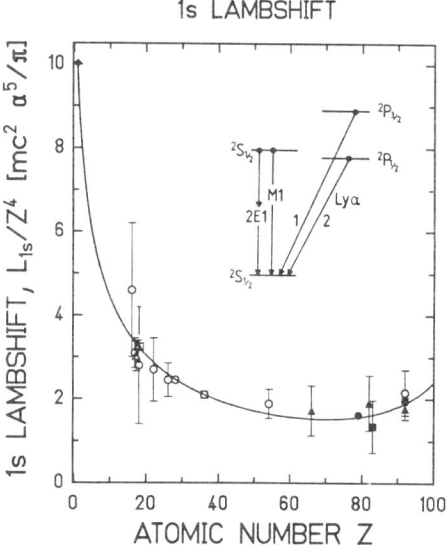

FIG. 35. Comparison of the measured ground state Lamb shift reduced by Z^4 for heavy, H-like ions with the predictions by Johnson and Soff (1985).

(1995a). Similar agreement has also been found for ground state transitions in He-like heavy systems (cf. Mokler *et al.*, 1994). However, because of the more complex decay pattern and the uncertainties from the blending contributions in the different lines, the situation is more complicated there and will not be discussed here. Anyhow, these measurements triggered sophisticated calculations, including higher order QED terms, in particular the electron–electron QED contributions (cf. Lindroth, 1995; Persson *et al.*, 1995; Indelicato, 1995). There is the hope that these effects can be tested soon by experiments.

B. Doubly Excited States

In Section IV, we emphasized that dielectronic recombination provides an ideal tool for atomic structure studies. In particular, it gives access to exploration of doubly excited states. For highly charged, very heavy ions, the structure of doubly excited states is governed by several fundamental effects. Besides relativistic effects, the current–current interaction, i.e., the Breit term, as well as QED contributions and nuclear size effects have to be included into calculations. As was pointed out in Section IV.A, fully relativistic calculations including all these effects not only describe the total recombination rates fairly well, they also give excellent agreement with the resonance positions within the measuring accuracies.

In Fig. 36, a recombination spectrum measured for Li-like Au^{76+} ions is displayed (Spies *et al.*, 1995). The test measurement was performed at the ESR with an energy uncertainty of about 0.5 eV. With some technical modifications already scheduled for the ESR cooler, the accuracy will be improved by an order of magnitude in the near future. For the present experiment, Li-like Au^{76+} ions have been bred in the ESR from He-like Au^{77+} ions originally injected from the heavy ion synchrotron SIS, cf. Section III.C.4 (Spies *et al.*, 1992). To detune the relative energies between electron and ion beams, high voltages up to ±5 kV have been applied to a

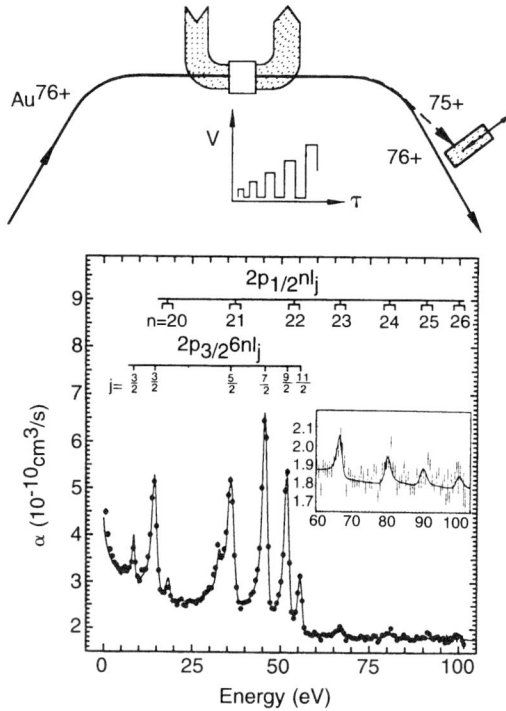

FIG. 36. Recombination rate for Li-like Au^{76+} ions as a function of the relative energy between electron and ion beam (Spies *et al.*, 1995). The full curve is a result of sophisticated calculations on dielectronic and radiative recombination convoluted with the experimental velocity distribution of the electrons over the whole interaction region (for details, see text). For clarity, the energy region beyond 60 eV is shown in the inset with an enlarged scale. The experimental arrangement is sketched at the top of the figure.

drift tube in the cooler. A fast periodic switching from cooling to measuring position was incorporated. The recombined Be-like Au^{75+} ions have been detected by the recombination detector behind the next downstream magnet as a function of the detuning voltage. In the figure, this voltage is already converted into the center-of-mass energy.

The recombination spectrum in Fig. 36 shows, on top of the rapidly decreasing radiative recombination continuum, sharp lines for the $\Delta n = 0$ dielectronic recombination resonances ($1s^2 2s_{1/2} + e \rightarrow 1s^2 2p_j n l_{j'}$). Excitations involving $2p_{3/2}$ electrons are the most prominent lines. Because of the large energy gaps between the atomic shells (n) at this high atomic number, we find within our energy range only the $2p_{3/2} 6l_j$ resonance group, which is split into several lines. To one j value of the electron with $n = 6$, two angular momenta l contribute, causing an additional small splitting. This is best visible for the $j = \frac{3}{2}$ states. All the resonances of the $2p_{3/2} 6l_j$ group cover an energy range of 50 eV because of the large fine structure splitting. Compare this with dielectronic recombination spectra for lighter ions, e.g., Li-like Cu^{26+}, shown previously in Fig. 24. There, no additional fine structure can be resolved at all within one series.

The fine structure seen for the $2p_{3/2} 6l_j$ resonances is well reproduced by fully relativistic calculations including Breit interaction, QED corrections, and nuclear size effect (Spies *et al.*, 1992). The curve shown in Fig. 36 together with the data is the result of such a calculation convoluted with the effective velocity distribution of the electrons over the interaction region of the cooler, including all the distortions caused by the drift tubes. For the $2p_{3/2}$ excitations, a full relativistic calculation of Zimmerer (1992) was applied; for the weak resonances involving $2p_{1/2}$ excitations (to be discussed), a semirelativistic approach of Pindzola and Badnell (1992) was used; and the radiative recombination continuum is based on the theory of Bethe and Salpeter (1957). An excellent agreement between the final result in an isolated resonance approximation (i.e., neglecting interferences between the various reaction channels) and the data is found.

In addition to the prominent resonances just discussed, we see the weak resonances of the $2p_{1/2}$ excitation series with an nl_j Rydberg electron. The positions for the resonances with $n = 20$ to 26 are indicated in Fig. 36. The resonances with $n = 20$ and 23 are clearly visible within the measured spectrum. For the higher n resonances, we display the energy region between 60 and 100 keV separately in the inset. There, the resonances up to $n = 26$ are visible. Although the rates are small for the $2p_{1/2} nl_j$ resonances and the present statistics is not adequate for a precise extrapolation of energies to higher n states, one can try to determine roughly the series limit. Using an effective charge (76.48) in the Rydberg formula or using correspondingly a quantum defect (0.098) formula, a $2s_{1/2} \rightarrow 2p_{1/2}$

energy splitting of 219.0 ± 3 eV and 218 ± 3 eV, respectively, has been reported for the limiting case that the second electron is just in the continuum, i.e., $n \to \infty$ or Li-like Au^{76+} ions (Spies et al., 1995).

A first calculation for this case gave a splitting of 217.5 eV (Zimmermann et al., 1994), in excellent agreement with the experimental result. Here, the one-electron Lamb shift contribution to the 2s level is on the order of 30 eV. This test experiment already gave an accuracy of 3 eV, i.e., a 10% measurement of the Lamb shift. As was pointed out, in future the accuracy of those experiments will be improved by at least an order of magnitude after introducing some technical modifications to the cooler. In the end, we expect that the accuracy of the DR experiments just described will be superior to those quoted for Li-like U^{89+} by Schweppe et al. (1991) using a Doppler-tuned spectroscopy technique for the $\Delta n = 0$ transition in the VUV range.

Moreover, there is the definite hope that, even for the heaviest ions, dielectronic recombination resonances with $\Delta n = 1$ transitions involving K shell electrons will be measured. Assuming similar accuracies to those already achieved for lighter ions or for $\Delta n = 0$ transitions, dielectronic recombination may be a very competitive tool for precision spectroscopy of the heaviest, highly charged ions.

C. RYDBERG STATES

By radiative recombination, highly excited projectile levels (n, l), i.e., Rydberg states, also can be populated in the cooler. According to Stobbe (1930), the population distribution (n, l) can be calculated; cf. Eq. (4). For small relative energies of the free electron compared with the emitted photon energy, the recombination rate is proportional to Z^2/n and decreases with the relative kinetic electron energy. Furthermore, (n, l) dependent coefficients have to be used. This spontaneous radiative recombination gives a very wide distribution for the Rydberg states. On the other hand, radiative recombination can also be induced by a corresponding photon field. This laser-induced radiative recombination (LIR) is a resonant process, selectively populating only one Rydberg state (Wolf, 1992). The induced recombination rate depends trivially on the laser intensity and, more importantly, on the third power of the photon wavelength. If we use the Rydberg formula for the binding energies, the induced rate is proportional to $(n/Z)^6$. Assuming a fixed laser wavelength means that very high Rydberg states can particularly be populated for high-Z projectiles. On the other hand, for the heavy ions the spontaneous rates also increase, masking increasingly stimulated transitions and the relative gain factor.

It is obvious that LIR is an ideal tool for probing the temperature distribution of the electrons in the cooler. On the other hand, the projectile Rydberg states also can be examined with high spectroscopic precision. For a low-Z heavy ion, laser-induced recombination was measured first at the TSR: By a dye laser, the recombination of a bare C^{6+} ion with a cooler electron into the $n = 14$ level of C^{5+} was stimulated; the recombined ions were detected behind the next downstream dipole magnet (for details, see Wolf et al., 1993). In particular, the electron density distribution around the continuum could be investigated. The Rydberg states for dressed, highly charged ions are very sensitive to the core potential of the ions, especially the core polarizability. This long range interaction has been studied for oxygen ions with a Li-like core (Schüssler et al., 1995a, b).

Loosely bound electrons in high Rydberg levels, with their relatively long lifetimes, may be field ionized by the motional electric field seen by the ion in the next downstream transverse magnetic dipole field. This is especially true for fast, i.e., heavy, ions, and the stimulated recombination cannot be observed because of field ionization. However, transferring the electron by an appropriately tuned second laser pulse to a deeper bound state, the recombination can be stabilized. Recently, this laser-stimulated, two-step recombination process was first seen for stored bare Ar^{18+} ions in the ESR (Borneis et al., 1994). More recently, this technique was also used at the TSR to study the fine structure of inter-Rydberg transitions, already mentioned (Schüssler et al., 1995a, b).

In Fig. 37, the principles for the two-step, laser-induced recombination experiment at the ESR are outlined in an example. Bare Ar^{18+} ions were stored at an energy of about 147 MeV/u ($\beta \approx 0.5$). In the overlap region of the cooler, an Nd:YAG laser was collinearly superimposed, stimulating an induced recombination into the $n_i = 84$ Rydberg level of Ar^{17+}. Here, the photon energy in the ion system is strongly red-shifted by the Doppler effect. Simultaneously, a tunable Ti:sapphire laser was superimposed antiparallel to the ion and electron beam direction, inducing a stimulated transition down to the $n_f = 36$ and $n_f = 37$ Rydberg state. Here, the blue-shift of the photons also was used to come down to Rydberg levels below the field ionization limit, which in the present case is at about $n_{ioniz} = 41$. The recombined Ar^{17+} ions are registered in the particle detector behind the next downstream dipole magnet, provided the spatial and time overlap of all the four beams are adjusted correctly. The recombination rate was measured as a function of the tunable wavelength of the second laser (Borneis et al., 1994).

With this two-step, laser-induced recombination technique, a precise Rydberg spectroscopy in an electron beam environment is now accessible (see also Schüssler et al., 1995a, b). In particular, using contramoving laser

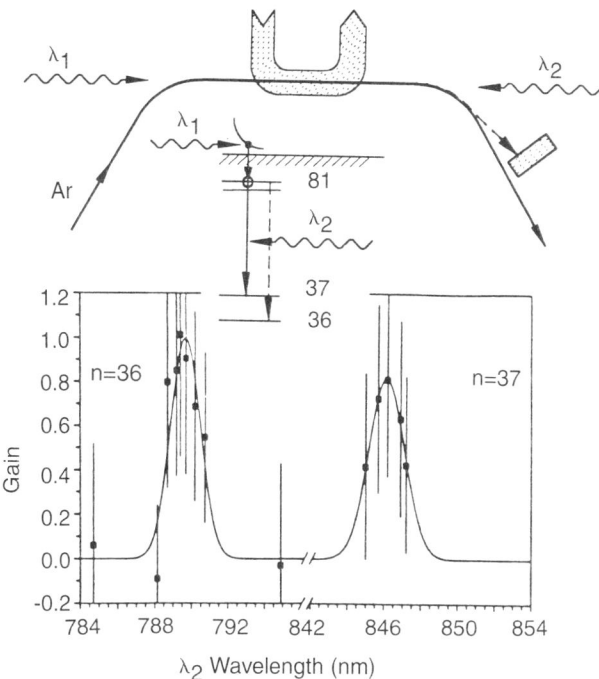

FIG. 37. Layout of the first two-step, laser-induced recombination experiment performed at the ESR with initially bare Ar^{18+} ions at 147 MeV/u. In the beam overlap region of the electron cooler, time-coincident co- and contramoving laser pulses are superimposed. Recombined Ar^{17+} ions are measured in the next downstream particle detector as a function of the time overlap and wavelength of the tunable contramoving laser pulses. For details, see text and Borneis et al. (1994).

beams, an almost Doppler-free spectroscopy of Rydberg states in fast beams might be possible. This is not only important for the physics involving Rydberg states, it gives a unique way to calibrate precisely ion velocities in storage rings. On the other hand, this technique may also be utilized to probe the critical structure of the continuum–bound state transition region in the cooler section.

D. HYPERFINE INTERACTIONS

Laser resonance fluorescence spectroscopy is a sensitive probe for atomic structure investigations. For high-Z ions, this method is well suited in particular for studying the hyperfine interaction where the angular momentum of the electron core J and the nuclear spin I couple to the total angular momentum F. In atomic hydrogen, this mechanism leads to the systems, around $Z = 90$, almost degenerate s and p levels in the L shell may mix, possibly leading to an atomic parity violation (Soff, 1991).

Beyond photon spectroscopy, dielectronic recombination gives equivalent access to fundamental questions of the structure of highly charged, very heavy atoms. In particular, $\Delta n = 1$, i.e., KLL, resonances will give detailed answers to QED contributions. With a cooler independent second electron target, these fundamental questions can be precisely studied in the near future. Here, too, the reaction dynamics governed by relativistic effects, particularly by magnetic interactions, will give novel insight into the important correlation contributions governing also the Auger effect at high-Z atoms.

Very subtle magnetic interactions lead to ground state hyperfine splitting. The QED coupling to the magnetic part of the virtual photon field can be probed uniquely here. Up to now, the uncertainties in the knowledge of the nuclear magnetization distribution (Bohr–Weisskopf effect) limited the accuracy for the QED part. In this case, as in general for all heavy ions, the nuclear part has to be known precisely in order to determine accurately the atomic structure effects.

A particular interplay between atomic and nuclear structure is the bound-state β decay, measured at the ESR already for two heavy species (Ho and Dy). These results will help fix the time scale for the age of the universe. On the other hand, by tuning the storage rings as high resolution mass spectrometers, the whole zoo of nuclear masses can be coupled precisely together via nuclear reactions, and their products stored in the rings. This may provide a new basis for understanding of the composition of our matter.

A crucial issue for heavy ion storage rings with the indispensible electron cooling devices is radiative recombination. On the one hand, colder and colder electron beams are provided, leading to high luminosity heavy ion beams with very low emittances (Danared et al., 1994). For instance, DR resonance structures can be measured with higher accuracies. On the other hand, the plasma physics in the cooler regime is of interest in itself. There, highly charged ions are in a temperature bath of extremely cold electrons. However, because of the recombination loss, the ionic Coulomb potential, and the magnetic guidance field, normal thermodynamics may get complicated. Moreover, RR seems to be an ideal tool to probe the energy and temperature distribution of the electrons around the highly charged ions.

Equivalent to RR, we have radiative electron capture, REC. This process tests the interaction of a quasi-free electron with the photon field of the Coulomb potential. In principle, the photoelectric effect has been studied at the ESR in the relativistic regime not accessible even at the third generation of synchrotron radiation facilities. Because of the high Z values of the ions and the relativistic velocities, higher-multiple transitions, in particular $M1$ transitions, contribute considerably to REC and, hence, to the photoelectric effect. Here, quite different emission patterns have been found. Normally forbidden 0° and 180° emission is allowed for K-REC. These effects have to be studied in more detail in the near future. Similar unrevealed relativistic effects can be expected for excitation and ionization processes, especially at high energies.

Nonradiative capture will be of particular interest at low projectile velocities. There, one- and multiple-electron capture into excited states dominates. A quite new aspect gives the multielectron capture populating predominately highly excited states: In the end, highly inverted systems may be created. Correlated decay may give information on the true emission volume (Hanbury-Brown and Twiss, 1954; see also Cindro et al., 1993). Not only are decelerated ions needed for the measurements, one may also benefit from extracting them to external experimental areas. First experimental tests with extracted low energy beams (e.g., 50-MeV/u U^{91+}) have already been performed at the ESR.

Extracted beams also give the possibility of studying classical atomic reactions at low collision energies—such as quasi-molecular collisions—but now with well-defined initial charge-state conditions. In particular, by $Pb^{81+} \rightarrow Pb(gas)$ collisions superheavy atoms can be probed, testing transiently the highest possible atomic fields. Interferences in the quasi-molecular radiation can measure the transient structure of these superheavy quasi-atoms (Schuch et al., 1988), near the diving into the negative continuum (Reinhardt and Greiner, 1977). Also, collision studies using the recoil momentum technique will be used at external and at in-ring experiments (Ullrich et al., 1994).

There is also the challenging field of forbidden transitions. Lifetimes can be studied for not too heavy ions in particular, revealing the crucial overlap of wavefunctions. For very heavy systems, for instance, the $2E1$ decay of the $1s2s\,^1S_0$ level in He-like ions reflects the complete structure of all the relevant bound and continuum states. All this information can be extracted from the continuum two-photon spectrum. These experiments will also benefit from decelerated extracted beams.

This brief summary of future possibilities of the physics with highly charged, very heavy ions is in no way complete. However, it demonstrates the unique power and the great potential of storage ring facilities in this challenging field of the fundamental physics of highly charged heavy ions. A flourishing future and a new understanding of the basics in this field are expected.

Acknowledgments

Both authors are deeply indebted to the support of their families during the writing of this chapter. It is a pleasure to devote this chapter to our wives. The fruitful collaboration with our colleagues at the rings in Darmstadt in particular, as well as in Aarhus, Heidelberg, and Stockholm, is highly acknowledged. We appreciate the data from their work appearing in the figures. One of the authors (PHM) benefited greatly from a stay at the University of Aarhus within a Humboldt professorship to Denmark.

References

Andersen, L. H., and Bolko, J. (1990). *Phys. Rev.* **A 42**, 1184.
Andersen, L. H., Hvelplund, P., Knudsen, H., and Kvistgaard, P. (1989). *Phys. Rev. Lett.* **62**, 2656.
Andersen, L. H., Bolko, J., and Kvistgaard, P. (1990). *Phys. Rev.* **A41**, 1293.
Andersen, L. H., Hvelplund, P., Lorents, D. C., Mathur, D., and Shen, H. (1995a). *In* "Biennal Report 1993/94," p. 49, Inst. Phys. Astron., University of Aarhus, Århus, Denmark.
Andersen, L. H., Mathur, D., Schmidt, H. T., and Vejby-Christensen, L. (1995b). *Phys. Rev. Lett.* **74**, 892.
Andersen, T., Andersen, L. H., Balling, P., Haugen, H. K., Hvelplund, P., Smith, W. W., and Taulbjerg, K. (1993). *Phys. Rev.* **A 47**, 890.
Anholt, R. (1979). *Phys. Rev.* **A 19**, 1004.
Bell, M., Chaney, J., Herr, H., Krienen, F., Möller-Petersen, P., and Petrucci, G. (1981). *Nucl. Instrum. Methods* **190**, 237.
Bethe, H. A., and Salpeter, E. E. (1957). *In* "Handbuch der Physik" (S. Flügge, ed.), Vol. 24, p. 1. Springer-Verlag, Berlin.
Beyer, H. F., Bollen, G., Bosch, F., Egelhof, P., Franzke, B., Hasse, R. W., Kluge, H. J., Kozhuharov, C., Kühl, T., Liesen, D., Mann, R., Mokler, P. H., Müller, A., Müller, R. W., Münzenberg, G., Poth, H., Schweikard, L., Schuch, R., and Werth, G. (1990). *Intern. Rep.* **GSI 90-20**.
Beyer, H. F., Finlayson, K. D., Liesen, D., Indelicato, P., Chantler, C. T., Deslattes, R. D., Schweppe, J., Bosch, F., Jung, M., Klepper, O., Koenig, W., Moshammer, R., Beckert, K., Eickhoff, H., Franzke, B., Gruber, A., Nolden, F., Spaedtke, P., and Steck, M. (1993). *J. Phys. B: At. Mol. Opt. Phys.* **26**, 1557.
Beyer, H. F., Liesen, D., Bosch, F., Finlayson, K. D., Jung, M., Klepper, O., Moshammer, R., Beckert, K., Eickhoff, H., Franzke, B., Nolden, F., Spaedkte, P., and Steck, M. (1994). *Phys. Lett. A* **184A**, 435.
Beyer, H. F., Menzel, G., Liesen, D., Gallus, A., Bosch, F., Deslattes, R., Indelicato, P., Stöhlker, Th., Klepper, O., Moshammer, R., Nolden, F., Eickhoff, H., Franzke, B., and Steck, M. (1995). *Z. Phys. D: At. Mol. Clusters* **35**, 169.
Bohr, A., and Weisskopf, V. F. (1950). *Phys. Rev.* **77**, 94.
Borneis, S., Bosch, F., Engel, T., Jung, M., Klaft, I., Klepper, O., Kühl, T., Marx, D., Moshammer, R., Neumann, R., Schröder, S., Seelig, P., and Völker, L. (1994). *Phys. Rev. Lett.* **72**, 209.

Bosch, F. (1987). *Nucl. Instrum. Methods Phys. Res., Sect. B* **23**, 190.
Bosch, F. (1989). *Nucl. Instrum. Methods Phys. Res., Sect. B* **52**, 945.
Bosch, F. (1993). *AIP Conf. Proc.* **295**, 3.
Bosch, F., Faestermann, Th., Friese, J., Heine, F., Kienle, P., Wefers, E., Zeitelhack, K., Beckert, K., Franzke, B., Klepper, O., Kozhuharov, C., Menzel, G., Moshammes, R., Nolden, F., Reich, H., Schlitt, B., Steck, M., Stöhlker, Th., Winkler, Th., and Takahashi, K. (1996). *Nature* (submitted).
Brinkman, H. C., and Kramers, H. A. (1930). *Proc. Acad. Sci.*, Amsterdam, Vol. 33, 973.
Burgess, A. (1964). *Astrophys. J.* **139**, 776.
Chen, M. H. (1990). *Phys. Rev. A* **41**, 4102.
Cindro. N., Karolija, M., and Shapira, D. (1993). *Prog. Part. Nucl. Phys.* **30**, 65.
Danared, H., Andler, G., Bagge, L., Herrlander, C. J., Hilke J., Jeansson, J., Källberg, A., Nilsson, A., Paal, A., Rensfelt, K.-G., Rosengard, U., Starker, J., and af Ugglas, M. (1994). *Phys. Rev. Lett.* **72**, 3775.
Drake, G. W. (1988). *Can. J. Phys.* **66**, 586.
Eichler, J. (1985). *Phys. Rev. A* **312**, 112.
Eichler, J. (1990). *Phys. Rep.* **193**, 165.
Eichler, J., Ichihara, A., and Shirai, T. (1995). *Phys. Rev. A* **51**, 3027.
Eickhoff, H., Beckert, K., Franczak, B., Franzke, B., Nolden, F., Poth, H., Schaaf, U., Schulte, H., Spädtke, P., and Steck, M. (1991). *In* "Cooler Rings and Their Applications" (T. Katayama and A. Noda, eds.), p. 11. World Scientific, Singapore.
Finkbeiner, M., Fricke, B., and Kühl, T. (1993). *Phys. Lett. A* **176A**, 113.
Franzke, B. (1987). *Nucl. Instrum. Methods Phys. Res., Sect. B* **24/25**, 18.
Franzke, B. (1988). *Phys. Scr.* **T22**, 41.
Graham, W. G., Berkner, K. H., Bernstein, E. M., Clark, M. W., Feinberg, B., McMahan, M. A., Morgan, T. J., Rathbun, W., Schlachter, A. S., and Tanis, J. A. (1990). *Phys. Rev. Lett.* **65**, 2773.
Grieser, R., Kühl, T., and Huber, G. (1995). *Am. J. Phys.* **63**, 665.
Gruber, A., Bourgeois, W., Franzke, B., Kritzer, A., and Treffert, C. (1989). *Nucl. Instrum. Methods Phys. Res., Sect. A* **282**, 87.
Hangst, J. S., Kristensen, M., Nielsen, J. S., Poulsen, O., Schiffer, J. P., and Shi, P. (1991). *Phys. Rev. Lett.* **67**, 1238.
Hanbury-Brown, R., and Twiss, R. Q. (1954). *Philos. Mag.* [7] **45**, 663.
Herlander, C. J. (1991). *In* "Cooler Rings and Their Applications" (T. Katayama and A. Noda, eds.), p. 15. World Scientific, Singapore.
Hino, K. I., and Watanabe, T. (1987). *Phys. Rev. A* **36**, 581.
Ichihara, A., Shirai, T., and Eichler, J. (1994). *Phys. Rev. A* **49**, 1875.
Indelicato, P. (1990). *AIP Conf. Proc.* **215**, 591.
Indelicato, P. (1995). *Phys. Rev. A* **51**, 1132.
Irnich, H., Geissel, H., Nolden, F., Beckert, K., Bosch, F., Eickhoff, H., Franzke, B., Fujita, Y., Hausmann, M., Jung, H. C., Kraus, G., Klepper, O., Kozhuharov, C., Magel, A., Münzenberg, G., Nickel, F., Radon, T., Reich, H., Schlitt, B., Schwab, W., Sümmerer, K., Suzuki, K., Steck, M., and Wollnik, H. (1995). *Phys. Rev. Lett.* **75**, 4182.
Jaeschke, E., Bisoffi, G., Blum, M., Friedrich, A., Geyer, C., Grieser, M., Holzer, B., Heyng, H. W., Habs, D., Jung, M., Krämer, D., Noda, A., Ott, W., Pollok, R. E., Repnow, R., Schmitt, F., and Steck, M. (1990). *Part. Accel.* **32**, 97.
Johnson, W. R., and Soff, G. (1985). *At. Data Nucl. Data Tables* **33**, 405.
Jung, M., Bosch, F., Beckert, K., Eickhoff, H., Folger, H., Franzke, B., Kienle, P., Klepper, O., Koenig, W., Kozhuharov, C., Mann, R., Moshammer, R., Nolden, F., Schaaf, U., Soff, G., Spädtke, P., Steck, M., Stöhlker, Th., and Sümmerer, K. (1992). *Phys. Rev. Lett.* **69**, 2164.

Kandler, T., Mokler, P. H. Stöhlker, Th., Geissel, H., Irnich, H., Kozhuharov, C., Kriessbach, A., Kucharski, M., Münzenberg, G., Nickel, F., Rymuza, P., Scheidenberger, C., Stachura, Z., Suzuki, T., Warczak, A., Dauvergene, D., and Dunford, R. W. (1995a). *Phys. Lett.* **204**, 274.

Kandler, T., Mokler, P. H., Geissel, H., Irnich, H., Kozhuharov, C., Kriessbach, A., Kucharski, M., Münzenberg, G., Nickel, F., Rymuza, P., Scheidenberger, C., Stachura, Z., Stöhlker, Th., Suzuki, T., Warczak, A., Dauvergne, D., and Dunford, R. W. (1995b) *Nucl. Instrum. Methods Phys. Res., Sect. B* **98**, 320.

Kandler, T., Mokler, P. H., and Stöhlker, Th. (1996). *Nucl. Instrum. Methods Phys. Res., Sect. B* **107**, 357.

Kienle, P. (1985). *Intern. Rep.* **GSI 85-16**.

Kilgus, G., Berger, J., Grieser, M., Habs, D., Hochadel, B., Jaeschke, E., Krämer, D., Neumann, R., Neureither, G., Ott, W., Schwalm, D., Steck, M., Stokstad, R., Szmola, E., Wolf, A., Schuch, R., Müller, A., and Wagner, M. (1990). *Phys. Rev. Lett.* **64**, 737.

Kilgus, G., Habs, D., Schwalm, D., Wolf, A., Badnell, N. R., and Müller, A. (1992). *Phys. Rev. A* **46**, 5730.

Kilgus, G., Habs, D., Schwalm, D., Wolf, A., Schuch, R., and Badnell, N. R. (1993). *Phys. Rev. A* **47**, 4859.

Klaft, I., Borneis, S., Engel, T., Fricke, B., Grieser, R., Huber, G., Kühl, T., Marx, D., Neumann, R., Schröder, S., Seelig, P., and Völker, L. (1994). *Phys. Rev. Lett.* **73**, 2425.

Kleber, M., and Jakubassa, D. H. (1974). *Nucl. Phys. A* **A252**, 152.

Klepper, O., Bosch, F., Daues, H. W., Eickhoff, H., Franczak, B., Franzke, B., Geissel, H., Gustafsson, O., Jung, M., Koenig, W., Kozhuharov, C., Magel, A., Münzenberg, G., Stelzer, H., Szerypo, J., and Wagner, M. (1992). *Nucl. Instrum. Methods Phys. Res., Sect. B* **70**, 427.

Kluge, H.-J. (1995). *Nucl. Instrum. Methods Phys. Res., Sect. B* **98**, 500.

Knapp, D. A., Marrs, R. E., Elliott, S. R., Magee, E. W., and Zasadzinski, R. (1993). *Nucl. Instrum. Methods Phys. Res., Sect. A* **334**, 305.

Kühl, T., Beckert, S., Borneis, S., Engel, T., Fricke, B., Grieser, M., Grieser, R., Habs, D., Huber, G., Klaft, I., Marx, D., Merz, P., Neumann, R., Schwalm, D., and Seelig, P. (1994). *At. Phys.* **14**, 30. AIP Conf. Proc. 323, New York.

Labzowsky, L. N., Klimchitskaya, G. L., and Dmitriev, Y. Y. (1993). "Relativistic Effects in the Spectra of Atomic Systems." IOP Publ., London.

Lamb, W. E., and Retherford, R. C. (1947). *Phys. Rev.* **72**, 241.

Larsson, M. (1996). *Rep. Prog. Phys.* (in press).

Levine, M. A., Marrs, R. E., Bardsley, J. N., Beiersdorfer, P., Bennett, C. L., Chen, M. H., Cowan, T., Dietrich, D., Henderson, J. R., Knapp, D. A., Osterheld, A., Penetrante, B. M., Schneider, D., and Scofield, J. H. (1989) *Nucl. Instrum. Methods Phys. Res., Sect. B* **43**, 431.

Liesen, D., Beyer, H. F., Finlayson, K. D., Bosch, F., Jung, M., Klepper, O., Moshammer, R., Beckert, K., Eickhoff, H., Franzke, B., Nolden, F., Spädtke, P., Steck, M., Menzel, G., and Deslattes, R. D. (1994). *Z. Phys. D: At. Mol. Clusters* **30**, 307.

Lindgren, I., Persson, H., Salomonson, S., and Labzowsky, L. N. (1995a). *Phys. Rev. A* **51**, 1167.

Lindgren, I., Persson, H., Salomonson, S., and Sunnergren, P. (1995b). *Phys. Sc.* **T59**, 179.

Lindroth, E. (1995). *Nucl. Instrum. Methods Phys. Res., Sect. B* **98**, 1.

Lotz, W. (1968). *Z. Phys.* **216**, 241.

Lupton, J. H., Dietrich, D. D., Hailey, C. J., Stewart, R. E., and Ziok, K. P. (1994). *Phys. Rev. A* **50**, 2150.

Madison, D. H., and Merzbacher, E. (1975). In "Atomic Inner-Shell Processes" (B. Crasemann, ed.), p. 1. Academic Press, London.
Marrs, R. E., Elliott, S. R., and Knapp, D. A. (1994a). Phys. Rev. Lett. **72**, 4082.
Marrs, R. E., Beiersdorfer, P., and Schneider, D. (1994b). Phys. Today **47**, 27.
Marrs, R. E., Elliott, S. E., and Stöhlker, Th. (1995). Phys. Rev. **A52**, 3577.
Marrus, R., and Mohr, P. (1978). Adv. At. Mol. Phys. **14**, 181.
Marrus, R., Siminiovici, A., Indelicato, P., Dietrich, D., Charles, P., Briand, J. P., Finlayson, K., Bosch, F., Liesen, D., and Parente, F. (1989). Phys. Rev. Lett. **63**, 502.
Marrus, R., Birkett, B., Simionovici, A., Indelicato, P., Beyer, H. F., Bosch, F., Gallus, A., Liesen, D., and Menzel, G. (1995). GSI Sci. Rep. 1994 **GSI 95-1**, 136.
Meyerhof, W. E., Anholt, R., Eichler, J., Gould, H., Munger, C., Alonso, J., Thieberger, P., and Wegner, H. E. (1985a). Phys. Rev. A **32**, 3291.
Meyerhof, W. E., Anholt, R., Eichler, J., Gould, H., Munger, C., Alonso, J., Thieberger, P., and Wegner, H. E. (1985b). Phys. Rev. A **49**, 1975.
Möhl, D. (1988). Phys. Sc. **T22**, 21.
Mohr, P. J. (1994). Nucl. Instrum. Methods Phys. Res., Sect. B **87**, 232.
Mokler, P. H., and Liesen, D. (1978). In "Progress-in Atomic Spectroscopy, Part C" (H. J. Beyer and H. Kleinpoppen, eds.), p. 321. Plenum, New York.
Mokler, P. H., and Reusch, S. (1988). Z. Phys. D: At. Mol. Clusters **8**, 393.
Mokler, P. H., Reusch, S., and Stöhlker, Th. (1989). Nucl. Instrum. Methods Phys. Res., Sect. A **278**, 93.
Mokler, P. H., Stöhlker, Th., Kozhuharov, C., Stachura, Z., and Warczak, A. (1991). Z. Phys. D: At. Mol. Clusters **21**, 197.
Mokler, P. H., Stöhlker, Th., Kozhuharov, C., Moshammer, R., Rymuza, P., Bosch, F., and Kandler, T. (1994). Phys. Sc. **T51**, 28.
Mokler, P. H., Stöhlker, Th., Kozuharov, C., Moshammer, R., Rymuza, P., Stachura, Z., and Warczak, A. (1995a). J. Phys. B: At. Mol. Opt. Phys. B **28**, 617.
Mokler, P. H., Stöhlker, Th., Dunford, R. W., Gallus, A., Kandler, T., Menzel, G., Prinz, H. T., Rymuza, P., Stachura, Z., Swiat, P., and Warczak, A. (1995b). Z. Phys. D: At. Mol. Clusters **35**, 97.
Möller, S. P. (1990). In "Proceedings of the First European Particle Accelerators Conference" (S. Tazzari, ed.), p. 328. World Scientific, Singapore.
Möller, S. P. (1994). In "Proceedings of the Fourth European Particle Accelerators Conference" (V. Suller and C. Petit-Jean-Genaz, eds.), p. 173, World Scientific, Singapore.
Müller, A. (1992). NATO ASI B, **296**, 155.
Müller, A. (1994). Nucl. Instrum. Methods Phys. Res., Sect. B **87**, 34.
Oppenheimer, J. R. (1928). Phys. Rev. **31**, 349.
Persson, H., Salomonson, S., Sunnergren, P., and Lindgren, I. (1996). Phys. Rev. Lett. **76**, 204.
Persson, H., Lindgren, I., Labzowsky, L. N., Plunien, G., Beier, T., and Soff, G. (1996). Phys. Rev. A (in press).
Pindzola, M. S., and Badnell, N. R. (1992). Communication in Spies et al. (1992).
Pollock, R. E. (1991). Annu. Rev. Part. Sci. **41**, 357.
Poth, H. (1990a). Phys. Rep. **196**, 135.
Poth, H. (1990b). Nature (London) **345**, 399.
Reinhardt, J., and Greiner, W. (1977). Rep. Prog. Phys. **40**, 219.
Rice, R., Basbas, G., and McDaniel, D. (1977). At. Data Nucl. Data Tables **20**, 503.
Rösel, F., Trautmann, D., and Baur, G. (1982). Nucl. Instrum. Methods Phys. Res., Sect. B **192**, 43.

Rymuza, P., Stöhlker, Th., Cocke, C. L., Geissel, H., Kozhuharov, C., Mokler, P. H., Moshammer, R., Nickel, F., Scheidenberger, C., Stachura, Z., Ullrich, J., and Warczak, A. (1993). *J. Phys. B: At. Mol. Opt. Phys.* **26**, 169.
Schennach, S., Müller, A., Wagner, M., Haselbauer, J., Uwira, O., Spies, W., Jennewein, E., Becker, R., Kleinod, M., Pröbstel, U., Angert, N., Klabunde, J., Mokler, P. H., Spädtke, P., and Wolf, B. (1991). *Z. Phys. D: At. Mol. Clusters* **21**, S205.
Schennach, S., Müller, A., Uwira, O., Haselbauer, J., Spies, W., Frank, A., Wagner, M., Becker, R., Kleinod, M., Jennewein, E., Angert, N., Mokler, P. H., Badnell, N. R., and Pindzola, M. S. (1994). *Z. Phys. D: At. Mol. Clusters* **30**, 291.
Schmidt, H. T., Forck, P., Grieser, M., Habs, D., Kenntner, J., Miersch, G., Repnow, R., Schramm, U., Schüssler, T., Schwalm, D., and Wolf, A. (1994). *Phys. Rev. Lett.* **72**, 1616.
Schmieder, R. W., and Marrus, R. (1973). *Nucl. Instrum. Methods* **110**, 459.
Schneider, D., Church, D., Weinberg, G., McDonald, J., Steiger, J., Beck, B., and Knapp, D. (1994). *EBIT Annu. Rep. LLNL. 1993* **UCRL-ID-118274**, 72.
Schneider, S. M., Schaffner, J., Soff, G., and Greiner, W. (1993). *J. Phys. B: At. Mol. Opt. Phys.* **26**, L581.
Schröder, S., Klein, R., Boos, N., Gerhard, M., Grieser, R., Huber, G., Karafillidis, A., Krieg, M., Schmidt, N., Kühl, T., Neumann, R., Balykin, V., Grieser, M., Habs, D., Jaeschke, E., Krämer, D., Kristensen, M., Music, M., Petrich, W., Schwalm, D., Sigray, P., Steck, M., Wanner, B., and Wolf, A. (1990). *Phys. Rev. Lett.* **64**, 2901.
Schuch, R., (1993). *In* "Review of Fundamental Processes and Applications of Atoms and Ions" (C. D. Lin, ed.), p. 169. World Scientific, Singapore.
Schuch, R., Meron, M., Johnson, B. M., Jones, K. W., Hoffmann, R., Schmidt-Böcking, H., and Tserruya, I. (1988). *Phys. Rev. A* **37**, 3313.
Schüssler, T., Schramm, U., Rüter, T., Broude, C., Grieser, M., Habs, D., Schwalm, D., and Wolf, A. (1995a). *Phys. Rev. Lett.* **75**, 802.
Schüssler, T., Schramm, U., Grieser, M., Habs, D., Rüter, T., Schwalm, D., and Wolf, A. (1995b). *Nucl. Instrum. Methods Phys. Res., Sect. B* **98**, 146.
Schweppe, J., Belkacem, A., Blumenfeld, L., Claytor, N., Feinberg, B., Gould, H., Kostroun, V. E., Levy, L., Misawa, S., Mowat, J. R., and Prior, M. H. (1991). *Phys. Rev. Lett.* **66**, 1434.
Soff, G. (1991). *Z. Phys. D: At. Mol. Clusters* **21**, S7.
Spädtke, P., Angert, N., Beckert, K., Bourgeois, W., Emig, H., Franzke, B., Langenbeck, B., Leible, K. D., Nolden, F., Odenweller, T., Poth, H., Schaaf, U., Schulte, H., Steck, M., and Wolf, B. H. (1991). *In* "Cooler Rings and Their Applications" (T. Katayama and A. Noda, eds.), p. 75. World Scientific, Singapore.
Spies, W., Müller, A., Linkemann, J., Frank, A., Wagner, M., Kozhuharov, C., Franzke, B., Beckert, K., Bosch, F., Eickhoff, H., Jung, M., Klepper, O., Koenig, W., Mokler, P. H., Moshammer, R., Nolden, F., Schaaf, U., Spädtke, P., Steck, M., Zimmerer, P., Grün, N., Scheid, W., Pindzola, M. S., and Badnell, N. R. (1992). *Phys. Rev. Lett.* **69**, 2768.
Spies, W., Uwira, O., Müller, A., Linkemann, J., Empacher, L., Frank, A., Kozhuharov, C., Mokler, P. H., Bosch, F., Klepper, O., Franzke, B., and Steck, M. (1995). *Nucl. Instrum. Methods* **98**, 158.
Stevefelt, J., Boulmer, J., and Delpech, J. F. (1975). *Phys. Rev. A* **12**, 1246.
Stobbe, M. (1930). *Ann. Phys.* (*Leipzig*) **7**, 661.
Stöhlker, Th. (1993). *Phys. Electron. At. Collisions, Abstr. Contrib. Pap. Int. Conf., 18th, 1993*, p. 615.
Stöhlker, Th. (1995a). *Intern. Rep.: Nachr.* **GSI 05-95**, 4.
Stöhlker, Th. (1995b). *Phys. Electron. At. Collisions*, AIP Conf. Proc. **360**, 525.

Stöhlker, Th., Geissel, H., Folger, H., Kozhuharov, C., Mokler, P. H., Münzenberg, G., Schardt, D., Schwab, Th., Steiner, M., Stelzer, H., and Sümmerer, K. (1991). *Nucl. Instrum. Methods Phys. Res.*, *Sect. B* **61**, 408.

Stöhlker, Th., Mokler, P. H., Geissel, H., Moshammer, R., Rymuza, P., Bernstein, E. M., Cocke, C. L., Kozhuharov, C., Münzenberg, G., Nickel, F., Scheidenberger, C., Stachura, Z., Ullrich, J., and Warczak, A. (1992). *Phys. Lett. A* **168A**, 285.

Stöhlker, Th., Mokler, P. H., Beckert, K., Bosch, F., Eickhoff, H., Franzke, B., Jung, M., Kandler, T., Klepper, O., Kozhuharov, C., Moshammer, R., Nolden, F., Reich, H., Rymuza, P., Spädtke, P., and Steck, M. (1993). *Phys. Rev. Lett.* **71**, 2184.

Stöhlker, Th., Geissel, H., Irnich, H., Kandler, T., Kozhuharov, C., Mokler, P. H., Münzenberg, G., Nickel, F., Scheidenberger, C., Suzuki, T., Kucharski, M., Warczak, A., Rymuza, P., Stachura, Z., Kriessbach, A., Dauvergene, D., Dunford, B., Eichler, J., Ichihara, A., and Shirai, T. (1994). *Phys. Rev. Lett.* **73**, 3520.

Stöhlker, Th., Kozhuharov, C., Mokler, P. H., Warczak, A., Bosch, F., Geissel, H., Eichler, J., Ichihara, A., Shirai, T., Stachura, Z., and Rymuza, P. (1995). *Phys. Rev. A* **51**, 2098.

Thompson, R. C. (1990). *Meas. Sci. Technol.* **1**, 93.

Ullrich, J., Dörner, R., Mergel, V., Jagutzki, O., Spielberger, L., and Schmidt-Böcking, H. (1994). *Comments At. Mol. Phys.* **30**, 285.

Uwira, O., Müller, A., Spies, W., Linkemann, J., Frank, A., Cramer, T., Empacher, L., Becker, R., Kleinod, M., Mokler, P. H., Kenntner, J., Wolf, A., Schramm, U., Schüssler, T., Schwalm, D., and Habs, D. (1995). *Hyperfine Interact.* **99**, 295.

Van der Meer, S. (1972). *Intern. Rep.* **CERN-ISR-PO/72-31**.

Weitz, M., Schmidt-Kahr, F., and Hänsch, T. W. (1992). *Phys. Rev. Lett.* **68**, 1120.

Wolf, A. (1988). *Phys. Scri.* **T22**, 55.

Wolf, A. (1992). *NATO ASI Ser.*, *Ser. B* **296**, 209.

Wolf, A., Habs, D., Lampert, A., Neumann, R., Schramm, U., Schüssler, T., and Schwalm, D. (1993). *At. Phys.* **13**, 228.

Zimmerer, P. (1992). Ph.D. Thesis, University of Giessen.

Zimmerer, P., Grün, N., and Scheid, W. (1990). *Phys. Lett. A* **148**, 457.

Zimmerer, P., Grün, N., and Scheid, W. (1991). *J. Phys. B: At. Mol. Opt. Phys.* **24**, 2633.

Zimmermann, M., Grün, N., and Scheid, W. (1994). *GSI Sci. Rep. 1993* **GSI-1994-1**, 167.

Index

A

ac magnetic trap, 216
Adiabatic compression, evaporative cooling, 213, 218
Adiabatic cooling, one-dimensional, 226–227
Adiabatic expansion, laser cooling in optical lattices, 128–130
Adiabaticity parameter, 341
Adiabatic passage
 Fock states, 258–260
 Schrödinger Cat states, 283
Alkali atoms
 Bose–Einstein condensation, 231
 evaporative cooling, 187, 192–193, 195, 204–206, 211, 217, 225–226, 228–230
  superconducting magnets, 213
 fermions, 208–209
 Feshbach resonance, 207
 laser cooling, 209
Anyons, 57
Atomic structure, highly charged heavy ions, 348–353
Atomic trampoline, 4
Atom laser, 23
Atom mirrors, 2–4, 8–10, 10–23
 evanescent-wave atom mirrors, 13–16, 18–21
 other electromagnetic mirrors, 21–23
 specular atom mirrors, 3–4
Atom optics
 atom mirrors, 2–4, 8–10, 10–23
 atom resonators, 5, 6, 23–39
 traps, *see* Atom traps
 waveguides, *see* Waveguides
Atom resonators, 6, 23
 cavities with two mirrors, 5, 23–29
 Fabry–Pérot type, 23–39
 gravitational cavity with parabolic mirror, 29–36
 red–blue pushme–pullyou resonator, 5, 36–39
Atom traps, 7–8
 ac magnetic trap, 216
 baseball trap, 212

blue-detuned concave atom traps, 59–60, 89
 conical gravitational trap, 68–72
 evanescent-wave cooling, 72–75
 particle-in-a-box with gravity, 6, 60–65
 pyramidal gravitational trap, 65–68
conservative traps, 212–217, 217–227
crossed dipole trap, 229
dark optical traps, 134, 182
dark SPOT trap, 211, 228–229
dipole trap, 201, 215–216
 history, 182
Ioffe–Pritchard trap, 198, 212–213
magnetic traps, 201, 209, 212–214, 217, 228
magneto-optical traps, 58, 211
microwave trap, 216
optical dipole traps, 201, 215–216, 217
optical lattices, 134–135
optically plugged dipole trap, 216, 226
Paul trap, 255
red-detuned convex atom traps, 75
 external solid fiber guide, 81–85
 inverted cone Yukawa-potential trap, 86–88
 microsphere whispering gallery trap, 76–81
time-averaged orbiting potential (TOP) trap, 214, 225, 226
VSCPT, 211–212, 226
yin–yang trap, 212
Atom waveguides, *see* Waveguides
Axial channeling, crystalline lattices, 146–152

B

Balmer transitions, highly charged heavy ions, 362
Baseball trap, 212
β decay, bound-state, 298, 360–362, 363
Bichromatic excitation, squeezed states, 275–278
Bloch oscillations, Stark ladder, 135–136

INDEX

Blue-detuned concave atom traps, 56–75, 89
Blue-detuned dipole trap, 215
Blue-detuned hollow-fiber wave guides, 45–55
Bohr–Weisskopf effect, 307, 360
Bose condensation, in atom traps, 206
Bose–Einstein condensation (BEC)
  alkali atoms, 231
  evaporative cooling, 182, 194, 195, 196, 200, 206–207, 231
Boser, 23
Bosons, evaporative cooling, 208
Bound-state β decay, 298, 360–362, 464
Bragg scattering, laser cooling in optical lattices, 130–133

C

Cavity quantum electrodynamics (CQED), 238–244, 262–264
  trapping states, 267–273
Centrifugal trap, red-detuned, 79–80
Cesium, evaporative cooling, 216
Charge changing collisions, crystalline lattices, 152–153
Collapses and revivals, 243, 283–290
Conical gravitational trap, 68–72
Conservative atom traps, 212–217, 217–227
Cooling
  cryogenic cooling, 209–210
  cyclic cooling, 217–218, 226
  evanescent-wave cooling, 72–75
  evaporative, see Evaporative cooling
  heavy ion storage/cooler rings, see Storage/cooler rings
  laser cooling, see Laser cooling
  one-dimensional adiabatic cooling, 226–227
  Raman cooling, 217–218, 226
  Raman sideband cooling, 257–258
  sideband cooling, 242, 252–258
  Sisyphus cooling, 99, 217
  sub-recoil cooling, 182, 211–212, 217, 231
Crossed dipole trap, 229
Cryogenic cooling, evaporative precooling, 209–210
Crystalline lattices
  channeling heavy ions, 139–144
    axial channeling, 146–152

    charge changing collisions, 152–153
    dielectronic excitation and recombination, 161–166
    electron impact ionization, 158–161
    hyperchanneling, 146–147
    interaction potentials, 145
    planar channeling, 145–146
    radiative electron capture, 153–158
    resonant coherent excitation, 166–167
    resonant transfer and excitation, 162–166
    trajectories, 144–145
Crystallography, optical lattices, 104–109
Cyclic cooling, 217–218, 226

D

Dark optical traps, 134, 182
Dark SPOT trap, 211, 228–229
Dielectric-waveguide-enhanced evanescent wave, 19–21
Dielectronic excitation, heavy ion channeling, 161–166
Dielectronic recombination, heavy ions, 161–166, 310, 312, 334, 353, 362–363
Dimension of evaporation, 197–199
Dipolar relaxation, 202, 205, 216
Dipole–dipole interactions, optical lattices, 134
Dipole trap, 201, 215–216
Doppler cooling, magnetic trap, 217, 228, 231
Doppler-shift-assisted ion spectroscopy, 349, 359
Doubly excited states, 353–356
Dysprosium, bound β decay, 360

E

Effective potential, atom mirrors, 16–18
Ehrenfest's theorem, 10
Elastic collision
  cross section, 199, 207–208
  evaporative cooling, 199, 201–206
Electron beam ion sources/traps (EBISs/EBITs), 308, 317
Electron capture
  highly charged ions, 308–313
  nonradiative capture, 310, 313, 363

radiative electron capture, 153–158, 308, 309–311, 340–344, 363
storage/cooler rings, 339–364
Electron impact ionization, channeling techniques, 158–161
Evanescent light-wave trapping
  atom mirrors, 2–4, 8–10, 10–23
  atom resonators, 6, 23
    cavities with two mirrors, 5, 23–29
    Fabry–Pérot type, 23–29
    gravitational cavity with parabolic mirror, 29–36
    red–blue pushme–pullyou resonator, 29–36
  evanescent-wave mirrors, 13–16, 18–21
  specular atom mirror, 3–4
  traps, 5, 7–8, 59–60
    conical gravitational trap, 68–72
    evanescent-wave cooling, 72–75
    external solid fiber guide, 81–85
    inverted cone Yukawa-potential trap, 86–88
    microsphere whispering-gallery trap, 76–81
    particle-in-a-box with gravity, 6, 60–65
    pyramidal gravitational trap, 65–68
  waveguides, 7
    hollow fiber guides, 39–55
Evanescent-wave atom mirrors, 8–10, 13–16
  dielectric-waveguide-enhanced, 19–21
  specular reflection, 13–15
  surface-plasmon-enhanced, 18–19
Evanescent-wave cooling, in gravitational traps, 72–75
Evanescent-wave hollow fiber guide, 45–55
Evaporation
  collisions, 183
    for Bose–Einstein condensation, 194, 195, 196, 200, 206–207
    elastic collision cross-section, 100, 207–208
    elastic and inelastic, 199, 201–206
    enhanced relaxation, 206–207
    for thermalization, 183, 199–200
  dimension of evaporation, 197–199
  models, 193–197, 200–201
  radiative evaporation, 220–226
  rf evaporation, 194–195, 197, 221–226
  runaway evaporation, 187–189, 192, 195
  theory, 183

  general scaling laws, 184–186, 196
  phase-space density, 189–192
  rate, 186–193
Evaporative cooling, 181–184, 231
  adiabatic compression, 213, 218
  alkali atoms, 187, 192–193, 195, 204–206, 211, 213, 217, 225–226, 228–230
  Bose–Einstein condensation, 182, 183, 194, 195, 196, 200, 206–207
  cesium, 216
  collisions, 183, 199–209
  dimension of evaporation, 197–199
  experimental techniques, 209, 218–227
    ac magnetic traps, 216–217
    adiabatic compression, 218
    conservative traps, 212–217, 217–227
    cryogenic cooling, 109–210
    laser cooling, 210–212, 217–227
    magnetic traps, 201, 209, 212–214
    microwave trap, 216
    optical dipole traps, 201, 215–216
    precooling, 199–212
  experiments, 183, 227–230
  fermions, 208–209
  hydrogen, 184, 192–193, 194, 195, 196, 204–206, 227–230
  lithium, 230
  models, 193–197, 200–201
  radiative evaporation, 220–226
  rf evaporation, 194–195, 197, 221–226
  rubidium, 230
  runaway evaporation, 187–189, 192, 195
  sodium, 204–206, 207, 217, 230
  theory, 183–192, 196
  thermalization, 199–200

F

Fabry–Pérot atom mirror, 23–29
Fast beam precision spectroscopy, 348–349
Fermions, evaporative cooling, 208–209
Feshbach resonances, 207
Fluorescence spectroscopy, optical lattices, 100–104, 120–122
Flux speaking, 147–150
Fock states, 243, 258–263, 258–273, 287, 288, 292

G

Gravitational traps, 29–36, 72–75
Gray optical lattice, 134
Ground state Lamb shift, 348–353
Grynberg-style lattices, 109–109, 119

H

Hänsch-style lattices, 107, 119
Harmonic oscillator, squeezed states, 274
Heavy ions
    channeling in crystalling lattices, 139–176
    highly charged, 297–365
        atomic structure studies, 300–308, 348–362
        charge changing processes, 332–347
        doubly excited states, 353–356
        future development, 362–364
        ground state Lamb shift, 348–353
        hyperfine interactions, 358–360
        physics, 297–316
        production, 316–319
        Rydberg states, 298, 356–358
        storage, 326–328
        storage/cooler rings, 299, 319–331
Helium
    cryogenic precooling, 210
    evaporative cooling, 204, 206
HITRAP project, 319
Hollow fiber waveguides
    blue-detuned, 45–55
    red-detuned, 39–44
Hydrogen
    Bose–Einstein condensation, 227–228
    cryogenic precooling, 210
    evaporative cooling, 184, 192–193, 194, 195, 196, 204–206, 216, 227–230
    magnetic trapping, 228
Hyperchanneling, 146–147
Hyperfine splitting, 359

I

Inverted cone Yukawa potential trap, 86–88
Ioffe–Pritchard trap, 198, 212–213
Ions
    highly charged heavy ions
        atomic structure, 300–308, 348–362

        physics, 297–316
        production, 316–319
        storage, 326–328
        storage/cooler rings, 299, 319–364
    trapped, nonclassical states of motion, 238–244, 292
Isoelectronic sequences, highly charged heavy ions, 300–308

J

Jaynes–Cumming model, 239, 262
    collapses and revivals, 243, 283–290

K

Kosterlitz–Thouless transition, 57
K-shell ionization, plane wave Born approximation, 313–314, 342–344, 346–348

L

Lamb–Dicke effect, 103
Lamb–Dicke limit (LDL), trapped ions, 241–242, 244–252, 253–255, 277
Lamb–Dicke parameter, 242, 244–245
Lamb shift, 301, 303, 348–353
Laser cooling, 96, 109, 109–119
    by adiabatic expansion, 128–130
    alkali atoms, 209
    band structure formalism, 115–116
    Bragg scattering, 130–133
    evanescent-wave cooling, 72–75
    evaporative precooling, 210–211
    formalism, 109–112
    Monte Carlo wavefunction technique, 113–115, 196
    recoil limit, 182
    secular approximation, 116–119
    semiclassical method, 99, 109–112
    trapped ions
        Lamb–Dicke limit, 253–255
        nonclassical states of motion, 238–292
Laser-induced radiative recombination, 356
Laser resonance fluorescence spectroscopy, 358

INDEX 375

Lasers
   evanescent light-wave devices
      atom mirrors, 2-4, 8-10, 10-23
      atom resonators, 5, 6, 23-29
      traps, 7-8, 59-88
      waveguides, 7, 39-58
   heavy ion production, 315-316
   optical lattices, 95-97, 133-136
      crystallography, 104-109
      laser cooling, 96, 109-119, 128-130, 130-133
      1D lin $\perp$ lin model system, 97-104
      spectroscopy, 119-128
Lattices, *see* Crystalline lattices; Optical lattices
Lithium, evaporative cooling, 230
Low excitation regime, 242, 248

M

Magnetic traps, 201, 209, 212-214, 217, 228
Magneto-optical traps, 58, 211
Magnets, evaporative cooling, 213
Master equation
   atomic motion, 109-116
   nonclassical states of motion, 250-252, 260, 262
Microsphere whispering gallery trap, 76-81, 89
Microwave trap, 216
Monte Carlo trajectory technique, evaporative cooling, 196, 199
Monte Carlo wavefunction simulation, laser cooling, 113-114, 196

N

Nonclassical states
   collapses and revials, 243, 283-290
   Fock states, 243, 258
      adiabatic passage, 258-260, 266
      quantum jumps, 260, 262-267, 290
      trapping states, 267-273
   ion traps, models, 242, 244-252
   pure and mixed states, 243-244, 290-292
   Schrödinger Cat states, 243-290, 278-283
   sideband cooliing, 242, 252-258
   squeezed states, 238-239, 243, 274-278

Nonradiative capture, heavy ions, 310, 313, 363

O

1D lin $\perp$ lin model system, 97, 100-104, 123
One-dimensional adiabatic cooling, 226-227
Optical dipole traps, 201, 215-216, 217
Optical heterodyne spectroscopy, optical lattices, 100-104, 121-122
Optical lattices, 95-97, 133-136
   crystallography, 104-105
      three-dimensional lattices, 107-109
      two-dimensional lattices, 105-107
   laser cooling, 109
      by adiabatic expansion, 128-130, 212, 213
      band structure formalism, 115-116
      Bragg scattering, 130-133
      formalism, 109-112
      Monte Carlo wavefunction technique, 113-115, 196
      secular approximation, 116-119
      semiclassical method, 99, 109-112
   1D lin $\perp$ lin model system, 97
      atomic motion, 97
      spectroscopy, 100-104, 123
Optically plugged dipole trap, 216, 226
Optical molasses, 96
Optical plug, 214
Optical shielding, 209

P

Parallel-mirror waveguides, 7, 55-58
Particle-in-a-box analysis, atom traps, 6, 60-65
Paul trap, 255
Permanent magnets, evaporative cooling, 213
Phase conjugation spectroscopy, optical lattices, 119
Phase-space density, evaporative cooling, 189-192, 200
Planar channeling, crystalline lattices, 145-146
Plane wave Born approximation (PWBA), K-shell ionization, 313-314, 342-344, 346-348

Population inversion, collapses and revivials, 284–289, 292
Probe transmission spectroscopy, optical lattices, 100–104, 122–123
Projectile ionization, 313, 342
Pushme–pullyou atom resonator, 5, 36–39
Pushme–pullyou trap, 80–81, 82
Pyramidal gravitational trap, 65–68

Q

Quantum jumps, Fock states, 260, 262–267, 290
Quantum logic gate, 292
Quantum mechanics
  coherent superposition states, 290
  evaporative cooling, 197
Quantum optics
  collapse and revival phenomenon, 243, 283–290
  nonclassical states of motion, 238–292

R

Radiative electron capture
  channeling techniques, 153–158
  heavy ions, 308, 309–311, 340–344, 363
Radiative evaporation, 220–226
Radiative recombination, heavy ions, 39–311, 308, 327, 332–333, 356
Raman cooling, 217–218, 226
Raman sideband cooling, 257–258
Raman sidebands, optical lattices, 124–126
Rate of evaporation, 186–193
Rayleigh scattering, optical lattices, 104, 127–128
Recoil-induced resonance, optical lattices, 127–128
Recombination process
  in cooler rings, 332–338
  dielectronic recombination, 161–166, 310, 312, 334, 353, 362–363
  laser-induced radiative recombination, 356
  radiative recombination, 308, 309–311, 327, 332–333, 356
  three-body recombination, 312–313, 336
Red–blue pushme–pullyou, solid fiber atom guide, 84–85

Red–blue pushme–pullyou resonator, 5, 36–39
Red–blue pushme–pullyou trap, 80–81, 82
Red-detuned centrifugal, solid fiber guide, 84–85
Red-detuned convex atom traps, 75–88
Red-detuned hollow fiber waveguides, 39–44
Red-detuned Yukawa potential trap, 86–88
Resonance fluorescence spectroscopy, optical lattices, 100–104, 121
Resonant coherent excitation, 166–167
Resonant transfer and excitation
  dielectronic recombination, 162–166
  heavy ions, 310, 312, 347–348
RETRAP project, 319
rf evaporation, 194–195, 197, 221–226
Rhenium, bound β decay, 362
Rubidium
  evaporative cooling, 230
  optical lattices, 133–134
Runaway evaporation, 187–189, 192, 195
Rydberg states, 298, 356–358

S

Saddle point, 220
Schrödinger Cat states, 243, 278–283, 290, 292
Schrödinger modal decomposition, cylindrical atom waveguide, 49–53
Secular approximation, laser cooling, 116–119
Sideband cooling, 242, 252–258
Sisyphus cooling, 99, 217
Sodium, evaporative cooling, 204–206, 207, 217, 230
Solid fiber waveguide, 81–85
Spectroscopy
  highly charged heavy ions, 362
    Balmer transitions, 362
    Doppler-shift-assisted ion spectroscopy, 349, 359
    fast beam precision spectroscopy, 348–349
    laser resonance fluorescence spectroscopy, 358
    VUV spectroscopy, 362
  optical lattices
    experimental results, 123–128

fluorescence spectroscopy, 100–104, 120–122
optical heterodyne spectroscopy, 100–104, 121–122
phase conjugation spectroscopy, 119
probe transmission spectroscopy, 100–104, 122–123
two-photon spectroscopy, 228
Specular atom mirror, 3–4
Specular reflection, evanescent-wave atom mirror, 13–15
Spontaneous force, 12, 14–15
Squeezed states, 239–239, 243, 274–278, 287, 288, 292
Storage/cooler rings, 317, 319–331
Strong coupling regine, 239–240
Strong excitation regime, 242, 278
Sub-recoil cooling, 182, 211–212, 217, 231
Superconducting magnets, evaporative cooling, 213
Surface-plasmon-enhanced evanescent wave, 18–19

T

Thermalization, evaporative cooling, 183, 199–200
Three-body recombination, 201–205, 312–313, 336
Three-dimensional evaporation, 197–199
Three-dimensional optical lattices, crystallography, 107–109
Time-averaged orbiting potential (TOP) trap, 214, 225, 226
Total trapping potential, 219–220
Trapped ions
  nonclassical states
    Fock states, 243, 258–273, 287, 288, 292
    models, 242, 244–252
    sideband cooling, 242, 252–258
  nonclassical states of motion
    collapses and revivals, 243, 283–290

pure and mixed states, 243–244, 290–292
Schrödinger Cat states, 243, 278–283, 290
squeezed states, 238–239, 243, 274–278
Trapping states, 267–273
Traps, see Atom traps
Tuning, evaporative cooling, 207–208
Tunneling, optical lattices, 135
Two-dimensional optical lattices, crystallography, 105–107
Two-photon spectroscopy, 228

V

Velocity-selective coherent populating trapping (VSCPT), 211–212, 226
VUV spectroscopy, 362

W

Wall-free confinement, evaporative cooling, 196
Wannier–Stark ladder, 135–136
Waveguides, 7
  cylindrical atom waveguide, 49–53
  hollow fiber guides
    blue-detuned, 45–55
    red-detuned, 39–44
  parallel-mirror waveguides, 7, 55–58
  pushme–pullyou guides, 5, 36–39
  red-detuned centrifugal solid fiber guide, 84–85
  solid fiber waveguides, 81–85
Whispering gallery trap, 36, 76–81

Y

Yin–yang trap, 212
Yukawa potential trap, 86–88

Contents of Volumes in This Serial

Volume 1

Molecular Orbital Theory of the Spin Properties of Conjugated Molecules, *G. G. Hall and A. T. Amos*

Electron Affinities of Atoms and Molecules, *B. L. Moiseiwitsch*

Atomic Rearrangement Collisions, *B. H. Bransden*

The Production of Rotational and Vibrational Transitions in Encounters between Molecules, *K. Takayanagi*

The Study of Intermolecular Potentials with Molecular Beams at Thermal Energies, *H. Pauly and J. P. Toennies*

High-Intensity and High-Energy Molecular Beams, *J. B. Anderson, R. P. Andres, and J. B. Fen*

Volume 2

The Calculation of van der Waals Interactions, *A. Dalgarno and W. D. Davison*

Thermal Diffusion in Gases, *E. A. Mason, R. J. Munn, and Francis J. Smith*

Spectroscopy in the Vacuum Ultraviolet, *W. R. S. Garton*

The Measurement of the Photoionization Cross Sections of the Atomic Gases, *James A. R. Samson*

The Theory of Electron–Atom Collisions, *R. Peterkop and V. Veldre*

Experimental Studies of Excitation in Collisions between Atomic and Ionic Systems, *F. J. de Heer*

Mass Spectrometry of Free Radicals, *S. N. Foner*

Volume 3

The Quantal Calculation of Photoionization Cross Sections, *A. L. Stewart*

Radiofrequency Spectroscopy of Stored Ions I: Storage, *H. G. Dehmelt*

Optical Pumping Methods in Atomic Spectroscopy, *B. Budick*

Energy Transfer in Organic Molecular Crystals: A Survey of Experiments, *H. C. Wolf*

Atomic and Molecular Scattering from Solid Surfaces, *Robert E. Stickney*

Quantum Mechanics in Gas Crystal-Surface van der Waals Scattering, *E. Chanoch Beder*

Reactive Collisions between Gas and Surface Atoms, *Henry Wise and Bernard J. Wood*

Volume 4

H. S. W. Massey—A Sixtieth Birthday Tribute, *E. H. S. Burhop*

Electronic Eigenenergies of the Hydrogen Molecular Ion, *D. R. Bates and R. H. G. Reid*

Applications of Quantum Theory to the Viscosity of Dilute Gases, *R. A. Buckingham and E. Gal*

Positrons and Positronium in Gases, *P. A. Fraser*

Classical Theory of Atomic Scattering, *A. Burgess and I. C. Percival*

Born Expansions, *A. R. Holt and B. L. Moiselwitsch*

Resonances in Electron Scattering by Atoms and Molecules, *P. G. Burke*

Relativistic Inner Shell Ionizations, *C. B. O. Mohr*

Recent Measurements on Charge Transfer, *J. B. Hasted*

Measurements of Electron Excitation Functions, *D. W. O. Heddle and R. G. W. Keesing*

Some New Experimental Methods in Collision Physics, *R. F. Stebbings*

Atomic Collision Processes in Gaseous Nebulae, *M. J. Seaton*

Collisions in the Ionosphere, *A. Dalgarno*

The Direct Study of Ionization in Space, *R. L. F. Boyd*

Volume 5

Flowing Afterglow Measurements of Ion-Neutral Reactions, *E. E. Ferguson, F. C. Fehsenfeld, and A. L. Schmeltekopf*

Experiments with Merging Beams, *Roy H. Neynaber*

Radiofrequency Spectroscopy of Stored Ions II: Spectroscopy, *H. G. Dehmelt*

The Spectra of Molecular Solids, *O. Schnepp*

The Meaning of Collision Broadening of Spectral Lines: The Classical Oscillator Analog, *A. Ben-Reuven*

The Calculation of Atomic Transition Probabilities, *R. J. S. Crossley*

Tables of One- and Two-Particle Coefficients of Fractional Parentage for Configurations $s^{\lambda}s'^{\mu}p^{q}$, *C. D. H. Chisholm, A. Dalgarno, and F. R. Innes*

Relativistic Z-Dependent Corrections to Atomic Energy Levels, *Holly Thomis Doyle*

Volume 6

Dissociative Recombination, *J. N. Bardsley and M. A. Biondi*

Analysis of the Velocity Field in Plasmas from the Doppler Broadening of Spectral Emission Lines, *A. S. Kaufman*

The Rotational Excitation of Molecules by Slow Electrons, *Kazuo Takayanagi and Yukikazu Itikawa*

The Diffusion of Atoms and Molecules, *E. A. Mason and T. R. Marrero*

Theory and Application of Sturmian Functions, *Manuel Rotenberg*

Use of Classical mechanics in the Treatment of Collisions between Massive Systems, *D. R. Bates and A. E. Kingston*

Volume 7

Physics of the Hydrogen Master, *C. Audoin, J. P. Schermann, and P. Grivet*

Molecular Wave Functions: Calculations and Use in Atomic and Molecular Processes, *J. C. Browne*

Localized Molecular Orbitals, *Harel Weinstein, Ruben Pauncz, and Maurice Cohen*

General Theory of Spin-Coupled Wave Functions for Atoms and Molecules, *J. Gerratt*

Diabatic States of Molecules—Quasi-Stationary Electronic States, *Thomas F. O'Malley*

Selection Rules within Atomic Shells, *B. R. Judd*

Green's Function Technique in Atomic and Molecular Physics, *Gy. Csanak, H. S. Taylor, and Robert Yaris*

A Review of Pseudo-Potentials with Emphasis on Their Application to Liquid Metals, *Nathan Wiser and A. J. Greenfield*

Volume 8

Interstellar Molecules: Their Formation and Destruction, *D. McNally*

Monte Carlo Trajectory Calculations of Atomic and Molecular Excitation in Thermal Systems, *James C. Keck*

Nonrelativistic Off-Shell Two-Body Coulomb Amplitudes, *Joseph C. Y. Chen and Augustine C. Chen*

Photoionization with Molecular Beams, *R.*

B. Cairns, Halstead Harrison, and R. I. Schoen

The Auger Effect, E. H. S. Burhop and W. N. Asaad

Volume 9

Correlation in Excited States of Atoms, A. W. Weiss

The Calculation of Electron–Atom Excitation Cross Sections, M. R. H. Rudge

Collision-Induced Transitions between Rotational Levels, Takeshi Oka

The Differential Cross Section of Low-Energy Electron–Atom Collisions, D. Andrick

Molecular Beam Electric Resonance Spectroscopy, Jens C. Zorn and Thomas C. English

Atomic and Molecular Processes in the Martian Atmosphere, Michael B. McElroy

Volume 10

Relativistic Effects in the Many-Electron Atom, Lloyd Armstrong, Jr. and Serge Feneuille

The First Born Approximation, K. L. Bell and A. E. Kingston

Photoelectron Spectroscopy, W. C. Price

Dye Lasers in Atomic Spectroscopy, W. Lange, J. Luther, and A. Steudel

Recent Progress in the Classification of the Spectra of Highly Ionized Atoms, B. C. Fawcett

A Review of Jovian Ionospheric Chemistry, Wesley T. Huntress, Jr.

Volume 11

The Theory of Collisions between Charged Particles and Highly Excited Atoms, I. C. Percival and D. Richards

Electron Impact Excitation of Positive Ions, M. J. Seaton

The R-Matrix Theory of Atomic Process, P. G. Burke and W. D. Robb

Role of Energy in Reactive Molecular Scattering: An Information-Theoretic Approach, R. B. Bernstein and R. D. Levine

Inner Shell Ionization by Incident Nuclei, Johannes M. Hansteen

Stark Broadening, Hans R. Griem

Chemiluminescence in Gases, M. F. Golde and B. A. Thrush

Volume 12

Nonadiabatic Transitions between Ionic and Covalent States, R. K. Janev

Recent Progress in the Theory of Atomic Isotope Shift, J. Bauche and R.-J. Champeau

Topics on Multiphoton Processes in Atoms, P. Lambropoulos

Optical Pumping of Molecules, M. Broyer, G. Goudedard, J. C. Lehmann, and J. Vigué

Highly Ionized Ions, Ivan A. Sellin

Time-of-Flight Scattering Spectroscopy, Wilhelm Raith

Ion Chemistry in the D Region, George C. Reid

Volume 13

Atomic and Molecular Polarizabilities—A Review of Recent Advances, Thomas M. Miller and Benjamin Bederson

Study of Collisions by Laser Spectroscopy, Paul R. Berman

Collision Experiments with Laser-Excited Atoms in Crossed Beams, I. V. Hertel and W. Stoll

Scattering Studies of Rotational and Vibrational Excitation of Molecules, Manfred Faubel and J. Peter Toennies

Low-Energy Electron Scattering by Complex Atoms: Theory and Calculations, R. K. Nesbet

Microwave Transitions of Interstellar Atoms and Molecules, *W. B. Somerville*

Volume 14

Resonances in Electron Atom and Molecule Scattering, *D. E. Golden*
The Accurate Calculation of Atomic Properties by Numerical Methods, *Brian C. Webster, Michael J. Jamieson, and Ronald F. Stewart*
(e, 2e) Collisions, *Erich Weigold and Ian E. McCarthy*
Forbidden Transitions in One- and Two-Electron Atoms, *Richard Marrus and Peter J. Mohr*
Semiclassical Effects in Heavy-Particle Collisions, *M. S. Child*
Atomic Physics Tests of the Basic Concepts in Quantum Mechanics, *Francis M. Pipkin*
Quasi-Molecular Interference Effects in Ion–Atom Collisions, *S. V. Bobashev*
Rydberg Atoms, *S. A. Edelstein and T. F. Gallagher*
UV and X-Ray Spectroscopy in Astrophysics, *A. K. Dupree*

Volume 15

Negative Ions, *H. S. W. Massey*
Atomic Physics from Atmospheric and Astrophysical Studies, *A. Dalgarno*
Collisions of Highly Excited Atoms, *R. F. Stebbings*
Theoretical Aspects of Positron Collisions in Gases, *J. W. Humberston*
Experimental Aspects of Positron Collisions in Gases, *T. C. Griffith*
Reactive Scattering: Recent Advances in Theory and Experiment, *Richard B. Bernstein*
Ion–Atom Charge Transfer Collisions at Low Energies, *J. B. Hasted*
Aspects of Recombination, *D. R. Bates*
The Theory of Fast Heavy Particle Collisions, *B. H. Bransden*
Atomic Collision Processes in Controlled Thermonuclear Fusion Research, *H. B. Gilbody*
Inner-Shell Ionization, *E. H. S. Burhop*
Excitation of Atoms by Electron Impact, *D. W. O. Heddle*
Coherence and Correlation in Atomic Collisions, *H. Kleinpoppen*
Theory of Low Energy Electron–Molecule Collisions, *P. G. Burke*

Volume 16

Atomic Hartree–Fock Theory, *M. Cohen and R. P. McEachran*
Experiments and Model Calculations to Determine Interatomic Potentials, *R. Düren*
Sources of Polarized Electrons, *R. J. Celotta and D. T. Pierce*
Theory of Atomic Processes in Strong Resonant Electromagnetic Fields, *S. Swain*
Spectroscopy of Laser-Produced Plasmas, *M. H. Key and R. J. Hutcheon*
Relativistic Effects in Atomic Collisions Theory, *B. L. Moiseiwitsch*
Parity Nonconservation in Atoms: Status of Theory and Experiment, *E. N. Fortson and L. Wilets*

Volume 17

Collective Effects in Photoionization of Atoms, *M. Ya. Amusia*
Nonadiabatic Charge Transfer, *D. S. F. Crothers*
Atomic Rydberg States, *Serge Feneuille and Pierre Jacquinot*
Superfluorescence, *M. F. H. Schuurmans, Q. H. F. Vrehen, D. Polder, and H. M. Gibbs*
Applications of Resonance Ionization Spectroscopy in Atomic and Molecular Physics, *M. G. Payne, C. H. Chen, G. S. Hurst, and G. W. Foltz*

Inner-Shell Vacancy Production in Ion–Atom Collisions, *C. D. Lin and Patrick Richard*

Atomic Processes in the Sun, *P. L. Dufton and A. E. Kingston*

Volume 18

Theory of Electron–Atom Scattering in a Radiation Field, *Leonard Rosenberg*

Positron–Gas Scattering Experiments, *Talbert S. Stein and Walter E. Kauppila*

Nonresonant Multiphoton Ionization of Atoms, *J. Morellec, D. Normand, and G. Petite*

Classical and Semiclassical Methods in Inelastic Heavy-Particle Collisions, *A. S. Dickinson and D. Richards*

Recent Computational Developments in the Use of Complex Scaling in Resonance Phenomena, *B. R. Junker*

Direct Excitation in Atomic Collisions: Studies of Quasi-One-Electron Systems, *N. Anderson and S. E. Nielsen*

Model Potentials in Atomic Structure, *A. Hibbert*

Recent Developments in the Theory of Electron Scattering by Highly Polar Molecules, *D. W. Norcross and L. A. Collins*

Quantum Electrodynamic Effects in Few-Electron Atomic Systems, *G. W. F. Drake*

Volume 19

Electron Capture in Collisions of Hydrogen Atoms with Fully Stripped Ions, *B. H. Bransden and R. K. Janev*

Interactions of Simple Ion–Atom Systems, *J. T. Park*

High-Resolution Spectroscopy of Stored Ions, *D. J. Wineland, Wayne M. Itano, and R. S. Van Dyck, Jr.*

Spin-Dependent Phenomena in Inelastic Electron–Atom Collisions, *K. Blum and H. Kleinpoppen*

The Reduced Potential Curve Method for Diatomic Molecules and Its Applications, *F. Jenč*

The Vibrational Excitation of Molecules by Electron Impact, *D. G. Thompson*

Vibrational and Rotational Excitation in Molecular Collisions, *Manfred Faubel*

Spin Polarization of Atomic and Molecular Photoelectrons, *N. A. Cherepkov*

Volume 20

Ion–Ion Recombination in an Ambient Gas, *D. R. Bates*

Atomic Charges within Molecules, *G. G. Hall*

Experimental Studies on Cluster Ions, *T. D. Mark and A. W. Castleman, Jr.*

Nuclear Reaction Effects on Atomic Inner-Shell Ionization, *W. E. Meyerhof and J.-F. Chemin*

Numerical Calculations on Electron-Impact Ionization, *Christopher Bottcher*

Electron and Ion Mobilities, *Gordon R. Freeman and David A. Armstrong*

On the Problem of Extreme UV and X-Ray Lasers, *I. I. Sobel'man and A. V. Vinogradov*

Radiative Properties of Rydberg States in Resonant Cavities, *S. Haroche and J. M. Ralmond*

Rydberg Atoms: High-Resolution Spectroscopy and Radiation Interaction—Rydberg Molecules, *J. A. C. Gallas, G. Leuchs, H. Walther, and H. Figger*

Volume 21

Subnatural Linewidths in Atomic Spectroscopy, *Dennis P. O'Brien, Pierre Meystre, and Herbert Walther*

Molecular Applications of Quantum Defect Theory, *Chris H. Greene and Ch. Jungen*

Theory of Dielectronic Recombination, *Yukap Hahn*

Recent Developments in Semiclassical Floquet Theories for Intense-Field Multiphoton Processes, *Shih-I Chu*

Scattering in Strong Magnetic Fields, *M. R. C. McDowell and M. Zarcone*

Pressure Ionization, Resonances, and the Continuity of Bound and Free States, *R. M. More*

Volume 22

Positronium—Its Formation and Interaction with Simple Systems, *J. W. Humberston*

Experimental Aspects of Positron and Positronium Physics, *T. C. Griffith*

Doubly Excited States, Including New Classification Schemes, *C. D. Lin*

Measurements of Charge Transfer and Ionization in Collisions Involving Hydrogen Atoms, *H. B. Gilbody*

Electron-Ion and Ion-Ion Collisions with Intersecting Beams, *K. Dolder and B. Pearl*

Electron Capture by Simple Ions, *Edward Pollack and Yukap Hahn*

Relativistic Heavy-Ion-Atom Collisions, *R. Anholt and Harvey Gould*

Continued-Fraction Methods in Atomic Physics, *S. Swain*

Volume 23

Vacuum Ultraviolet Laser Spectroscopy of Small Molecules, *C. R. Vidal*

Foundations of the Relativistic Theory of Atomic and Molecular Structure, *Ian P. Grant and Harry M. Quiney*

Point-Charge Models for Molecules Derived from Least-Squares Fitting of the Electric Potential, *D. E. Williams and Ji-Min Yan*

Transition Arrays in the Spectra of Ionized Atoms, *J. Bauche, C. Bauche-Arnoult, and M. Klapisch*

Photoionization and Collisional Ionization of Excited Atoms Using Synchroton and Laser Radiation, *F. J. Wuilleumier, D. L. Ederer, and J.L. Picqué*

Volume 24

The Selected Ion Flow Tube (SIDT): Studies of Ion-Neutral Reactions, *D. Smith and N. G. Adams*

Near-Threshold Electron-Molecule Scattering, *Michael A. Morrison*

Angular Correlation in Multiphoton Ionization of Atoms, *S. J. Smith and G. Leuchs*

Optical Pumping and Spin Exchange in Gas Cells, *R. J. Knize, Z. Wu, and W. Happer*

Correlations in Electron-Atom Scattering, *A. Crowe*

Volume 25

Alexander Dalgarno: Life and Personality, *David R. Bates and George A. Victor*

Alexander Dalgarno: Contributions to Atomic and Molecular Physics, *Neal Lane*

Alexander Dalgarno: Contributions to Aeronomy, *Michael B. McElroy*

Alexander Dalgarno: Contributions to Astrophysics, *David A. Williams*

Dipole Polarizability Measurements. *Thomas M. Miller and Benjamin Bederson*

Flow Tube Studies of Ion-Molecule Reactions, *Eldon Ferguson*

Differential Scattering in He-He and He^+-He Collisions at KeV Energies, *R. F. Stebbings*

Atomic Excitation in Dense Plasmas, *Jon C. Weisheit*

Pressure Broadening and Laser-Induced Spectral Line Shapes, *Kenneth M. Sando and Shih-I Chu*

Model-Potential Methods, *G. Laughlin and G. A. Victor*

Z-Expansion Methods, *M. Cohen*

Schwinger Variational Methods, *Deborah Kay Watson*

Fine-Structure Transitions in Proton-Ion Collisions, *R. H. G. Reid*

Electron Impact Excitation, *R. J. W. Henry and A. E. Kingston*

Recent Advances in the Numerical Calculation of Ionization Amplitudes, *Christopher Bottcher*

The Numerical Solution of the Equations of Molecular Scattering, *A. C. Allison*

High Energy Charge Transfer, *B. H. Bransden and D. P. Dewangan*

Relativistic Random-Phase Approximation, *W. R. Johnson*

Relativistic Sturmian and Finite Basis Set Methods in Atomic Physics, *G. W. F. Drake and S. P. Goldman*

Dissociation Dynamics of Polyatomic Molecules, *T. Uzer*

Photodissociation Processes in Diatomic Molecules of Astrophysical Interest, *Kate P. Kirby and Ewine F. van Dishoeck*

The Abundances and Excitation of Interstellar Molecules, *John H. Black*

Volume 26

Comparisons of Positrons and Electron Scattering by Gases, *Walter E. Kauppila and Talbert S. Stein*

Electron Capture at Relativistic Energies, *B. L. Moiseiwitsch*

The Low-Energy, Heavy Particle Collisions—A Close-Coupling Treatment, *Mineo Kimura and Neal F. Lane*

Vibronic Phenomena in Collisions of Atomic and Molecular Species, *V. Sidis*

Associative Ionization: Experiments, Potentials, and Dynamics, *John Weiner, Françoise Masnou-Sweeuws, and Annick Giusti-Suzor*

On the β Decay of ^{187}Re: An Interface of Atomic and Nuclear Physics and Cosmochronology, *Zonghau Chen, Leonard Rosenberg, and Larry Spruch*

Progress in Low Pressure Mercury-Rare Gas Discharge Research, *J. Maya and R. Lagushenko*

Volume 27

Negative Ions: Structure and Spectra, *David R. Bates*

Electron Polarization Phenomena in Electron–Atom Collisions, *Joachim Kessler*

Electron–Atom Scattering, *I. E. McCarthy and E. Weigold*

Electron–Atom Ionization, *I. E. McCarthy and E. Weigold*

Role of Autoionizing States in Multiphoton Ionization of Complex Atoms, *V. I. Lengyel and M. I. Haysak*

Multiphoton Ionization of Atomic Hydrogen Using Perturbation Theory, *E. Karule*

Volume 28

The Theory of Fast Ion–Atom Collisions, *J. S. Briggs and J. H. Macek*

Some Recent Developments in the Fundamental Theory of Light, *Peter W. Milonni and Surendra Singh*

Squeezed States of the Radiation Field, *Khalid Zaheer and M. Suhail Zubairy*

Cavity Quantum Electrodynamics, *E. A. Hinds*

Volume 29

Studies of Electron Excitation of Rare-Gas Atoms into and out of Metastable Levels Using Optical and Laser Techniques, *Chun C. Lin and L. W. Anderson*

Cross Sections for Direct Multiphoton Ionization of Atoms, *M. V. Ammosov, N. B. Delone, M. Yu. Ivanov, I. I. Bondar, and A. V. Masalov*

Collision-Induced Coherences in Optical Physics, *G. S. Agarwal*

Muon-Catalyzed Fusion, *Johann Rafelski and Helga E. Rafelski*

Cooperative Effects in Atomic Physics, *J. P. Connerade*

Multiple Electron Excitation, Ionization, and Transfer in High-Velocity Atomic and Molecular Collisions, *J. H. McGuire*

Volume 30

Differential Cross Sections for Excitation of Helium Atoms and Helium-Like Ions by Electron Impact, *Shinobu Nakazaki*

Cross-Section Measurements for Electron Impact on Excited Atomic Species, *S. Trajmar and J. C. Nickel*

The Dissociative Ionization of Simple, Molecules by Fast Ions, *Colin J. Latimer*

Theory of Collisions between Laser Cooled Atoms, *P. S. Julienne, A. M. Smith, and K. Burnett*

Light-Induced Drift, *E. R. Eliel*

Continuum Distorted Wave Methods in Ion-Atom Collisions, *Derrick S. F. Crothers and Louis J. Dubé*

Volume 31

Energies and Asymptotic Analysis for Helium Rydberg States, *G. W. F. Drake*

Spectroscopy of Trapped Ions, *R. C. Thompson*

Phase Transitions of Stored Laser-Cooled Ions, *H. Walther*

Selection of Electronic States in Atomic Beams with Lasers, *Jacques Baudon, Rudolf Düren, and Jacques Robert*

Atomic Physics and Non-Maxwellian Plasmas, *Michèle Lamoureux*

Volume 32

Photoionization of Atomic Oxygen and Atomic Nitrogen, *K. L. Bell and A. E. Kingston*

Positronium Formation by Positron Impact on Atoms at Intermediate Energies, *B. H. Bransden and C. J. Noble*

Electron-Atom Scattering Theory and Calculations, *P. G. Burke*

Terrestrial and Extraterrestrial H_3^+, *Alexander Dalgarno*

Indirect Ionization of Positive Atomic Ions, *K. Dolder*

Quantum Defect Theory and Analysis of High-Precision Helium Term Energies, *G. W. F. Drake*

Electron-Ion and Ion-Ion Recombination Processes, *M. R. Flannery*

Studies of State-Selective Electron Capture in Atomic Hydrogen by Translational Energy Spectroscopy, *H. B. Gilbody*

Relativistic Electronic Structure of Atoms and Molecules, *I. P. Grant*

The Chemistry of Stellar Environments, *D. A. Howe, J. M. C. Rawlings, and D. A. Williams*

Positron and Positronium Scattering at Low Energies, *J. W. Humberston*

How Perfect are Complete Atomic Collision Experiments?, *H. Kleinpoppen and H. Hamdy*

Adiabatic Expansions and Nonadiabatic Effects, *R. McCarroll and D. S. F. Crothers*

Electron Capture to the Continuum, *B. L. Moiseiwitsch*

How Opaque Is a Star? *M. J. Seaton*

Studies of Electron Attachment at Thermal Energies Using the Flowing Afterglow–Langmuir Technique, *David Smith and Patrik Španěl*

Exact and Approximate Rate Equations in Atom-Field Interactions, *S. Swain*

Atoms in Cavities and Traps, *H. Walther*

Some Recent Advances in Electron-Impact Excitation of $n = 3$ States of Atomic Hydrogen and Helium, *J. F. Williams and J. B. Wang*

Volume 33

Principles and Methods for Measurement of Electron Impact Excitation Cross Sections for Atoms and Molecules by Optical Techniques, *A. R. Filippelli, Chun C. Lin, L. W. Andersen, and J. W. McConkey*

Benchmark Measurements of Cross Sections for Electron Collisions: Analysis of Scattered Electrons, *S. Trajmar and J. W. McConkey*

Benchmark Measurements of Cross Sections for Electron Collisions: Electron Swarm Methods, *R. W. Crompton*

Some Benchmark Measurements of Cross Sections for Collisions of Simple Heavy Particles, *H. B. Gilbody*

The Role of Theory in the Evaluation and Interpretation of Cross-Section Data, *Barry I. Schneider*

Analytic Representation of Cross-Section Data, *Mitio Inokuti, Mineo Kimura, M. A. Dillon, Isao Shimamura*

Electron Collisions with N_2, O_2 and O: What We Do and Do Not Know, *Yukikazu Itikawa*

Need for Cross Sections in Fusion Plasma Research, *Hugh P. Summers*

Need for Cross Sections in Plasma Chemistry, *M. Capitelli, R. Celiberto, and M. Cacciatore*

Guide for Users of Data Resources, *Jean W. Gallagher*

Guide to Bibliographies, Books, Reviews, and Compendia of Data on Atomic Collisions, *E. W. McDaniel and E. J. Mansky*

Volume 34

Atom Interferometry, *C. S. Adams, O. Carnal, and J. Mlynek*

Optical Tests of Quantum Mechanics, *R. Y. Chiao, P. G. Kwiat, and A. M. Steinberg*

Classical and Quantum Chaos in Atomic Systems, *Dominique Delande and Andreas Buchleitner*

Measurements of Collisions between Laser-Cooled Atoms, *Thad Walker and Paul Feng*

The Measurement and Analysis of Electric Fields in Glow Discharge Plasmas, *J. E. Lawler and D. A. Doughty*

Polarization and Orientation Phenomena in Photoionization of Molecules, *N. A. Cherepkov*

Role of Two-Center Electron–Electron Interaction in Projectile Electron Excitation and Loss, *E. C. Montenegro, W. E. Meyerhof, and J. H. McGuire*

Indirect Processes in Electron Impact Ionization of Positive Ions, *D. L. Moores and K. J. Reed*

Dissociative Recombination: Crossing and Tunneling Modes, *David R. Bates*

Volume 35

Laser Manipulation of Atoms, *K. Sengstock and W. Ertmer*

Advances in Ultracold Collisions: Experiment and Theory, *J. Weiner*

Ionization Dynamics in Strong Laser Fields, *L. F. DiMauro and P. Agostini*

Infrared Spectroscopy of Size Selected Molecular Clusters, *U. Buck*

Femtosecond Spectroscopy of Molecules and Clusters, *T. Baumer and G. Gerber*

Calculation of Electron Scattering on Hydrogenic Targets, *I. Bray and A. T. Stelbovics*

Relativistic Calculations of Transition Amplitudes in the Helium Isoelectronic Sequence, *W. R. Johnson, D. R. Plante, and J. Sapirstein*

Rotational Energy Transfer in Small Polyatomic Molecules, *H. O. Everitt and F. C. De Lucia*

Volume 36

Complete Experiments in Electron–Atom Collisions, *Nils Overgaard Andersen, and Klaus Bartschat*

Stimulated Rayleigh Resonances and Recoil-Induced Effects, *J.-Y. Courtois and G. Grynberg*

Precision Laser Spectroscopy Using Acousto-Optic Modulators, *W. A. van Wijngaarden*

Highly Parallel Computational Techniques for Electron–Molecule Collisions, *Carl Winstead and Vincent McKoy*

Quantum Field Theory of Atoms and Photons, *Maciej Lewenstein and Li You*

Volume 37

Evanescent Light-Wave Atom Mirrors, Resonators, Waveguides, and Traps, *Jonathan P. Dowling and Julio Gea-Banacloche*

Optical Lattices, *P. S. Jessen and I. H. Deutsch*

Channeling Heavy Ions through Crystalline Lattices, *Herbert F. Krause and Sheldon Datz*

Evaporative Cooling of Trapped Atoms, *Wolfgang Ketterle and N. J. van Druten*

Nonclassical States of Motion in Ion Traps, *J. I. Cirac, A. S. Parkins, R. Blatt, and P. Zoller*

The Physics of Highly-Charged Heavy Ions Revealed by Storage/Cooler Rings, *P. H. Mokler and Th. Stöhlker*

ISBN 0-12-003837-4